2nd International Conference on Mechatronics and Electrical Systems (ICMES 2017)

IOP Conference Series: Materials Science and Engineering
Volume 339

Wuhan, China
15 – 17 December 2017

ISBN: 978-1-5108-6115-2
ISSN: 1757-8981

Printed from e-media with permission by:

Curran Associates, Inc.
57 Morehouse Lane
Red Hook, NY 12571

Some format issues inherent in the e-media version may also appear in this print version.

Printed by Curran Associates, Inc. (2018)

Institute of Physics
Dirac House, Temple Back
Bristol BS1 6BE UK

Phone: 44 1 17 929 7481
Fax: 44 1 17 920 0979

techtracking@iop.org

Additional copies of this publication are available from:

Curran Associates, Inc.
57 Morehouse Lane
Red Hook, NY 12571 USA
Phone: 845-758-0400
Fax: 845-758-2633
Email: curran@proceedings.com
Web: www.proceedings.com

Table of contents

Volume 339

2017 2nd International Conference on Mechatronics and Electrical Systems (ICMES 2017)

15–17 December 2017, Wuhan, China

Accepted papers received: 15 March 2018
Published online: 5 April 2018

Preface

Papers

Modern Electronic Information Technology and Applications

IOP Conf. Series: Materials Science and Engineering **339** (2018) 011001 doi:10.1088/1757-899X/339/1/011001

Preface

It was a great pleasure for us to organize 2018 2nd International Conference on Mechatronics and Electrical Systems (ICMES 2018) from December 22-25, 2017 at CUG in Wuhan, China. This conference is supported by China University of Geosciences (Wuhan).

ICMES is a highly selective, single-track meeting based on technical submissions presenting significant, original, and previously unpublished scientific research and technical developments for industrial application. Many researchers and practitioners present and discuss their advancements in the fields of Mechatronics and Electrical Systems, on topics like Electrical and Electronic System, Mechatronics Applications in Agriculture and Regional Industries and Intelligent Systems and Information Technology etc.

ICMES 2017 received more than 70 submitted papers whose authors are from over 6 countries, and 39 excellent papers had been selected. The papers are accepted after a selective peer-review based on the coherence of the topic with the Conference themes and on the content and quality of the work contained. This Conference Proceedings compile the accepted papers into ICMES 2017 conference proceedings, which represent an important collection of information on the latest international achievements and developments in the fields of Mechatronics and Electrical Systems: these should serve as a good base for inspiration for further advancements and developments.

On behalf of the organizing committee, we express my heartfelt gratitude to the participants for their outstanding contributions, to committee members and reviewers for their great professionalism and efforts and to the sponsors for their valuable contributions support to this successful conference.

Finally, we are grateful to all those who have contributed to the success of ICMES 2017.

Huafeng Ding
Dean, Professor
China University of Geosciences, China
ICMES 2017 Organizing Committees

IOP Conf. Series: Materials Science and Engineering **339** (2018) 011001 doi:10.1088/1757-899X/339/1/011001

Statement of Peer Review

All papers published in this volume of IOP Conference Series: Materials Science and Engineering has been peer reviewed through processes administered by the proceedings Editors. Reviews were conducted by expert referees to the professional and scientific standards expected of a proceedings journal published by IOP Publishing.

IOP Conf. Series: Materials Science and Engineering **339** (2018) 011002 doi:10.1088/1757-899X/339/1/011002

Conference Committee

General Conference Chair
Prof. Huafeng Ding, China University of Geosciences, China

International Advisory Committee
Prof. Dan Zhang, York University, Canada

Conference Committee Chair
Prof. Jamaluddin Mahmud, Universiti Teknologi MARA, Malaysia
Prof. Shishir Kumar Sahu, National Institute of Technology, Rourkela, India

Conference Program Committee
Prof. Chaolong Song, China University of Geosciences (Wuhan), China
Prof. Tianye Huang, China University of Geosciences (Wuhan), China
Prof. Jie Luo, China University of Geosciences (Wuhan), China
Prof. Yiyong Yang, China University of Geosciences (Beijing), China
Prof. Christos Spitas, Delft University of Technology, Netherlands/ Nazarbayev University, Kazakhstan

Local Organizing Chair
Prof. Bo Li, China University of Geosciences (Wuhan), China

Conference Technical Committee
Prof. Sanjay Deokar, Zeal College of Engineering and Research, India
Assoc. Prof. Chong Wen Tong, University of Malaya, Malaysia
Prof. Md Alamgir Hossain, Military Institute of Science and Technology, Bangladesh
Dr. Sanjeet Dwivedi, Danfoss Power Electronics, Denmark
Dr. S A Mohan Krishna, Vidyavardhaka College of Engineering, India
Prof. Wen-Tsai Sung, National Chin-Yi University of Technology, China
Assoc. Prof. QingSong Xu, University of Macau, China
Assoc. Prof. Joseph Antony, University of Leeds, UK
Assoc. Prof. Darko Brodic, University of Belgrade, Serbia
Dr. Xinggang Yan, University of Kent, UK
Dr. Ching-Tai Ng, The University of Adelaide, Australia
Dr. Paulo NobreBalbis dos Reis, University of Beira Interior, Portugal
Prof. Eyhab EL-Kharashi, Ain Shams University, Egypt
Assoc. Prof. Saeed Tavakoli, University of Sistan and Baluchestan, Iran
Dr. Jia-xing Guo, University of Virginia, USA
Assoc. Prof. Ray Hwa Wong, Hwh Hsia University of Technology, Taiwan
Prof. Dan Dobrota, Constantin Brancusi University of Targu Jiu, Romania
Assoc. Prof. Xiaonlin Wang, University of Tasmania, Australia
Dr. Dianzi Liu, University of East Anglia, UK

IOP Conf. Series: Materials Science and Engineering **339** (2018) 011002 doi:10.1088/1757-899X/339/1/011002

Prof. Rafał Grzejda, West Pomeranian University of Technology, Poland
Prof. Elena Cerro Prada, Universidad Politécnica de Madrid, Spain
Prof. Ahmad Fakharian, Islamic Azad University, Iran
Assoc. Prof. Mohd Azlan MohdIshak, UniversitiTeknologi MARA, Malaysia
Dr. Yongli Shi, China Academy of Space Technology, China
Prof. Ziliang An, Shanghai Institute of Technology, China
Assoc. Prof. Mohammad Salah, Hashemite University, Jordan
Dr. Kemal UÇAK, Muğla Sıtkı Koçman University, Turkey

IOP Conf. Series: Materials Science and Engineering **339** (2018) 011003 doi:10.1088/1757-899X/339/1/011003

Peer review statement

All papers published in this volume of *IOP Conference Series: Materials Science and Engineering* have been peer reviewed through processes administered by the proceedings Editors. Reviews were conducted by expert referees to the professional and scientific standards expected of a proceedings journal published by IOP Publishing.

IOP Conf. Series: Materials Science and Engineering **339** (2018) 012001 doi:10.1088/1757-899X/339/1/012001

A Power Transformers Fault Diagnosis Model Based on Three DGA Ratios and PSO Optimization SVM

Hongzhe Ma, Wei Zhang, Rongrong Wu, Chunyan Yang

Electric Power Scientific Research Institute of Guangxi Power Grid, Guangxi Power Grid Corporation, Nanning, China

Email: 15235030639@163.com

Abstract. In order to make up for the shortcomings of existing transformer fault diagnosis methods in dissolved gas-in-oil analysis (DGA) feature selection and parameter optimization, a transformer fault diagnosis model based on the three DGA ratios and particle swarm optimization (PSO) optimize support vector machine (SVM) is proposed. Using transforming support vector machine to the nonlinear and multi-classification SVM, establishing the particle swarm optimization to optimize the SVM multi classification model, and conducting transformer fault diagnosis combined with the cross validation principle. The fault diagnosis results show that the average accuracy of test method is better than the standard support vector machine and genetic algorithm support vector machine, and the proposed method can effectively improve the accuracy of transformer fault diagnosis is proved.

1. Introduction

An oil-immersed power transformer is the core equipment for electric energy conversion in the electric power system. A fault in a transformer may result in not only substantial repair cost but also power interruptions to thousands of customers, therefore it is essential to assess its working condition online and detect the potential fault as soon as possible [1].

In the past few years, some monitoring methods, including windings displacement [2] and hot spot temperature [3], were applied to detect faults of oil-immersed power transformers, however DGA is still a more convenient and effective online monitoring method comparing to the above methods [4]. To dig the potential fault law from the gases content, some new fault diagnosis methods, including fuzzy logic [5] and support vector machine (SVM) [6], have been proposed in the recent years. In these methods, SVM is an effective method and is wildly applied in classification problems (fault diagnosis of transformers based on DGA is often considered as a classification problem) [7] because its advantages of small sample learning, global optimization, and structural risk minimization [8]. In general, some parameters have great influence on the performance of support vector machines (SVM), including kernel function, penalty coefficient C and kernel parameter gamma. These parameters are uncertain and should be optimized by artificial intelligence (AI) method. Some AI methods such as ICA, GA, were used to optimize the two parameters in the classification problems and receive a good performance. PSO, firstly put forward by Eberhart and Kennedy in 1995 [9], is an effective AI optimization algorithm for optimizing the SVM parameters because this algorithm has good global convergence solutions and efficient local search ability. Therefore, in this paper, a new support vector machine model based on particle swarm optimization was proposed to find out whether the approach can detect faults of transformers accurately and effectively.

IOP Conf. Series: Materials Science and Engineering **339** (2018) 012001 doi:10.1088/1757-899X/339/1/012001

2. Transform SVM and Select Kernel Function

2.1. Selection of DGA Feature

At present, the main methods of transformer fault diagnosis are using all the data of DGA (H_2, CH_4, C_2H_2, C_2H_4, C_2H_6, CO, CO_2) or part of DGA data as the characteristic quantity to diagnose faults. Literature [10] indicates that the three ratios of C_2H_2/C_2H_4, CH_4/H_2 and C_2H_4/C_2H_6 gas ratios are more closely associated with the operation of the transformer. These three ratios are the most representative ratio characteristics. Therefore, this paper uses the three gas ratios as DGA feature pairs for fault diagnosis.

2.2. Transform the Linear SVM to the Nonlinear SVM

In order to solve the following quadratic programming in (1), the concept of SVM is introduced.

$$\min \Phi(\omega,\xi)=\frac{1}{2}\|\omega\|^2+C\sum_{i=1}^{l}\xi_i$$
$$s.t.\begin{cases} y_i\left[\omega^T\varphi(x_i)+b\right]\geq 1-\xi_i \\ \xi_i\geq 0,\ i=1,2,\ldots,l \end{cases} \tag{1}$$

In the above formula, ξ_i is the slack variable and C is the penalty parameter. Therefore, formula (2) shows the Lagrange function simulation.

$$L(\omega,b,\xi,a,\beta)=\Phi(\omega,\xi)-\sum_{i=1}^{l}\alpha_i\left\{y_i[\omega^T\varphi(x_i)+b]-1+\xi_i\right\}-\sum_{i=1}^{l}\beta_i\xi_i \tag{2}$$

In the formula, the Lagrange multiplier is α_i and β_i. The Lagrangian function is transformed to the following quadratic programming (QP) problem (3) with the constraints in (4).

$$\max \Psi(\alpha)=-\frac{1}{2}\sum_{i,j=1}^{l}\alpha_i\alpha_j y_i y_j(x_i x_j)+\sum_{i=1}^{l}\alpha \tag{3}$$

$$f(x)=sign\left[\sum_{i=1}^{l}\alpha_i y_i K(x,x_i)+b\right] \tag{4}$$

In the formula (3), $K(x_i,x_j)=\varphi(x_i)\varphi(x_j)$ represents a kernel function.

At last, the formula (5) represents the nonlinear two-class classifier.

$$f(x)=sign\left[\sum_{i=1}^{l}\alpha_i y_i K(x,x_i)+b\right] \tag{5}$$

2.3. Transform the Two-Class SVM to the Multiple-Class SVM

The standard SVM is a typical two classification classifier, namely "either this or that, but the transformer fault classification is a multiple-class classification problem. Therefore, it needs to extend the two classification support vector machine to multi classification support vector machine for solving multiple-class classification problems. The OAO (one against one) scheme is a good method [11] for transforming and extending, so the schemes are adopted this method in the next section.

2.4. Establishment of the Optimal Objective Function

After obtaining the model of the nonlinear-multiple classification SVM, the selection of kernel function of SVM is important for establishing the optimal objective function. The RBF is a real valued function that depends only on the specific value of the distance. The RBF kernel function, shown in equation (6), is effective and needs less parameters setting, so it is used in the paper.

$$K(x_i, x_j) = \exp(-\gamma \|x_i - x_j\|^2), \gamma > 0 \quad (6)$$

In formula (6), γ is a free parameter in the RBF kernel function and it is inversely proportional to the width of RBF. This parameter needs to be optimized, and the punishment coefficient C for support vector machines also needs to be optimized. This paper intends to optimize these two parameters through the particle swarm optimization. Combining the RBF kernel functions with SVM, we can build the RBF-SVR model. The structure of RBF-SVR is shown in Fig. 1.

3. The Particle Swarm Optimization Optimizes the SVM Parameter

The particle swarm optimization (PSO) is a popular population-based heuristic algorithm that simulates the social behavior of individuals such as birds flocking, a school of fish swimming or a colony of ants moving to a potential position to achieve particular objectives in a multidimensional space. PSO is found to have the extensive capability of global optimization for its simple concept, easy implementation, scalability, robustness, and fast convergence [12].
PSO is mathematically modeled as follows:

$$v_{sd}(t+1) = v_{sd}(t) + c_1(t) r_1(t)\left(p_{sd}(t) - x_{sd}(t)\right) + c_2(t) r_2(t)\left(p_{gd}(t) - x_{sd}(t)\right) \quad (7)$$

$$x_{sd}(t+1) = x_{sd}(t) + v_{sd}(t+1) \quad (8)$$

In Equations (7) and (8) above, t is the number of generations; The random variable t is subject to a uniform distribution in the (0, 1) interval; $c_1(t)$ and $c_2(t)$ is the acceleration constant; $x_{sd}(t)$ is the position of the s particle for the t generation; p_{sd} is the optimal location for all particles in the t generation.

The flow of PSO for parameters optimization can described as follows [13]:

(1) Initialize the parameters of SVM;

(2) A group of initial particles are randomly generated within the range of parameters, and the prediction error of SVM is chosen as the fitness value;

(3) Update the individual fitness values and global fitness values of particles;

(4) Update the position and speed of particles;

(5) Determine if the termination condition is satisfied. If meeting the accuracy requirements or achieving the maximum number of iterations, end the operation, otherwise, continue to perform. Fig. 2 is the flow chart of PSO-SVM fault diagnosis.

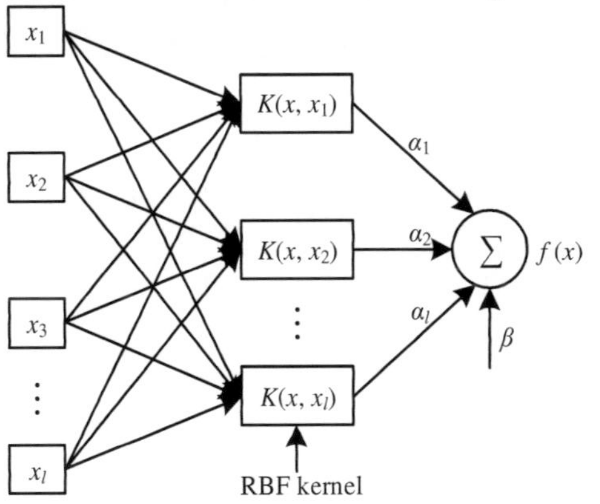

Figure 1. Structure diagram of RBF-SVR model.

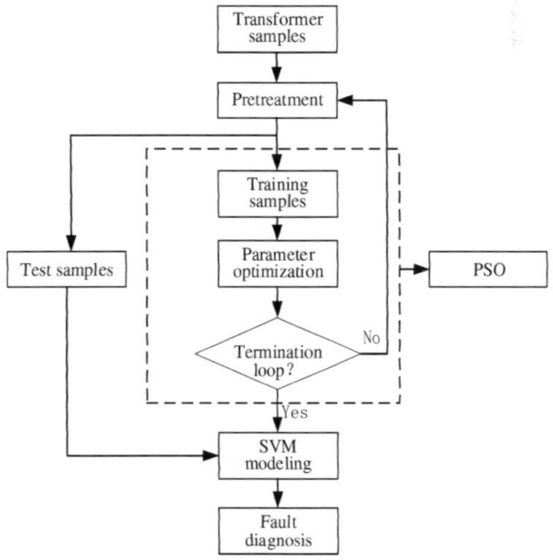

Figure 2. Flow chart of PSO-SVM fault diagnosis.

IOP Conf. Series: Materials Science and Engineering **339** (2018) 012001 doi:10.1088/1757-899X/339/1/012001

4. Fault Diagnosis of Power Transformers Based on PSOSVM

This paper collects DGA data from IEC TC 10. The data contains 118 samples, which matched with five diagnosis types, i.e. low-energy discharge, high-energy discharge, thermal fault of low and medium temperature, thermal fault of high temperature and normal condition.

The 118 samples were randomly divided into the training sample set containing 90 samples and the test sample set containing 28 samples. The input of the classifier is the three ratios of C_2H_2/C_2H_4, CH_4/H_2 and C_2H_4/C_2H_6 gas ratios. All input data were normalized in the range of [0 1] before training to improve the generalization performance of SVM.

The target of PSO is to realize the process of the optimization of the parameters. When the candidate parameters C and γ are set, they vary in the range of two fixed ranges, $[10^0\ 10^3]$ and $[10^{-2}\ 10^2]$, respectively. The parameters of the PSO algorithm used in this paper are as follows: the number of populations and the number of initial particles were fixed to 30 and 8, respectively; the dimension of the optimized function was set to 2; the maximum number of generations was 10.

The Variation curves of PSO fitness are shown in Fig. 3. At last, the optimal parameters of the multi class support vector machine classifier based on PSO algorithm are $C = 504.0727$, $\gamma = 4.2139$.

According to the selected two parameters, the OAO-nonlinear multiple classification SVM classifier is trained and applied for fault diagnosis of transformers. Fig. 4 shows the testing results of 28 samples for fault diagnosis. Apparently, only 4 samples were classified incorrectly and the classification accuracy of test samples hit 85.71% (24/28). The possible reason of the error may be that the four cases are bad data with measuring error.

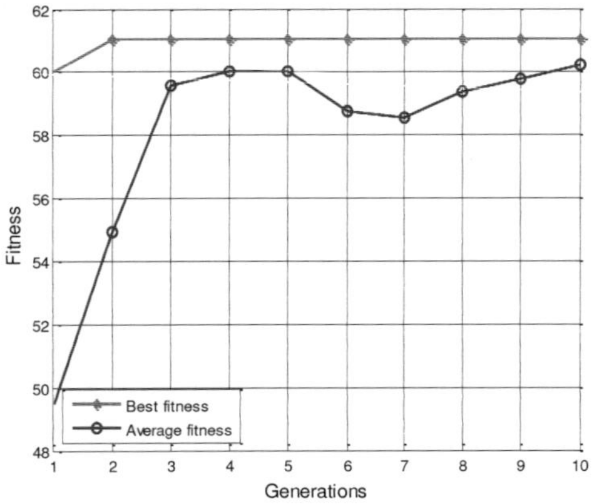

Figure 3. Variation curves of PSO fitness.

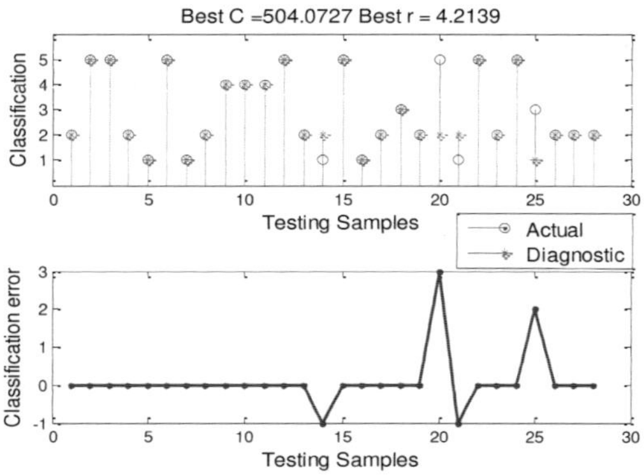

Figure 4. Classification accuracy of testing samples.

In order to compare with other fault diagnosis methods, the standard SVM, GA-SVM and the method of this paper are used to do the fault diagnosis experiments respectively.

Table 1. Cases Studies with Different Approches

Methods	Accuracy	
	Training	Testing
SVM	72.22%	57.14%
GA-SVM	80.00%	60.71%
This paper	81.15%	85.71%

In order to compare with other fault diagnosis methods, the standard SVM, particle swarm optimization support vector machines (SVM) and the method of this paper are used to do the fault diagnosis experiments respectively.

5. Conclusion

A novel method combing particle swarm optimization with SVM for fault diagnosis of oil-immersed transformers is proposed in this paper. It is shown that the proposed method can provide the high accuracy fault diagnosis model of the transformer oil-dissolved gases on the basis of the field test results. The method could be served as an effective tool for the assessments of transformer condition.
Conclusions can be summarized as follows:
Three classification approaches were studied for fault diagnosis of transformers. The classification accuracy of the three approaches, including SVM, GASVM and PSOSVM, reaches 57.14%, 60.71%, 85.71%, respectively. The comparing results demonstrate that the proposed PSOSVM approach has better performance than the other two approaches.

References

[1] Liu X, Liao R J, Lv Y 2015 Study on Influences and Elimination of Test Temperature on PDC Characteristic Spectroscopy of Oil-Paper Insulation System *J. Journal of Electrical Engineering & Technology* vol 10 no 3 pp 1107-1113
[2] Shon C H, Yi S H, Lee H J 2012 Study on the Transfer Functions for Detecting Windings Displacement of Power Transformers with Impulse Method *J. Journal of Electrical Engineering & Technology* vol 7 no 6 pp 876-883
[3] Kweon D J, Koo K S, Woo J W, Kwak J S 2012 A Study on the Hot Spot Temperature in 154kV Power Transformers *J. Journal of Electrical Engineering & Technology*vol 7 no 3 pp 312-319
[4] Khan S, Equbal M D, Islam T 2015 A Comprehensive Comparative Study of DGA Based Transformer Fault Diagnosis Using Fuzzy Logic and ANFIS Models *J. IEEE Transactions on Dielectrics and Electrical Insulation* vol 22 no 1 pp 590-596
[5] Aghaei J, Gholami A, Shayanfar H A 2010 Dissolved gas analysis of transformers using fuzzy logic approach *J. European Transactions on Electrical Power* vol 20 no 5 pp 630-638
[6] Wu H Y, Hsu C Y, Lee T F 2009 Improved SVM and ANN in incipient fault diagnosis of power transformers using clonal selection algorithms *J. International Journal of Innovative Computing Information and Control* vol 5 no 7 pp 1959-1974
[7] David M S, Dictino C, Jesus M C 2013 Identification of a surface marine vessel using LS-SVM *J. Journal of Applied Mathematics* vol 2013 pp 1–11
[8] Bazi Y and Melgani F 2006 Toward an optimal SVM classification system for hyperspectral remote sensing images *J. IEEE Transactions on Geoscience and Remote Sensing* vol 44 no 11 pp 3374–3385
[9] Kennedy J and Eberhart R 2002 Particle swarm optimization *IEEE International Conference on Neural Networks* vol 4 pp 1942-1948
[10] *EC Publication 599* Interpretation for the Analysis of Gases in transformers and other Oil-Filled Electrical Equipment in Service
[11] Hsu C W, Lin C J 2002 A comparison of methods for multiclass support vector machines *J. IEEE Transactions on Neural Networks* vol 13 no 1 pp 415–425
[12] Qie H, Wang L, Liu B 2007 Parameter estimation for chaotic systems by particle swarm optimization *J. Chaos Solitons & Fractals 34.2* pp 654-661
[13] Zheng H Q, Zhang S L, Sun X Y 2017 Classification recognition of anchor rod based on PSO-SVM *Control And Decision Conference (CCDC), 2017 29th Chinese*

Arrester Resistive Current Measuring System Based on Heterogeneous Network

Yun Hua Zhang[1,2], Zai Lin Li[1,2], Feng Yuan[1,2], Feng Hou Pan[1,2], Zhan Nan Guo[1,2] and Yue Han[1]

[1] Research and Development Department, State Grid Liaoning Electric Power Co., Ltd. Electric Power Research Institute, Heping District, Shenyang 110000, China
[2] Research and Development Department, Shenyang Kekai Electric Power Technology Co., Ltd., Hunnan District, Shenyang 110000, China

Email: zhangyunhua@dbdky.com

Abstract. Metal Oxide Arrester (MOA) suffers from aging and poor insulation due to long-term impulse voltage and environmental impact, and the value and variation tendency of resistive current can reflect the health conditions of MOA. The common wired MOA detection need to use long cables, which is complicated to operate, and that wireless measurement methods are facing the problems of poor data synchronization and instability. Therefore a novel synchronous measurement system of arrester current resistive based on heterogeneous network is proposed, which simplifies the calculation process and improves synchronization, accuracy and stability and of the measuring system. This system combines LoRa wireless network, high speed wireless personal area network and the process layer communication, and realizes the detection of arrester working condition. Field test data shows that the system has the characteristics of high accuracy, strong anti-interference ability and good synchronization, which plays an important role in ensuring the stable operation of the power grid.

1. Introduction

The metal Oxide arrester (MOA) is widely used in power system, and its condition under the continuous voltage plays a vital role in safe operation of power grid system. After a long term operation, the arrester will appear the problems of internal damp, surface contamination, resistors aging and insulation deterioration, resulting in the lightning protection performance of the arrester become worse [1]. As the resistive current can reflect the operation of the arrester, the resistive current detection is listed as a standard experimental project.

At present, the common method of MOA resistive current measurement requires to layout long cables, which has increased the workload and insecurity of field testers. And the wireless method can simplify the complexity of field wiring and easy to operate[2-4], however the anti-interference ability of ordinary wireless communication is poor[5]. In view of the above problems, combined with the existing process layer communication network[6] that confirming to I0 interface specification in the substation, and based on LoRa wireless network[7], high speed wireless personal area network and heterogeneous network[8-9], this paper has done some research, and then a novel synchronous measurement has been designed, through the wireless communication in two frequency bands, which not only could solve the field complex wiring problems and simplify the calculation process, but also could greatly improve the synchronization performance and the measurement accuracy of the system.

2. Measurement method of arrester resistive current based on heterogeneous network

The heterogeneous network[10-11] is a new network composed of different manufacturers of computers, network equipment and systems, in order to meet the diversity of terminal services, different types of networks are integrated[12]. In this system, the heterogeneous network mainly refers to high speed wireless personal area network and LoRa wireless network. And the full name of LoRa is “Long Range”, and it mainly aims at internet of things, and is applied to battery powered wireless equipment. LoRa signals are also powerful for building penetration and large scale internet of things deployment, and the high speed wireless personal area network is used at the end of the whole network chain to realize the communication among different terminals.

Based on the advantages of the heterogeneous network, the arrester resistive current detection system has been designed. The smallest cell of the system is shown in figure 1, which uses the voltage and current measuring device respectively to collect the secondary voltage signal and the leakage current signal of the arrester on the same phase[13], then voltage and current signals are handled by using the Hanning window with FFT, which decreases the time of calculation, and the simplified window function is shown as formula (1).And then the handled data is transmitted to the gateway.

$$\omega(n)=1/2-1/2\cos(2\pi n/N),\ n=0,\ 1,\ \ldots,\ N-1 \tag{1}$$

The gateway is used to receive voltage and current signals of A, B and C three phase arresters, and analyze and calculate the measured signals, and obtain the resistive current value of three phase arresters. The system can be used as both on line monitoring equipment and charged detection equipment in the substation. When used as charged detection equipment, it is necessary to configure with handheld device in LoRa wireless network. And when the system is used as on line monitoring device, data is interacted with IED through the LoRa wireless network or the process layer communication network, the overall structure of the system is shown in figure 2.

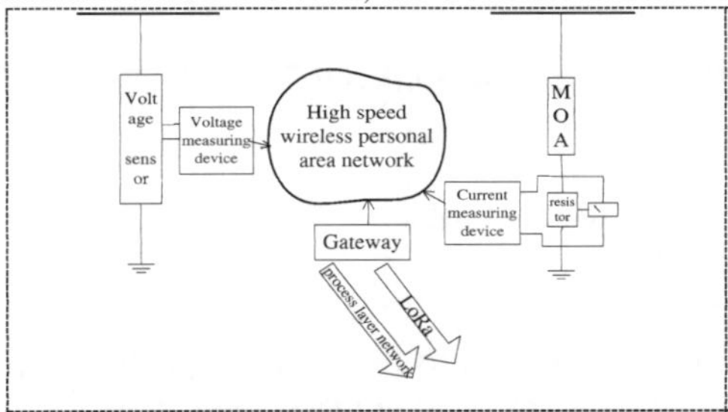

Figure 1. Single phase arrester resistive current detection system.

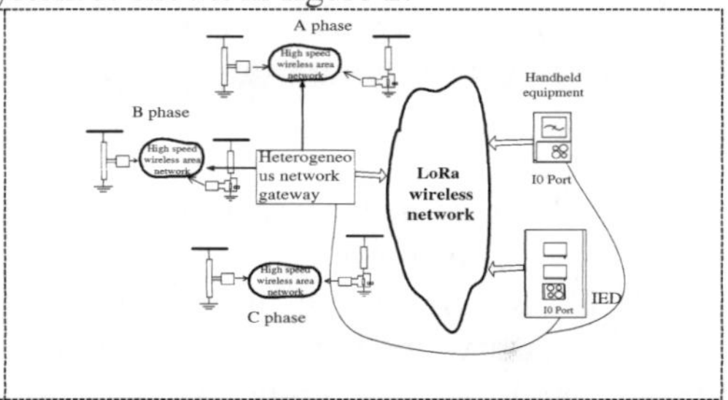

Figure 2. The overall structure of detection system of three phase arresters.

The heterogeneous network gateway can communicate well with the field handheld device and the IED through LoRa network, mainly relying on the characteristics of high reliability, strong penetration, long distance and low power consumption. As the flexibility and reconfiguration of heterogeneous network, the system can be extended to the on line monitoring and charged detection of all arresters or other equipment in the station. The LoRa network brings together information of all kinds of arresters at different voltage levels, and carries out the comprehensive analysis to get the health status of each arrester, and finally realize the automation and intelligentization of power grid[14].

3. System design and implementation

The heterogeneous network transmission architecture of this system mainly adopts the idea of layered structure design, the network structure layer is divided into three parts: high speed wireless personal area network layer, LoRa wireless network layer and the process layer network, and the data in different network layers can be exchanged through the data information processing and conversion unit of the heterogeneous network gateway, and the network block diagram of resistive current measurement system for MOA based on heterogeneous network is shown in figure 3. As a data transmission layer of current and voltage of three phase arresters for A, B and C, the high speed wireless personal area network is used to transmit the data of arrester collected by acquisition terminal to heterogeneous network gateway, and after calculated and analyzed by the gateway, the data is transmitted to handheld device and IED through LoRa wireless network and the process layer network.

3.1. Implementation principle of the system

The data acquisition unit includes A, B, C three phase voltage acquisition devices and A, B, C three phase current acquisition devices, as shown in figure 4, a simplified block diagram of a single phase current or voltage data acquisition device is presented, which mainly includes current and voltage sensors, analog to digital conversion unit, central processing unit, high speed wireless personal area network, non-volatile memory, etc.

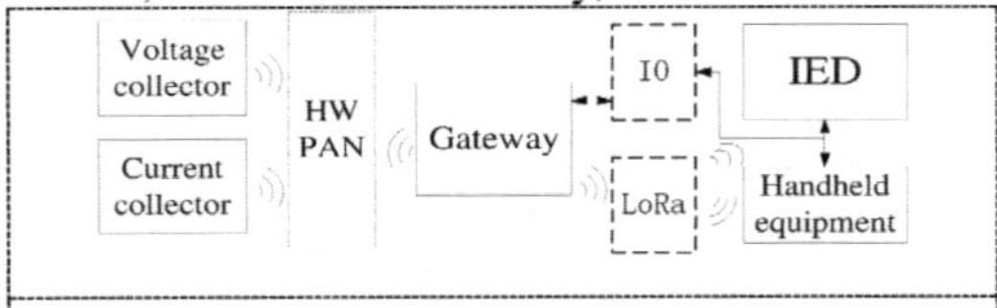

Figure 3. Block diagram of network structure of resistive current measuring instrument.

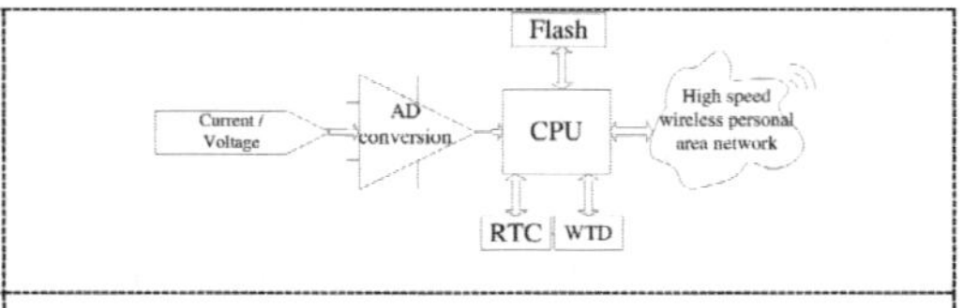

Figure 4. The simplified block diagram of data acquisition device.

The current sensor is used to measure the leakage current of arrester; and the voltage sensor is used to measure the voltage of the secondary side of the voltage transformer; the central processor unit is integrated with the DSP kernel to calculate and analyze the data; the analog to digital conversion unit uses a 16 bit high precision analog to digital conversion chip; the high speed wireless personal area network adopts 2.4G wireless module for communication, which has the advantages of high communication rate and strong anti-interference ability; the nonvolatile memory is used to store dynamic configuration instructions and measurement data.

The heterogeneous network gateway mainly includes high speed wireless person area network, central processing unit, LoRa wireless network, process layer communication and nonvolatile memory, as shown in figure 5. LoRa wireless network has the characteristics of long distance wireless transmission, low power consumption, long battery life, multi node and low cost, which provides great convenience for field measurement of arrester; the process layer communication can be achieved by Ethernet or 485 field bus for data communication with IED.

Handheld equipment mainly includes LoRa wireless network, process layer, central processing unit, liquid crystal display and nonvolatile memory as shown in figure 6. The central processing unit adopts 800M industrial level microprocessor AM3359, and the large size LCD provides a good man-computer interaction function to store and display the resistive current data of the arrester.

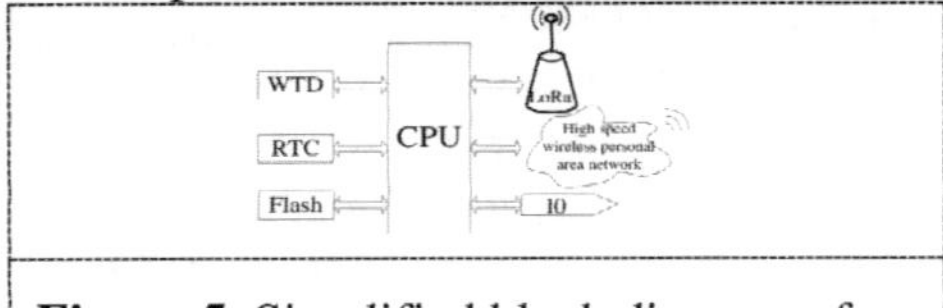

Figure 5. Simplified block diagram of gateway.

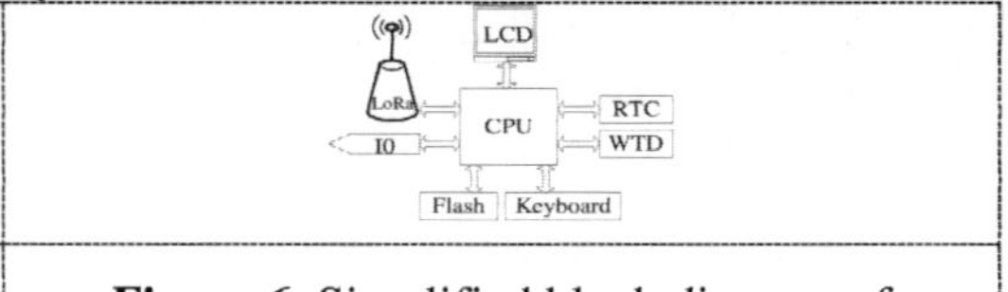

Figure 6. Simplified block diagram of handheld device.

3.2. System function implementation

The system work flow chart is shown in figure 7. The arrester current measurement system mainly includes four parts: arrester voltage and current collector, heterogeneous network gateway, handheld device and IED. The collector is used to measure the resistive current of the arrester and collect the voltage signal of the voltage transformer synchronously, and then send the data to heterogeneous network gateway; and the gateway is used to receive the data collected from the collector, then the data is analyzed and calculated, and the calculated results are passed to handheld and IED through LoRa wireless network and process layer network to complete a data collection, communication and display.

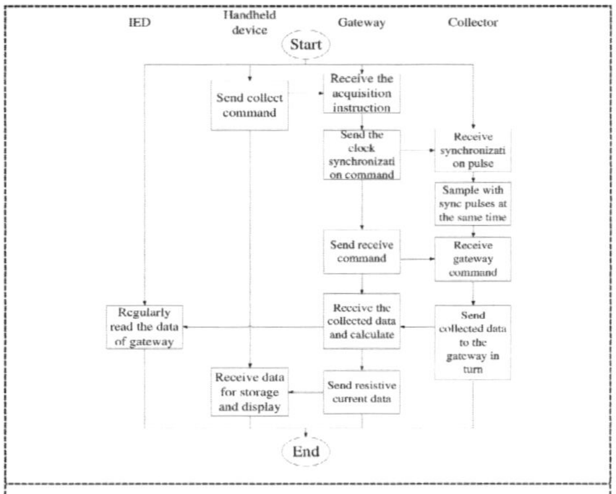

Figure 7. System workflow diagram.

4. System Verification

First of all, the accuracy test of the system was carried out by using the DK-34B1 three phase current source in the laboratory, when the angular phase difference Φ between the fundamental voltage and the fundamental full current was set to 85 °, the measurement results of different full current values were shown in table 1, and the unit of all data is mA in this table. Then the system was verified at 220kV 50HZ test field in Shenyang tiger stone station. The packet loss rate with different communication distance was tested.

Table 1. The variation of resistive current of arrester at different full current.

Set the full current rms	Theoretical resistive peak current	Measured value 1	Measured value 2	Measured value 3	Measured value 4
0.1	0.012	0.012	0.012	0.012	0.012
1.0	0.123	0.123	0.124	0.122	0.123
4.0	0.492	0.494	0.495	0.496	0.495
6.7	0.826	0.828	0.829	0.826	0.827
10	1.232	1.243	1.242	1.239	1.241

The MOA related data was transmitted with 1000 packets once, the result was shown in table 2.

Table2. The quality of network communication varies with distance.

Number of group	Distance (Km)	Receive packets	Packet loss rate
1	0.2	1000	0.0%
2	0.5	999	0.1%
3	1.2	995	0.5%
4	2.5	987	1.3%

The three phase arresters of the 220kV 17L group was measured by this system at the same time, the measurement result was shown in table 3.

Table 3. The actual measurement situation of the field.

Test items	A phase	B phase	C phase
Voltage RMS (V)	236.19	236.19	236.19
Three harmonic (%)	0.72	0.7	0.71
Five harmonic (%)	0.12	0.13	0.12
Seven harmonic (%)	0.04	0.03	0.04
Full current(mA)	0.586	0.596	0.59.
Resistance current fundamental peak (mA)	0.119	0.073	0.088
Phase angle difference (°)	81.7	85	83.9

5. References

[1] Fan M, Xie J, Jiang F C and Wang J 2017 *Insulators and Surge Arresters*. **2017**(2):64-68
[2] Yao X T, Wang J R and Jiang Z J 2016 *High Voltage Apparatus*. **2016**(11):95-101
[3] Duan D P, Jiang X C, Sun C X, Sheng G H and ZengY 2008 *Transactions of China Electrotechnical Society*. **23**(7):56–61
[4] Zhou S B, Liang W M and Yong M C 2017 *Insulators and Surge Arresters*. **2017**(2):44-48
[5] Song H, Sheng G H and Wang C Q 2012 *High Voltage Apparatus*. **48**(5):49–55
[6] Gao Q, Cheng D W and Geng B H 2013 *Electric Power Information and Communication Technology*. **11**(10):11–15
[7] Sun M, Zhang N Q, Jin L B and Yu S B 2016 *Video Engineering*. **40**(10):77-81
[8] Liu Shen, Shao Zhen 2016 *Telecommunications Technology*. **1**(5):43–46
[9] Han W, Wu C H and Gao W 2016 *Power System and Clean Energy*. **32**(11):43-50
[10] Tang Q W and Li B 2013 *Computer Engineering and Design*. **34**(6):1906–11
[11] Wei X H and Wang H 2015 *Journal of System Simulation*. **27**(7):1541–47
[12] Teng S H, Jiang Z J and Chen J J 2016 *Microeletronics & Computer*. **2016**(10):78-81
[13] Wang S, Niu B, Guo J, Hu P F and Li Z Z 2013 *Power System Technology*. **37**(1):201–5
[14] Zhu B S, Song Y and Wei X L 2012 *Proceedings of the CSEE*. **32**(16):165–172

IOP Conf. Series: Materials Science and Engineering **339** (2018) 012003 doi:10.1088/1757-899X/339/1/012003

High-accuracy resolver-to-digital conversion via phase locked loop based on PID controller

Yaoling Li and Zhong Wu

School of Instrumentation Science and Opto-electronics Engineering, Beihang University, Beijing 100191, China

Email: liyaoling@buaa.edu.cn

Abstract. The problem of resolver-to-digital conversion (RDC) is transformed into the problem of angle tracking control, and a phase locked loop (PLL) method based on PID controller is proposed in this paper. This controller comprises a typical PI controller plus an incomplete differential which can avoid the amplification of higher-frequency noise components by filtering the phase detection error with a low-pass filter. Compared with conventional ones, the proposed PLL method makes the converter a system of type III and thus the conversion accuracy can be improved. Experimental results demonstrate the effectiveness of the proposed method.

1. Introduction

Shaft sensors are often used to obtain the information of angular position and velocity for servo control systems [1]. Compared with the other shaft sensors, resolver is widely used owing to its superior properties in ruggedness, reliability and high accuracy [2]. However, resolvers only can output amplitude-modulated orthogonal pair signals containing angular position information. High-accuracy resolver-to-digital converters (RDCs) should be developed to obtain the information of angular position and velocity for servo applications.

The most convenient solution to RDC is the special integrated circuit (IC), such as AD2S80A, AD2S1210. RDC ICs can offer simple interfaces between resolvers and computers, but the bandwidth of RDC ICs can not be adjusted flexibly as it depends on the external resistances and capacitances.

Software-based RDC can avoid the problems of RDC ICs. Software-based RDC methods can be classified into two kinds: arctangent method and phase locked loop (PLL) method [3][4][5][6]. Arctangent method is the most straightforward one, but the angular speed is usually derived by differential operation which will amplify higher-frequency noise. PLL method is essentially a closed-loop system and has been widely used in RDC ICs and software-based RDC. Most of RDC ICs adopt type II tracking loop as PLL structure, and it has been extended to software-based RDC. Besides, angle tracking observer (ATO) proposed in [7] is also a PLL system. Both type II tracking loop and ATO have better performance in disturbance attenuation. However, the conversion accuracy is still restricted by the system structure of conventional PLL methods.

In order to improve RDC accuracy, the problem of RDC is transformed into the problem of angle tracking control, and a PLL method based on PID controller is proposed for software-based RDC. This controller consists of a typical PI controller and an incomplete differential with a low-pass filter. Since the low-pass filter can remove higher-frequency noise components from signals, the incomplete differential often be implemented in engineering. The proposed PLL method makes RDC become a system of type III so that the demodulation accuracy can be improved.

2. Review of typical PLL-based RDCs

The ideal outputs of resolver are two orthogonal amplitude-modulated signals [8]. In software-based RDCs, the envelope detection is always implemented by the peripheral hardware circuits to relieve the pressure on sampling. The envelopes of resolver after detection can be written as:

$$y_s = \sin\theta,\ y_c = \cos\theta \tag{1}$$

The information of shaft angle and speed can be obtained from (1) by PLL-based RDCs [9]. Figure 1(a) shows the typical structure of a PLL-based RDC which includes a phase detector (PD), a low-pass filter (LPF), and a voltage-controlled oscillator(VCO). PD is in charge of detecting the phase error $\tilde{\theta}$ between the shaft angle θ and its estimate $\hat{\theta}$. The phase error $\tilde{\theta}$ is given in a nonlinear form as:

$$\varepsilon = \sin\theta\cos\hat{\theta} - \cos\theta\sin\hat{\theta} = \sin(\theta - \hat{\theta}) = \sin\tilde{\theta} \tag{2}$$

For very small $\tilde{\theta}$, $\sin\tilde{\theta} \approx \tilde{\theta}$. Therefore, figure 1(a) can be simplified into figure 1(b).

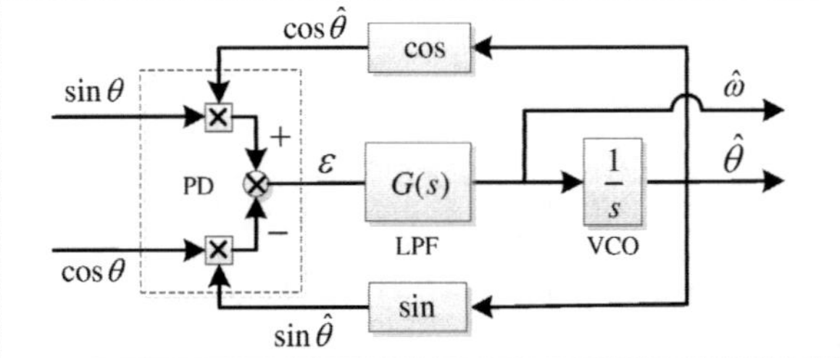

(a) Actual form.

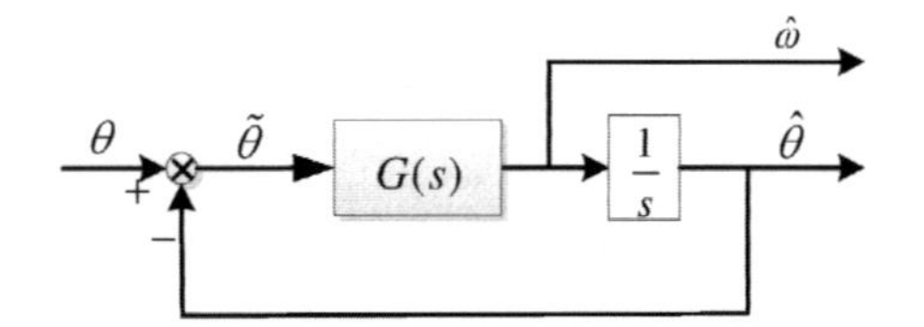

(b) Simplified form.

Figure 1. Block diagram of conventional PLL based RDC.

Type II tracking loop method and angle tracking observer (ATO) method are two main kinds of PLL-based RDCs. Their performance will be analyzed as follows.

2.1. Type II tracking loop based RDC

According to the datasheet of AD2S1210 [10], AD2S1210 is a type II tracking loop based PLL in fact. Two error transfer functions concerning shaft angle and speed can be derived as:

$$\Phi^1_{\tilde{\theta}/\theta}(s) = \frac{\tilde{\theta}(s)}{\theta(s)} = \frac{\tau_2 s^3 + s^2}{\tau_2 s^3 + s^2 + k_a\tau_1 s + k_a},\ \Phi^1_{\tilde{\omega}/\omega}(s) = \frac{\tilde{\omega}(s)}{\omega(s)} = \frac{\tau_2 s^3 + s^2}{\tau_2 s^3 + s^2 + k_a\tau_1 s + k_a} \tag{3}$$

where $k_a > 0$ is the open-loop gain, and $\tau_1 > \tau_2$ are compensation time constants.

When the resolver rotates at a constant speed of C rad/s, from the final-value theorem, the steady state errors of shaft angle and speed estimation are both equal to 0. And when the resolver rotates at an acceleration of A rad/s^2, from the final-value theorem, the steady state error of shaft speed estimation is also equal to 0, but the estimate of shaft angle contains steady state error which is equal to A / K_a.

2.2. ATO based RDC

ATO method based on nonlinear state observer for RDC has been proposed in [11]. Two error transfer functions concerning shaft angle and speed also can be obtained as:

$$\Phi^2_{\tilde{\theta}/\theta}(s) = \frac{\tilde{\theta}(s)}{\theta(s)} = \frac{s^2}{s^2 + k_\theta s + k_\omega},\ \Phi^2_{\tilde{\omega}/\omega}(s) = \frac{\tilde{\omega}(s)}{\omega(s)} = \frac{s^2 + k_\theta s}{s^2 + k_\theta s + k_\omega} \tag{4}$$

where k_θ, k_ω are positive observer gains.

Similarly, when the resolver rotates at a constant speed of C rad/s, from the final-value theorem, the steady state error of shaft angle and speed estimation are equal to 0. When the resolver rotates at a constant acceleration of A rad/s^2, the estimates of shaft angle and speed contain steady state errors which are equal to A / k_ω and Ak_θ / k_ω, respectively.

3. Improved PID controller for RDC

To improve the conversion accuracy of typical PLL based RDC, the problem of RDC is transformed into the problem of angular tracking control and a PID controller based RDC is designed as follows.

3.1. Design of PID controller based RDC

Define an extended state as $\theta_I = \int \theta dt$, then an augmented system for RDC can be constructed as:

$$\dot{\hat{\theta}}_I = \hat{\theta},\ \dot{\hat{\theta}} = \hat{\omega},\dot{\hat{\omega}} = u \tag{5}$$

The problem of RDC is to design the control u subject to $\hat{\theta} \to \theta$ as $t \to \infty$. Actually, an error state feedback can accomplish the above aim. We can design the controller as:

$$u = K_P\tilde{\theta} + K_I\tilde{\theta}_I + K_D\tilde{\omega} \tag{6}$$

where K_P , K_I , K_D are positive controller gains, $\tilde{\theta}_I = \theta_I - \hat{\theta}_I$, $\tilde{\theta} = \theta - \hat{\theta}$,$\tilde{\omega} = \omega - \hat{\omega}$. It is noted that system (5) is controllable, the poles of the closed-loop system can be configured by choosing suitable controller gains. Therefore, the demodulation error $\tilde{\theta}$ can be asymptotically bounded owing to the uncertainties in acceleration. Substituting (6) into (5) gives the PID form of RDC as:

$$\dot{\hat{\theta}}_I = \hat{\theta},\ \dot{\hat{\theta}} = \hat{\omega},\ \dot{\hat{\omega}} = K_P\tilde{\theta} + K_I\tilde{\theta}_I + K_D\tilde{\omega} \tag{7}$$

3.2. Realization of PID based RDC

To minimize the influence of the differential computation in (6), a low-pass filter $f(s) = 1/(1+Ts)$ can be used to realize an incomplete differential, as shown in Figure 2(a). Considering the phase detection, the actual form of the new RDC can be implemented as shown in Figure 2(b).

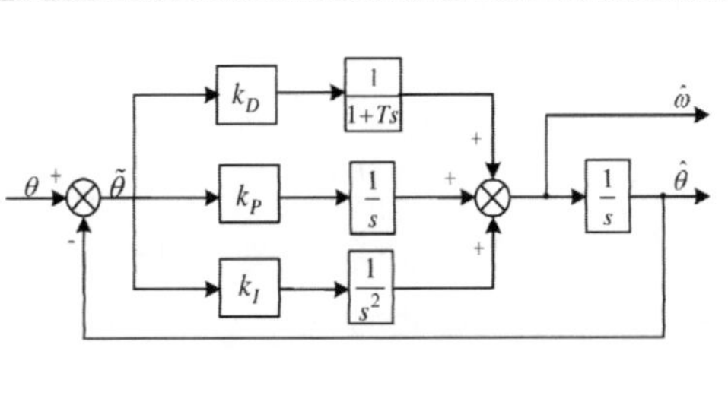

(a) Simplified form.

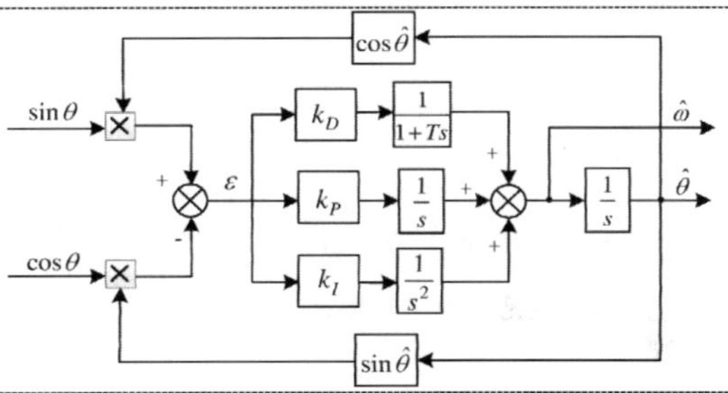

(b) Actual form.

Figure 2. Block diagram of Improved PID Controller for RDC.

According to Figure 2(a), the error transfer functions concerning shaft angle and speed of the improved PID controller based RDC can be written as:

$$\Phi^3_{\tilde{\theta}/\theta}(s) = \frac{\tilde{\theta}(s)}{\theta(s)} = \frac{Ts^4 + s^3}{Ts^4 + s^3 + (k_D + k_PT)s^2 + (k_P + k_IT)s + k_I} \tag{8}$$

$$\Phi^3_{\tilde{\omega}/\omega}(s) = \frac{\tilde{\omega}(s)}{\omega(s)} = \frac{Ts^4 + s^3}{Ts^4 + s^3 + (k_D + k_PT)s^2 + (k_P + k_IT)s + k_I} \tag{9}$$

According to the error transfer functions (8) and (9), when the resolver rotates at a constant speed of C rad/s, from the final-value theorem, the steady state errors of shaft angle and speed estimation are both equal to 0. When the resolver rotates at a constant acceleration of A rad/s^2, the steady state error of shaft angle and speed estimation are also equal to 0.

From the above analysis, as the proposed PLL method makes RDC become a system of type III, it can be seen that the conversion accuracy can be improved by the proposed PLL method compared with the typical PLL method.

4. Experimental results

To evaluate the effectiveness of the proposed RDC, experiments have been carried out on a platform including a resolver simulator and a DSP-based detection board. By using the resolver simulator, the actual values of the shaft angle and speed can be preset which is convenient to evaluate the performance of the proposed RDC by comparing the estimated values with the preset actual ones.

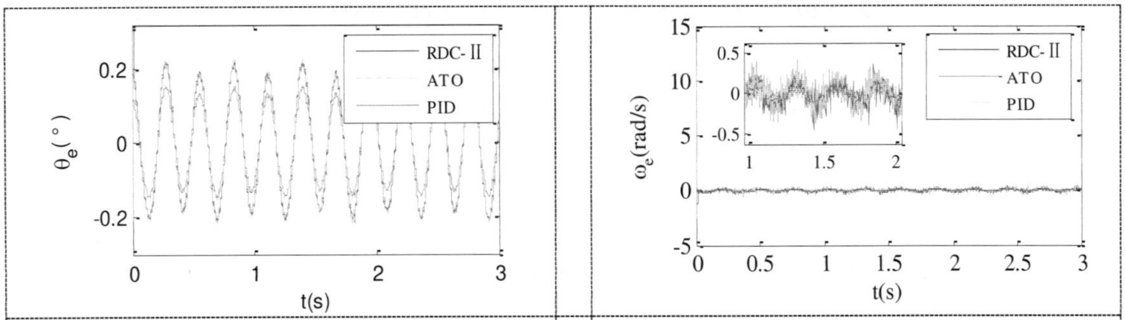

(**a**) Shaft angle estimation errors. (**b**) Shaft speed estimation errors.

Figure 3. Shaft angle and speed estimation errors at a constant speed of 4π rad/s.

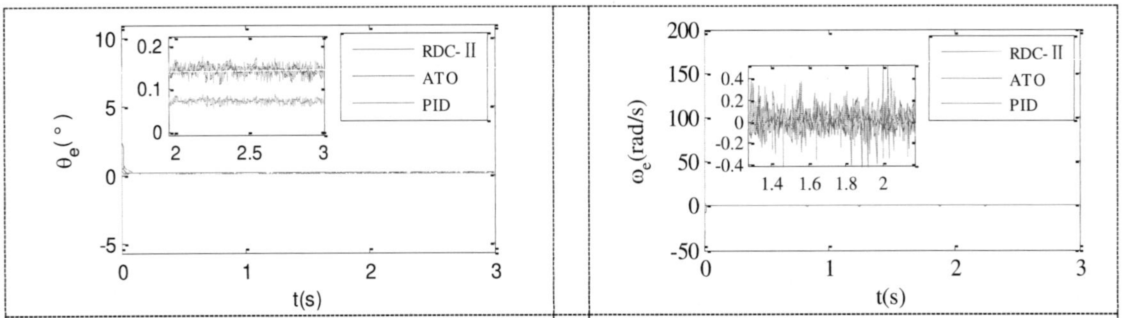

(**a**) Shaft angle estimation errors. (**b**) Shaft speed estimation errors.

Figure 4. Shaft angle and speed estimation errors at a constant acceleration of 4π rad/s^2.

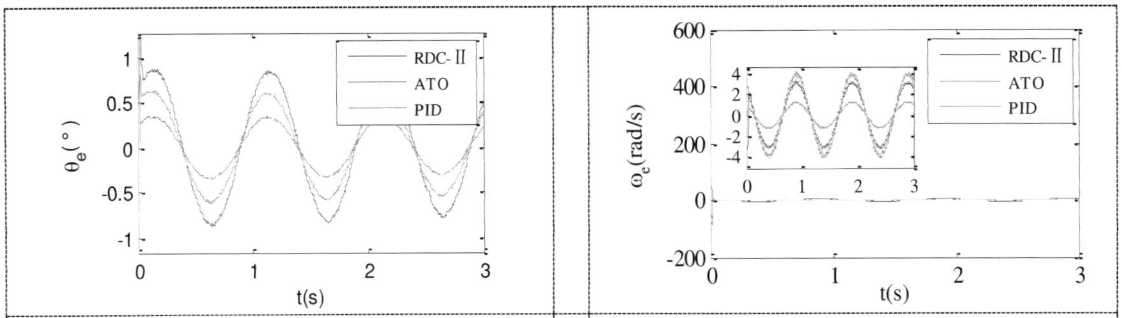

(**a**) Shaft angle estimation errors. (**b**) Shaft speed estimation errors.

Figure 5. Shaft angle and speed estimation errors at a sine speed of $(8\pi + 4\pi \sin 2\pi t)$ rad/s.

To compare the performance of type II tracking loop, ATO and the proposed method, the parameters of ATO and the proposed method are adjusted according to the bandwidth of shaft speed given in [10]. Thus, the parameters of type II tracking loop are $k_a = 92.7\times10^3$, $t_1 = 8\times10^{-3}$ s, $t_2 = 728\times10^{-6}$ s; the parameters of ATO are chosen to be $k_\theta = 1200$, $k_\omega = 1000000$; the parameters of the proposed method are adjusted as $k_p = 180000$, $k_D = 700$, $k_I = 3000000$, $T = 7\times10^{-4}$ s. By using the above parameters of three methods, experimental results in three cases can be obtained, as shown in Figures (3), (4), and (5). It needs to be explained that type II tracking loop, ATO and the proposed method are denoted as RDC-II, ATO, and PID in the figures respectively.

Case 1: When the resolver rotates at a constant speed of 4π rad/s, it can be seen from figure 3(a) that the estimation error of shaft angle by the proposed method is about 0.18° which is smaller than RDC-II and ATO method whose shaft angle errors are both about 0.21°. And from figure 3(b), the performance of the shaft speed estimation by the proposed method is also much better than that by the other two methods.

Case 2: When the resolver rotates at a constant acceleration of 4π rad/ s^2, as we can see from figure 4(a), the shaft angle estimation errors of these three methods both contain steady state errors, while the shaft angle steady state errors of improved PID method is about 0.08° which is much smaller than that of the other two methods whose shaft angle errors are both about 0.16°. Besides, from figure 4(b), it is clear that the shaft speed error of improved PID method is also much smaller than that of RDC-II and ATO method.

Case 3: When the resolver rotates at a sine speed of $(8\pi + 4\pi\sin 2\pi t)$ rad/s, from figure 5(a), it can be seen that the shaft angle error of RDC-II method is slightly larger than that of ATO method and improved PID method. But from figure 5(b), the shaft speed errors of improved PID method are much smaller than that of RDC-II and ATO method.

To sum up, as the proposed PLL method makes RDC become a system of type III, the conversion accuracy can be improved by the proposed PLL method compared with the typical PLL method.

5. Conclusion

In this paper, a PLL method based on improved PID controller has been designed to estimate the shaft angle and speed of a resolver. This method makes RDC become a system of type III, thus the conversion accuracy can be improved. Theoretical analysis and experimental results show that the proposed method has higher steady-state accuracy compared with the typical ones.

References

[1] Hou B, Zhou B, Song M L, Lin Z H and Zhong R 2016 A Novel Single-Excitation Capacitive Angular Position Sensor Design *Sensors.* **16** 1196

[2] Hou C C, Chiang Y H and Lo C P 2014 DSP- Based Resolver- to- Digital Conversion System Designed in Time Domain *Power Electron.* **7** 2227

[3] Khaburi D A 2012 Software-Based Resolver-to-Digital Converter for DSP-Based Drives Using an Improved Angle-Tracking Observer *IEEE Trans Instrum. Meas.* **61** 922

[4] Bergas J J, Ferrater S C, Gross G, Ramyrez P R and Galceran A S 2012 High-Accuracy All-Digital Resolver-to-Digital Conversion *IEEE Trans Ind Electron.* **59** 326

[5] Hoseinnezhad R and Harding P 2005 A Novel Hybrid Angle Tracking Observer for Resolver to Digital Conversion *Proc. Int. IEEE Conf. Decis Control Eur Control.* (Seville, Spain) p 7020

[6] Park S and Kim K C 2014 Study on the Optimal Design of a Novel Slotless Resolver by FEM *IEEE Trans. Magnetics* **50** 1

[7] Yim C H, Ha I J and Ko M S 1992 A Resolver-to-Digital Conversion Method for Fast Tracking *IEEE Trans. Ind. Electron* **39** 369

[8] Aung N L H, Chao B, Mamun A A, Cheng S S and Yu Y Q 2013 A demodulation technique for spindle rotor position detection with resolver *IEEE Trans. Magnetics* **49** 2614

[9] Sivappagari C M R and Konduru N R 2014 Modified ATO algorithm based high accuracy RDC using pulse excitation *Autom. Control, Energy and Syst.* 1

[10] Variable Resolution, 10-Bit to 16-Bit R/D Converter with Reference Oscillator, AD2S1210 [Internet] 2008 Available from: www.analog.com

[11] Zhang J and Wu Z 2017 Composite state observer for resolver-to-digital conversion *Meas. Sci. & Technol* **28(6)** 065103

IOP Conf. Series: Materials Science and Engineering **339** (2018) 012004 doi:10.1088/1757-899X/339/1/012004

A simple approach to detect and correct signal faults of Hall position sensors for brushless DC motors at steady speed

Yongli Shi[1], Zhong Wu[2], Kangyi Zhi [1] and Jun Xiong[1]

[1]Beijing Institute of Control Engineering, China Academy of Space Science, Beijing 100190, China
[2]School of Instrumentation Science and Optoelectronics Engineering, Beihang University, Beijing 100191, China

Email: yonglishi@163.com

Abstract. In order to realize reliable commutation of brushless DC motors (BLDCMs), a simple approach is proposed to detect and correct signal faults of Hall position sensors in this paper. First, the time instant of the next jumping edge for Hall signals is predicted by using prior information of pulse intervals in the last electrical period. Considering the possible errors between the predicted instant and the real one, a confidence interval is set by using the predicted value and a suitable tolerance for the next pulse edge. According to the relationship between the real pulse edge and the confidence interval, Hall signals can be judged and the signal faults can be corrected. Experimental results of a BLDCM at steady speed demonstrate the effectiveness of the approach.

1. Introduction

Brushless DC motors (BLDCMs) have been used in many fields from aerospace, military, industrial to domestic applications owing to their advantages on smaller size, higher efficiency, simpler structure, easier control, and higher performance[1]. Switch-mode Hall position sensors are often mounted in BLDCMs to detect the information of the rotor position and instruct the current commutation. However, the reliability of Hall position sensors together with signal processing and transmission circuits maybe degraded with long-time running in the harsh environment. Once if Hall signals become abnormal due to the faults in Hall position sensors or the signal channels, false commutation will lead to improper running of the motor, even with large peak current which can harm the health of the motor[2].

For the detection and correction of Hall signal faults in BLDCMs, many research results have been reported in literatures. Ma and Zhou proposed a method to diagnosis and reconstruct Hall signals for doubly salient permanent magnet motor by using the last pulse edge in [3], but the pulse errors were not taken into account. Fu and his collaborators proposed a linear extrapolation based method to judge Hall signals and reconstruct the faulty signals in [4], but the mounting errors of Hall sensors were required *a prior*. Tashakori and Ektesabi designed a simple fault tolerant control system for Hall-effect sensors failure of BLDC motors in [5], but the FFT based failure diagnosis method required large amount calculations and the remedial strategy of failure signals also required the information of the mounting errors of Hall sensors.

The back electromotive force (EMF) method is an alternative to detect and correct faulty Hall signals for BLDCMs without the information of the mounting errors of Hall sensors in [6] and [7]. Since novel measurement information is introduced in [6] and [7], EMF method can be applied to not only the detection and correction of the mounting errors of Hall sensors, but also the situation of three breakdown Hall sensors. However, this method may increase the complexity of hardware and can not be used at low speed.

In order to improve the reliability of BLDCMs running at steady speed, this paper proposes a simple approach to detect and correct signal faults of Hall position sensors. This approach only adopts prior information of pulse intervals in the last electrical period and the mounting errors of Hall sensors are not necessary. Compared with other ones, this approach is easy to implement and the other information except for the pulse intervals in the last electrical period is not needed.

2. BLDCM principles and problem description

2.1. BLDCM principles

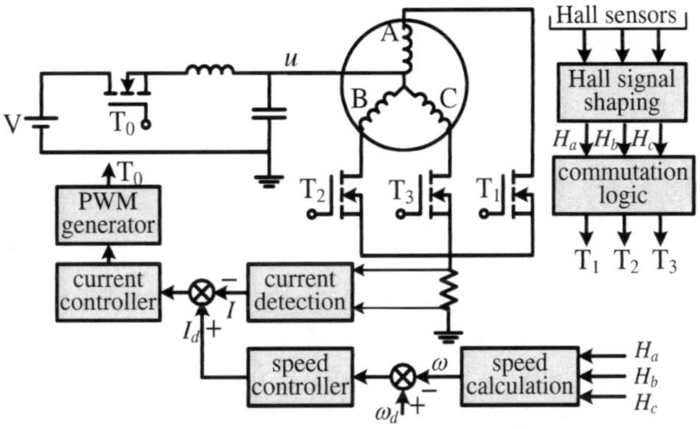

Figure 1. Schematics of BLDCM speed control system.

Figure 1 shows the principles of a BLDCM speed control system commutated by three Hall-effect position sensors and supplied by a DC/DC buck converter plus a three-phase half-bridge converter. Hall position sensors are mounted inside BLDCM with 120° phase difference and can output switch-mode signals representing rotor positions. The buck converter is in charge of regulating the motor current by chopping the supplied voltage, while the half-bridge converter is in charge of commutating the phase currents according to Hall commutation signals. For any time instant, only one phase is energized by turning on the correspondent power switch. If the rising edge of Hall signal for some phase is detected, this phase will be energized after turning off the conducting phase by manipulating correspondent switches, i.e., commutation.

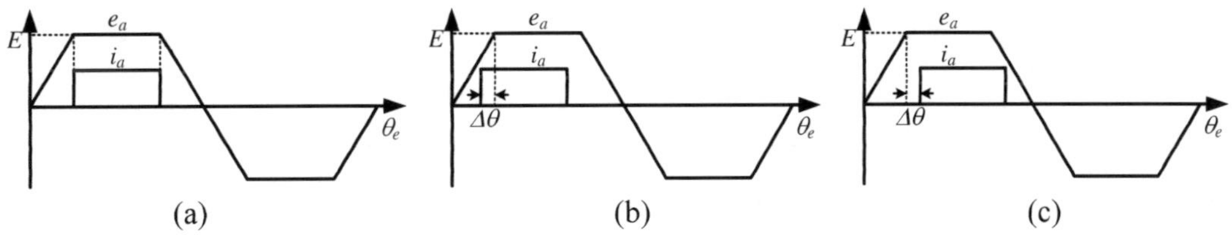

Figure 2. Diagrams of commutation (a) accurate (b) advanced (c) retarded.

Under ideal conditions, the commutation instant happens at 30° and the conduction mode goes on during the whole 120° constant plateau of EMF for any phase, as shown in Figure 2(a). For any phase in conduction mode, the voltage equation can be written as:

$$u = Ri_k + L di_k / dt + e_k \quad (1)$$

where u, R, and L denote the chopping voltage, winding resistor, and inductance respectively; i_k and e_k

denote the current and EMF of phase *k* respectively, *k*=*a*, *b*, *c*. If the commutation is accurate, EMF e_k will be its amplitude *E* in (1) and the desired current can be obtained by current loop. The ideal stator current has rectangular form with width of 120°. Assume the amplitude of the stator current is *I*, then the electromagnetic torque can be written as:

$$T_e = \frac{1}{\omega}(e_a i_a + e_b i_b + e_c i_c) = \frac{EI}{\omega} = K_T I \tag{2}$$

where ω and K_T denote the motor speed and torque coefficient respectively; K_T=E/ω. By using electromagnetic torque in (2), the motor speed can be regulated smoothly according to the dynamical equation as follows:

$$J\frac{d\omega}{dt} = T_e - D\omega - T_L \tag{3}$$

where *J*, *D*, and T_L denote inertia, damping coefficient, and load torque of the motor respectively.

2.2. Problem description

However, Hall position sensors and signal transmission channels may fail during long-term operations in the harsh environment. As a result, Hall signals used as commutation instructions will become abnormal. For example, the edge of Hall signal may arrive in advance or late. In severe cases, Hall signal may remain high or low all the time [4]. Abnormal Hall signals will lead to improper commutations. Figure 2(b) shows an advanced commutation due to an early rising edge, while Figure 2(c) shows a retarded commutation due to a late rising edge of Hall signal in phase A. Obviously, the EMF e_k in (1) will not be its amplitude *E*, it may be less or far less than *E*. Thus, (u–e_k) will become larger and peak current will arise according to (1). Especially, when e_k is negative during commutation interval, the peak current may be larger than the permitted value and it will endanger the safe operation of the motor.

Therefore, a detection and correction method for Hall signal faults will be proposed to improve reliability of BLDCMs in this paper. Even when the commutation information becomes incomplete owing to the failure in the partial Hall sensors or signal transmission channels, BLDCM still can work properly by using the proposed method.

3. Detection and correction of Hall signal faults

In this section, a simple approach is proposed to detect and correct Hall signal faults. The main idea of the approach is to detect and correct the next edge according to the prior information of Hall signals in the last electrical period. For the next edge of Hall signal, the approach includes three steps: prediction of the time instant, configuration of the confidence interval, and design of detection and correction algorithm.

3.1. Prediction of next edge instant

The core of the detection and correction method is to predict the time instant of the next edge for Hall signals. As we know, the electrical degree between two contiguous pulse edges (rising or falling) of Hall signals should be 60° if many kinds of unexpected factors are not taken into account, such as installing errors of Hall sensors, mounting errors of rotor magnetic poles, and delay of processing circuits, etc. However, the electrical degree of adjacent pulse edges will not be 60° owing to the effects of the above factors. Even if the stability of the motor speed is high enough, the pulse width of Hall signals will vary at a certain extent. Therefore, the time instant of the next pulse edge can not be predicted by using the width of the last pulse when the effects of the above unexpected factors are taken into account.

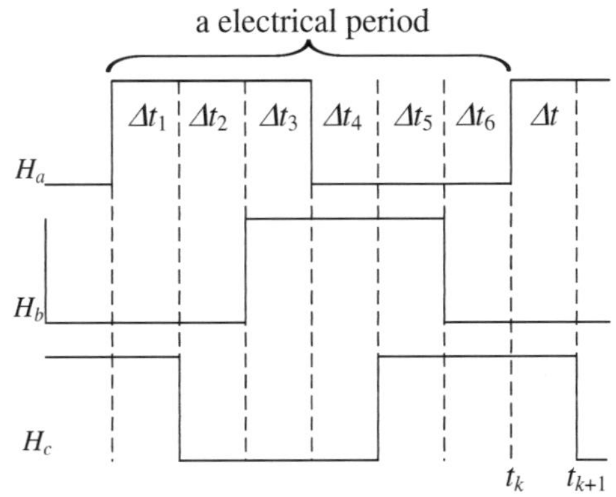

Figure 3. Relationship of 3-phase Hall signals in an electrical period.

If only installing errors of Hall sensors are taken into account, the time instant of the next pulse edge can be predicted by using the prior information of the pulse width in the last electrical period for Hall signals in BLDCMs at steady speed. As shown in Figure 3, the time intervals between contiguous pulse edges (rising or falling) in the last electrical period can be denoted by Δt_1, Δt_2, ··, Δt_6. If the last edge of the last electrical period is assumed to be the rising edge of phase A at instant t_k, then the next edge needs to be detected is the falling edge of phase C. Assume the time instant of the next edge is t_{k+1}, then the time interval Δt between the rising edge of phase A and the falling edge of phase C needs to be predicted before the prediction of t_{k+1}.

Examine the time intervals Δt_1, Δt_2, ··, Δt_6 in the last electrical period, it can be found that Δt_1 is the mirror before 360° electrical degree of the time interval Δt to be estimated. Even if installing errors of Hall sensors exist, Δt equals to Δt_1 approximately when the stability of the motor speed is high enough. Therefore, the next time interval Δt can be estimated as:

$$\Delta\hat{t} = \Delta t_1 \tag{4}$$

In practical engineering, only 6 registers are needed to store 6 time intervals between two contiguous pulse edges in the last electrical period for Hall signals. When estimating the next time interval, only the earliest one is needed. Compared with the other methods to predict the next time interval, this one does not require large amount of calculations and excessive register resources.

3.2. Design of confidence interval for next edge instant

However, Hall signals can not be judged only by using the predicted time instant of the next edge since the real signals are disturbed by various errors and disturbances. Therefore, a confidence interval should be set to exclude the effects of possible errors and disturbances according to certain standard. If the next edge of real Hall signals falls into the confidence interval, Hall signals can be judged to be normal. Otherwise, Hall signals can be judged to be abnormal and needs to be corrected.

Since the time instant of the next edge can be affected by the speed ripple of the motor, installing errors of Hall sensors, mounting errors of rotor magnetic poles, and delay of signal processing circuits, the confidence interval should be designed taking these factors into account. Assume the ripple ratio of the pulse is denoted by $\alpha < 1$, then the minimal width can be set as:

$$\Delta\hat{t}_{\min} = (1-\alpha)\Delta\hat{t} \tag{5}$$

The maximal width can be set as:

$$\Delta\hat{t}_{\max} = (1+\alpha)\Delta\hat{t} \tag{6}$$

Then, the confidence interval of the time instant for the next pulse edge can be derived as:

$$\hat{t}_{k+1} \in [t_k + \Delta\hat{t}_{\min}, t_k + \Delta\hat{t}_{\max}] \tag{7}$$

IOP Conf. Series: Materials Science and Engineering **339** (2018) 012004 doi:10.1088/1757-899X/339/1/012004

3.3. Design of detection and correction algorithm

If the time instant t_k of the next real pulse edge falls into the confidence interval (7), Hall signals can be judged to be normal. If t_k is earlier than $(t_k + \Delta\hat{t}_{\min})$, Hall signals can be judged to be advanced or a pulse is added. If t_k is later than $(t_k + \Delta\hat{t}_{\max})$, Hall signals can be judged to be retarded or a pulse is missed. The detection and correction algorithm of Hall signals is shown in Figure 4.

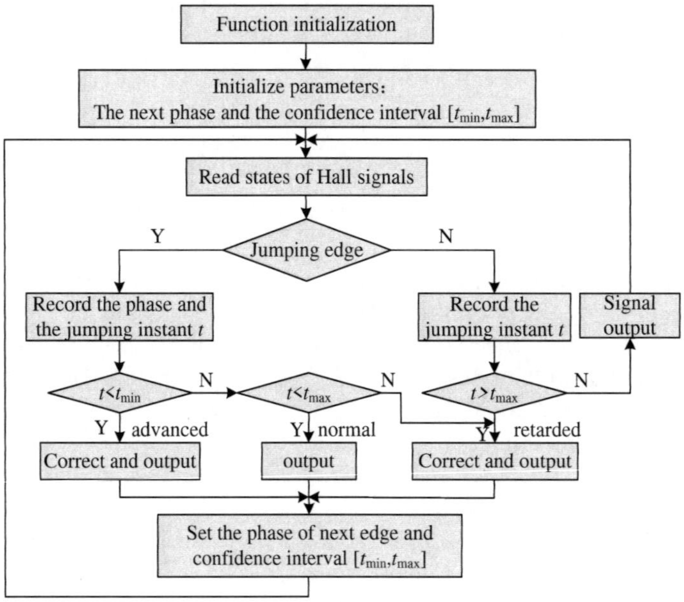

Figure 4. Detection and correction algorithm of Hall signals.

4. Experimental results

To verify the effectiveness of the proposed approach in this paper, experiments have been carried out by using a BLDCM with pole pairs of 6, rated voltage of 50V, and rated current of 3A. In the experiment, BLDCM is controlled by a FPGA control unit plus 3-phase half bridge and runs at a speed of 3000r/min. To simulate the faults of Hall signals, three switches are used in the signal transmission channels. Opening the switch can cut off the signal transmission to simulate the break faults of Hall signals. The detection and correction algorithm is implemented in FPGA and experimental results are derived when switching off Hall signal of phase B, as shown in Figure 5 and Figure 6. Since Hall signals are reversed by the open collector circuits, high level denotes break fault in the figures.

Switching off H_b without fault detection and correction, experimental results are shown in Figure 5. From Figure 5(a), it can be seen that H_b turns into high level due to the break fault and the commutation of phase A becomes abnormal. After the break fault happens, the current i_a of phase A has large peaks which is harmful to the health of the motor. The current I of DC link also exceeds the normal value. As a result, the speed in Figure 5(b) drops suddenly.

When Hall signals are detected and corrected by the proposed approach, experimental results are shown in Figure 6. From Figure 6(a), it is known that the faulty Hall signal H_b is corrected to be H_b' and the effect of the signal fault on BLDCM is attenuated. Both the current i_a of phase A and the current I of DC link have no obvious variations, As a result, the speed in Figure 6(b) also has enough stability. From these results, it can be concluded that the proposed approach is effective.

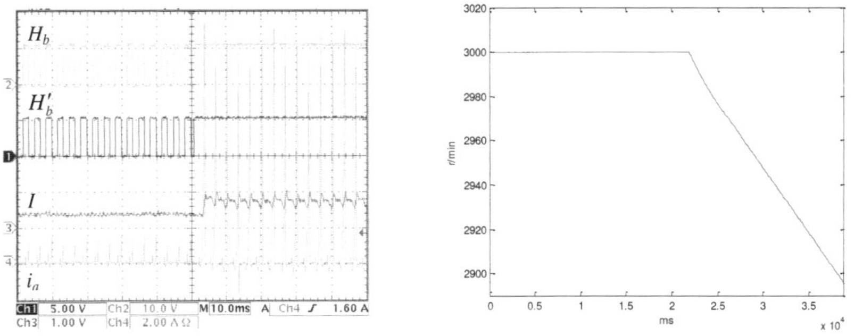

(a) (b)

Figure 5. Uncorrected case (a) Hall signals and currents (b) speed.

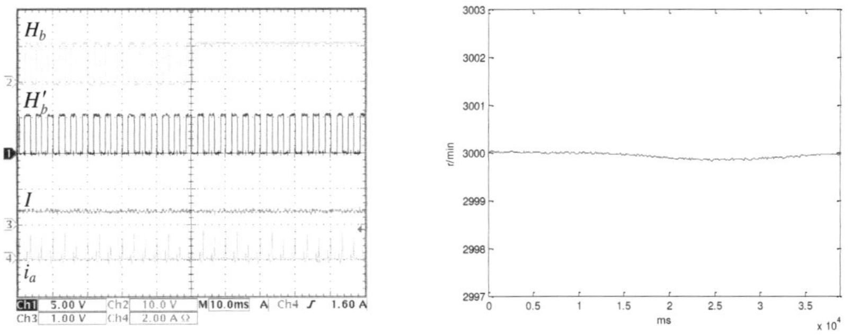

(a) (b)

Figure 6. Corrected case (a) Hall signals and currents (b) speed.

5. Conclusion

A simple approach is proposed to detect and correct signal faults of Hall position sensors for BLDCMs running at steady speed. This approach only adopts prior information of pulse intervals in the last electrical period and is easy to implement. Experimental results indicate that this approach is feasible.

References

[1] Chau KT, Chan CC and Liu CH 2008 Overview of permanent-magnet brushless drives for electric and hybrid electric vehicles *IEEE Transactions on Industrial Electronics* **55** 2246-2257

[2] Liu WJ, Liu CR, Wang NH and Wang DY 2011 Quantitative and Qualitative Model Based Fault Diagnosability Evaluation of Momentum Wheel *Chinese Space Science and Technology* **31** 54-63

[3] Ma CS and Zhou B 2008 Fault diagnosis and fault-tolerant control of position signal for doubly salient motor *Proceedings of the CSEE* **28** 73-78

[4] Fu AQ, Wu Z and Lyu HT 2013 Fault detection and correction of switch-mode Hall position sensors for brushless DC motors *Micromotors* **46** 62-67

[5] Tashakori A and Ektesabi M 2013 A simple fault tolerant control system for Hall effect sensors failure of BLDC motor *Proceedings of the 8th IEEE Conference on Industrial Electronics and Applications* (IEEE) p1011-1016

[6] Choi JH, Park JS, Gu BG and Won CY 2012 Position estimation and control of BLDC motor for VVA module with unbalanced hall sensors *Proceedings of 2012 IEEE International Conference on Power and Energy (PECon)* (IEEE) p390 -395

[7] Hung CW, Chen JH and Huang KC 2012 A correction circuit of Hall-sensor-signal-based speed measurement for BLDC motors *Artificial Life & Robotics* **17** 80-85

IOP Conf. Series: Materials Science and Engineering **339** (2018) 012005 doi:10.1088/1757-899X/339/1/012005

Control method of Three-phase Four-leg converter based on repetitive control

Wang Hui

School of Electrical Engineering, Shandong University, Jinan250061

Email: 262341218@qq.com

Abstract. The research chose the magnetic levitation force of wind power generation system as the object. In order to improve the power quality problem caused by unbalanced load in power supply system, we combined the characteristics and repetitive control principle of magnetic levitation wind power generation system, and then an independent control strategy for three-phase four-leg converter was proposed. In this paper, based on the symmetric component method, the second order generalized integrator was used to generate the positive and negative sequence of signals, and the decoupling control was carried out under the synchronous rotating reference frame, in which the positive and negative sequence voltage is PI double closed loop, and a PI regulator with repetitive control was introduced to eliminate the static error regarding the fundamental frequency fluctuation characteristic of zero sequence component. The simulation results based on Matlab/Simulink show that the proposed control project can effectively suppress the disturbance caused by unbalanced loads and maintain the load voltage balance. The project is easy to be achieved and remarkably improves the quality of the independent power supply system.

1. Introduction

Maglev wind power generator is often used in Wind-photovoltaic Hybrid street lamps, garden lighting, landscape lighting, traffic lights and other small off-grid power supply system and the large wind farm because of its advantages such as breeze starting, efficient power generation and stable operation, and it also plays an important role in the researches of power transport, use and control of large magnetic levitation wind power system.

Three-phase Four-leg inverter provides assess for nonlinear and unbalanced load with zero sequence current by adding the fourth bridge arm, making sure that the inverter can still provide a balanced three-phase sinusoidal voltage to load end under various severe load conditions. [1] Due to the addition of a bridge arm, the switch condition changes from the original 8 to 16, so the modulation mode becomes more complex comparing the traditional Three-leg inverter. At present, there are many strategies aiming at four-leg inverter such as pulse width modulation, space vector modulation technology and saddle pulse width modulation and so on[2-5]. SAPWM modulation is equivalent to common SVPWM modulation wave, which substantially both mean injecting 3 harmonics in a sinusoidal modulated signal. In order to realize independent control of three-phase voltage, different decoupling control strategies have been developed in different coordinate systems by scholars both at home and abroad. PIR-P dual-loop control in dq0 rotating coordinate system was proposed [6], which can suppress unbalanced loads through the reasonable resonant controller. A study suggests [7-8] that the load voltage is controlled by the PI regulator under the synchronous rotating coordinate of positive and negative sequence, and because the positive and negative sequence separation are not carried out, so the PI regulator is unable to track the resulting two octave voltage component without static difference. Therefore, based on positive and negative sequence separation, a closed-loop control algorithm in double synchronous rotating coordinate system is proposed [9], and the zero sequence voltage component is controlled by proportional resonant controller. On this foundation [9], an improved derated generalized integrator for fast sorting was introduced [10]. The study [11] suggests that Repetitive control is introduced under two coordinate systems to achieve stable output voltage with unbalanced loads.

Based on the above literature, this paper presents an improved positive sequence, negative sequence and zero sequence separation control strategy: the load voltage is divided by Second-order generalized integrator in order to increase the separation speed of positive and negative sequence, and the positive and negative sequence voltage components are respectively controlled in the positive and negative sequence synchronous coordinate system. Then, zero axis control is simplified, combining with repetitive control dual-closed loop of PI regulator, and the effectiveness of the control scheme has been proved by simulation.

2. VSC - HVDC transmission system structure

Fig. 1. It shows the topology of the three-phase four-leg inverter with load operation. The fourth leg is consist of switch tube *T7* and *T8*, in which the point *n* is connected with the load common point *g* through the inductance *Ln*. The output of the switch tube S1-S6 is connected with the load by the LC filter.

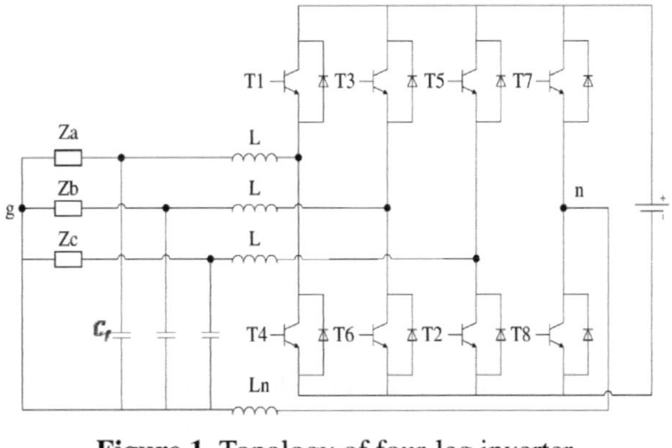

Figure 1. Topology of four-leg inverter

Since the DC side of inverter is supplied by a DC voltage source, so the voltage is constant and the switching frequency is 12.8 KHZ which is much higher than the fundamental frequency of the output voltage (50Hz), and therefore the converter average switching model can be used to simplify the analysis.

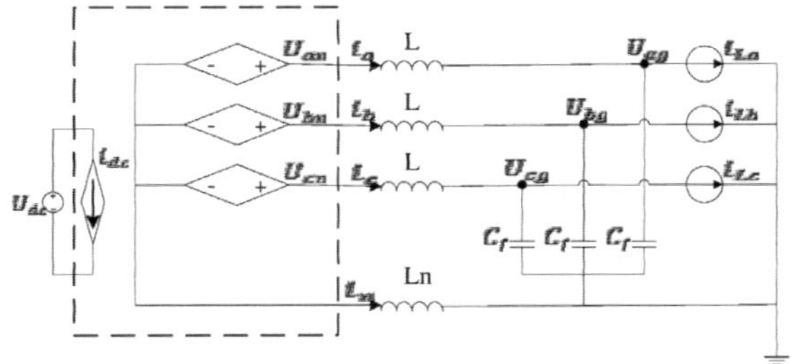

Figure 2. Average switching model of four-leg inverter

$$u_{xn} = U_{dc} * d_{xn} \tag{1}$$

$$Lpi_x = u_{xn} - u_{xg} + L_n pi_n \tag{2}$$

$$i_a + i_b + i_c + i_n = 0 \tag{3}$$

IOP Conf. Series: Materials Science and Engineering **339** (2018) 012005 doi:10.1088/1757-899X/339/1/012005

$$C_f pu_{xg} = i_x - i_{Lx} \tag{4}$$

In the above formula, x=a, b, c; u_{xg} is the load voltage, u_{xn} is the output voltage of the inverter, d_{xn} is the duty cycle of the phase voltage, p is the differential operator d/dt. In order to decouple the traffic flow control, the average switching model is transformed to the *d-q* synchronous rotating coordinate system by using the transform matrix *Tabc/dq0* just as shown in formula (5). In a synchronous rotating coordinate system, the formula (2) - (4) can change to formula (5) and formula (6). Among them, W inverter output fundamental angular frequency. Subscript *dq0* represent the corresponding *dq0* component of the physical quantity respectively.

$$T_{abc/dq0} = \frac{2}{3}\begin{bmatrix} \cos\omega t & \cos(\omega t - 2\pi/3) & \cos(\omega t + 2\pi/3) \\ -\sin\omega t & -\sin(\omega t - 2\pi/3) & -\sin(\omega t + 2\pi/3) \\ 1/2 & 1/2 & 1/2 \end{bmatrix} \tag{5}$$

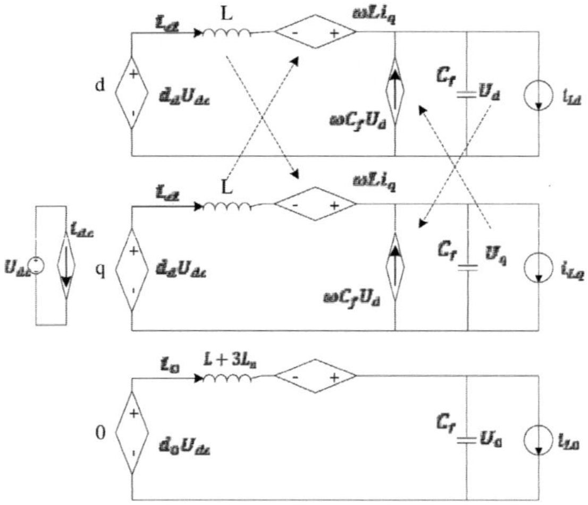

Figure 3. The inverter switch model under rotating coordinate system

$$\begin{cases} Lpi_d = d_d U_{dc} + \omega L i_q - u_d \\ Lpi_d = d_q U_{dc} - \omega L i_d - u_q \\ (L+3L_n)pi_0 = d_0 U_{dc} - u_0 \end{cases} \tag{6}$$

$$\begin{cases} C_f pu_d = \omega C_f u_q + i_d - i_{Ld} \\ C_f pu_q = -\omega C_f u_d + i_q - i_{Lq} \\ C_f pu_0 = i_0 - i_{L0} \end{cases} \tag{7}$$

3. The controller design

3.1. Repetitive controller

Based on internal model principle, a repetitive controller makes an internal model *1/ (1-e-ts)* which contains all the fundamental harmonic sub internal mode, inserting it into the controller, so that the system can realize the static tracking of the periodic AC signals [12]. It can be seen from its equation, the eigenvalues of roots *0, W* and *2W* are all distributed on the imaginary axis, and the gain of system at these characteristic frequency points is infinitely large. Therefore, the closed-loop feedback system can realize perfect tracking of AC signals. The sampling points of each cycle are *N*, and their internal modes are represented in the *Z* domain as follows:

$$G_R(z) = \frac{1}{1-z^{-N}} \quad (8)$$

From the *Z* domain structure of internal model, the repetitive controller is a positive feedback loop with periodic delay. Its function is accumulating the input signal periodically, similar to the integral link. When the given and feedback are inconsistent, the output of controller will continue to increase until the error is eliminated. However, the controller has *N* poles located on the unit circle, which makes the system be in critical undefined state. Therefore, the low pass filter Q (z) is introduced to obtain the improved repetitive controller, as shown in Fig. 4. Meanwhile, taking a constant less than 1, usually 0.95. Because the repetitive controller has a slow dynamic response, a repetitive controller combine with PI control is adopted.

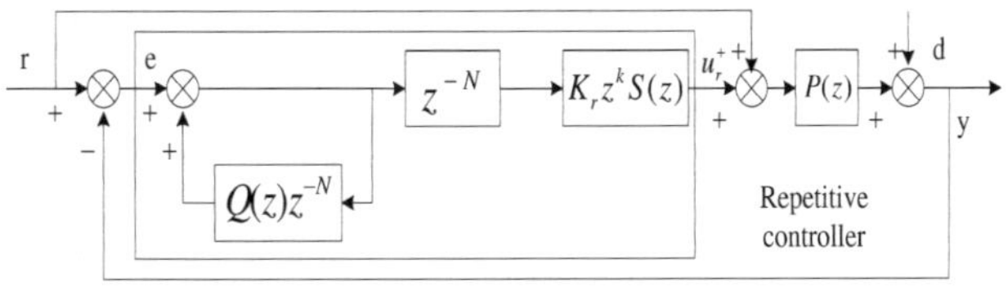

Figure 4. The improved repetitive controller

The repetitive controller is composed of a delayed positive feedback link and a compensator. The output error of inverter is accumulated by the fundamental period according to the delayed positive feedback. The function of compensator is to realize medium low frequency cancellation and high frequency attenuation with the inverter object, so that the compensation signal given by the internal model can correct the amplitude and phase correctly against to the disturbance, so as to realize the difference of the implicit waveform. S (z) is designed to eliminate the peak value of the amplitude frequency characteristic curve of P (z) and attenuate the high frequency, and we take the value of S (z) as a cut-off frequency, and we also take S (z) as a first order filter approximating to P (z) in order to achieve high frequency attenuation and lead to phase compensation realized by leading link.

3.2 Symmetrical components method

According to the symmetrical component theory [13], the three-phase unbalanced vector system can be decomposed into three symmetrical vector systems. The symmetric vectors in each system are called positive sequence components, negative sequence components and zero sequence components in turn. Therefore, under unbalanced load, the three-phase load voltage and inductance current can be expressed as:

$$u_{xg} = u_{xg+} + u_{xg-} + u_0 \quad (9)$$

$$i_x = i_{x+} + i_{x-} + i_0 \quad (10)$$

$$u_{xn} = u_{xn+} + u_{xn-} + u_{xn0} \quad (11)$$

$$\begin{cases} u_{xn+} = Lpi_{x+} + u_{xg+} \\ u_{xn-} = Lpi_{x-} + u_{xg-} \\ u_{xn0} = (L+3L_n)pi_0 + u_0 \end{cases} \quad (12)$$

Through the coordinate transformation, the physical quantity in the abc coordinate system is transformed into the ab0 coordinate system.

$$T_{abc/\alpha\beta 0} = \frac{2}{3}\begin{bmatrix} 1 & -1/2 & -1/2 \\ 0 & \sqrt{3}/2 & -\sqrt{3}/2 \\ 1/2 & 1/2 & 1/2 \end{bmatrix} \quad (13)$$

In this coordinate system, the positive and negative sequence components have nothing to do with the zero sequence component, and f_0^+ , f_0^- is 0. The phase shift operator composed of second order generalized integrator SOGI is $S = e^{-j90^{\circ}}$.

$$\begin{cases} f_{\alpha\beta}^{+} = \dfrac{1}{2}\begin{bmatrix} 1 & -S \\ S & 1 \end{bmatrix} f_{\alpha\beta} \\ f_{\alpha\beta}^{-} = \dfrac{1}{2}\begin{bmatrix} 1 & S \\ -S & 1 \end{bmatrix} f_{\alpha\beta} \end{cases} \tag{14}$$

The following is a SOGI shifted circuit diagram.

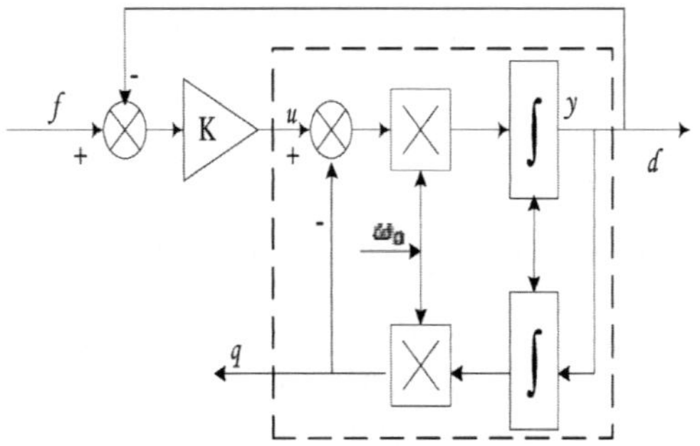

Figure 5. The phase shift operator is composed of second order generalized integrator

The transfer function of SOGI:

$$G_s = \frac{y(s)}{u(s)} = \frac{s\omega_0}{s^2 + \omega_0^2} \tag{15}$$

According to the above Fig.5, we can see:

$$\begin{cases} u_{\alpha\beta+} = \dfrac{k\omega_0 s}{s^2 + k\omega_0 s + \omega_0^2} u_{\alpha\beta} \\ u_{\alpha\beta-} = \dfrac{k\omega_0{}^2}{s^2 + k\omega_0 s + \omega_0^2} u_{\alpha\beta} \end{cases} \tag{16}$$

Based on SOGI phase-shift circuit of positive and negative sequence component, the extraction implementation flow chart is just as shown in Fig.6

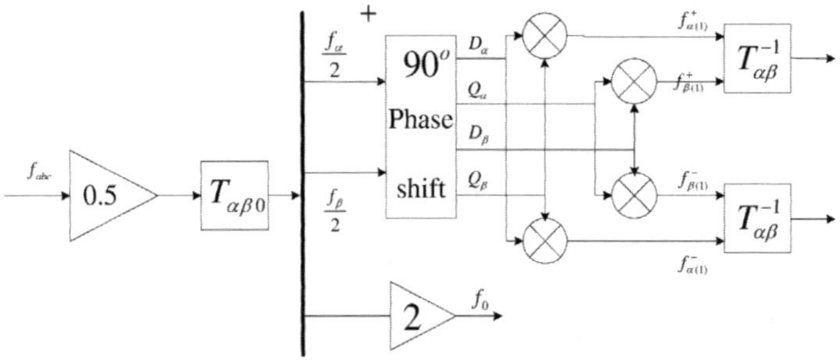

Figure 6. The fundamental principle

3.3 Dividing-sequence control

In order to achieve the voltage balance control, the control objectives should be set as that positive sequence component for a given value, negative sequence and zero sequence voltage components are zero. In the forward rotating coordinate system, the positive sequence *d-q* component is the direct flow, and in the negative rotation coordinate system, the negative sequence *d-q* component is also the direct flow. Therefore, the *PI* controller can achieve the control effect. When the load is unbalanced, the zero sequence component is a set of alternating currents with equal amplitude and phase, and the traditional *PI* controller is difficult to achieve good control effect, so the PI controller with repetitive control is used to track the zero sequence voltage current component. The repetitive controller is used to eliminate static error, PI controller is used to adjust the system dynamic performance. The system control diagram is as shown. There is no coupling between zero line and d-q axis, so the zero sequence component exists only on the zero axis and can be independently controlled.

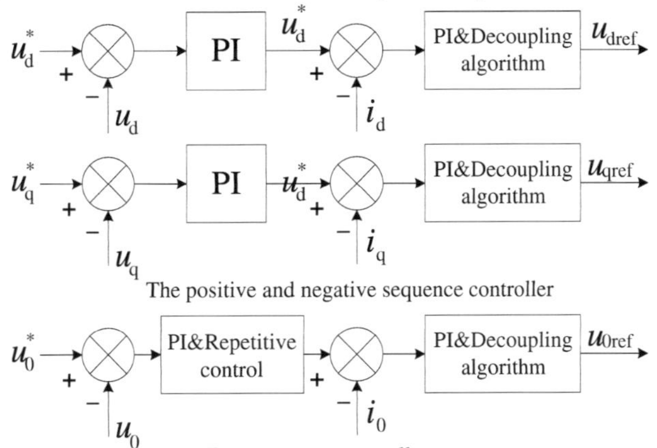

Figure 7. Dividing-sequence controller structure

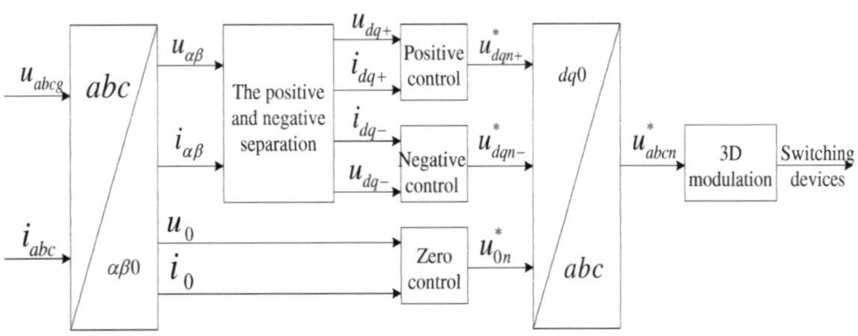

Figure 8. System control block diagram

4. The simulation and conclusion

In order to verify the effectiveness of control strategy proposed in this paper, a simulation model is built on the Matlab/Simulink platform as shown in Fig. 1. The component parameters are shown in Table 1, and first the three-phase unbalanced load experiment is carried out, in which the three-phase load of ABC is 10, 10, 30, and the capacitance voltage and inductance current are measured just as shown in Figure.

Then, the single-phase load experiment is carried out, in which the A load is 10 and the B C both no-load. The capacitance voltage and inductance current output are measured in the experiment just as shown in Figure.

Table 1. The simulation parameters

parameters	parameters
DC voltage /V	600
Filtering inductance /L	5
Filter capacitor /uf	30
The midline inductor /mH	2
Switching frequency /kHz	12. 8
Output voltage instruction value /V	220

The output voltage frequency/Hz	50

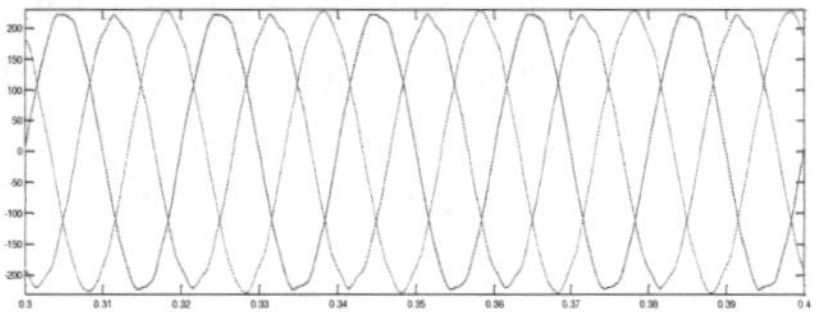

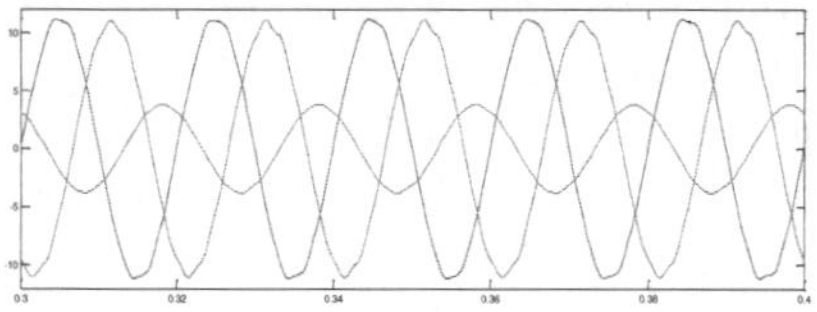

A. Improved points sequence control output voltage waveform
B. Improved points sequence control output current waveform

The Figure shows under the improved points sequence control strategy, the three-phase output voltage is 223.1V, 218.7V, 223.4V, the harmonic content is 1.72%, 1.46%, 1.56%, and the maximum voltage imbalance is 1.54%. The calculation shows that the current negative sequence imbalance is 0.8%, and the zero sequence imbalance is 0.7%.

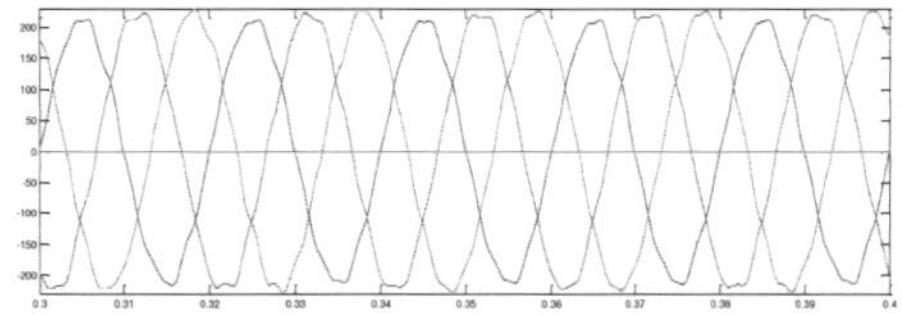

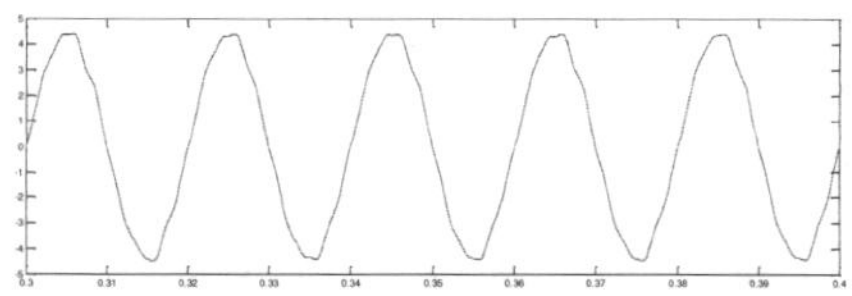

A. Improved points sequence control output voltage waveform
B. Improved points sequence control output current waveform

The Figure shows under the improvement points sequence control strategy, The three-phase output voltage is 213.8V, 223.3V, 224.1V, the harmonic content is 2.79%, 3.17%, 2.76%, and the maximum voltage imbalance is 2.81%. The calculation shows that the current negative sequence imbalance is 0.7%, and the zero sequence imbalance is 1.4%.

5. Conclusion

The simulation results show that the improved sequence control strategy can suppress unbalanced loads effectively and has a good effect on improving the power quality of the maglev wind power grid-connected system.

Reference

[1] Steigerwald R L. Current fed inverter with commutation independent of load inductance: US, US4039926[P]. 1977.
[2] Zhang R, Prasad V H, Boroyevich D, et al. Three-dimensional space vector modulation for four-leg voltage-source converters[J]. IEEE Transactions on Power Electronics, 2002, 17(3):314-326.
[3] Perales M A, Prats M M, Portillo R, et al. Three-dimensional space vector modulation in abc coordinates for four-leg voltage source converters[J]. IEEE Power Electronics Letters, 2003, 1(4):104-109.
[4] Yao C. Study of the Unification of SPWM and SVPWM in Three-Phase Four-Wire Systems[J]. Transactions of China Electrotechnical Society, 2007, 22(12):122-127.
[5] Zhou W P, Zheng-Guo W U, Tang J S, et al. A Novel Algorithm of SVPWM and the Study on the Essential Relationship Between SVPWM and SPWM[J]. Proceedings of the Csee, 2006, 26(2):133-137.
[6] Fei L L, Zhang K, Cai Y L. The Control of Three-Phase Four-Bridge Arm Inverter under the Unbalanced Load Conditions[J]. Telecom Power Technology, 2011.
[7] Sun J, Hou Z, Su Y. Study of control method of three-phase four-wire inverter power supply[J]. Transactions of China Electrotechnical Society, 2004, 19(4):61-65.
[8] Gannett R A, Sozio J C, Boroyevich D. Application of synchronous and stationary frame controllers for unbalanced and nonlinear load compensation in 4-leg inverters[C]// Applied Power

Electronics Conference and Exposition, 2002. Apec 2002. Seventeenth IEEE. IEEE, 2002:1038-1043 vol.2.

[9] Xiao Z, Jin X M, Tang F, et al. Islanded-mode control and implementation of a three-phase four-leg based microgrid converter under unbalanced load conditions[J]. Power System Protection & Control, 2013.

[10] Zhou X, Jin X, Tang F, et al. Improved Sequence Decomposition Control Strategy for Three-phase Four-leg Inverter with Unbalanced Load[J]. High Voltage Engineering, 2014, 40(8):2504-2512.

[11] Zheng Y N, Han C. Researches on the Control of Three-phase Four-leg Inverter based on Repetitive Control[J]. Power Electronics, 2009, 43(11):68-70.

[12] Fortescue C L. Method of symmetrical co-ordinates applied to the solution of polyphase networks[J]. American Institute of Electrical Engineers Proceedings of the, 1918, 37(6):629-716.

[13] Xue S, Cai J. Detection of fundamental positive and negative sequence components based on second-order generalized integrator[J]. Electric Power Automation Equipment, 2011, 31(11):69-73.

IOP Conf. Series: Materials Science and Engineering **339** (2018) 012006 doi:10.1088/1757-899X/339/1/012006

Adaptive control of servo system based on LuGre model

Wang Jin, Liu Niancong, Chen Jianlong and Geng Weitao
School of nuclear technology and automation engineering, Chengdu university of technology, Chengdu 610059,China

Email: 1179036999@qq.com

Abstract. This paper established a mechanical model of feed system based on LuGre model. In order to solve the influence of nonlinear factors on the system running stability, a nonlinear single observer is designed to estimate the parameter z in the LuGre model and an adaptive friction compensation controller is designed. Simulink simulation results show that the control method can effectively suppress the adverse effects of friction and external disturbances. The simulation show that the adaptive parameter k_z is between 0.11-0.13, and the value of gamma1 is between 1.9-2.1. Position tracking error reaches level 10^{-3} and is stabilized near 0 values within 0.3 seconds, the compensation method has better tracking accuracy and robustness.

1. Introduction

With the progress of science and technology, more and more machining accuracy requirements are put forward for NC machine tools [1-2]. However, due to the nonlinear friction disturbance, the ball screw feed system in the traditional control methods cannot meet the requirements of high precision control[3].

The key of friction compensation is to establish accurate friction model, and calculate the friction force of the system according to the speed, position etc. The friction model can be divided into static and dynamic[4]. At present, the LuGre model is the most widely used, the model accurately describes the dynamic and static characteristics of frictional, and has a good compensation effect. Kamalzadeh[5] proposed for the adaptive sliding mode controller for axial vibration characteristics of ball screw feed system model; Choi J[6] establish dynamic friction model according to LuGre model, show the friction hysteresis characteristic of the system, and apply it to the system sliding mode controller; Zhou Jinzhu[7] has designed a nonlinear observer according to the LuGre model, and the integral backstepping adaptive control algorithm is used.

This paper set up mechanics model of the NC system based on LuGre model, an adaptive friction compensation method is designed, analyses the stability of the adaptive control method and asymptotic convergence, and the validity of the compensation method is verified by simulation.

2. Feed system dynamics modeling

The simplified physical model of the feeding system of NC machine tools is shown in Figure 1：

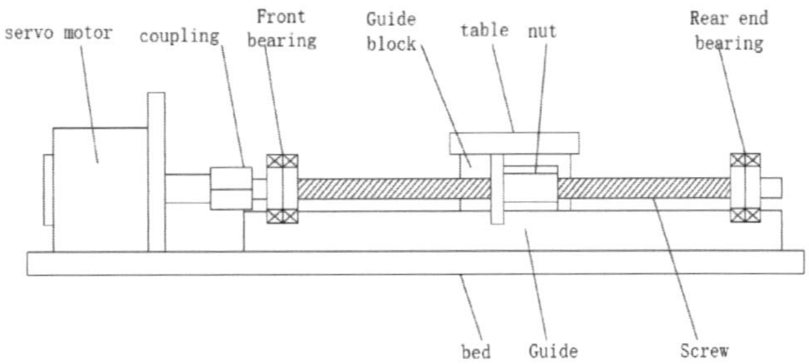

Figure 1. servo system structure

The equation of motion of the work table:

$$m\ddot{x} = F_D - F_f - F_r \tag{1}$$

Where m- table quality; x- axial displacement of table; F_D- driving force of the table; F_f- friction force of the work table; F_r- The disturbance force of the worktable.

Equations of motion of the motor screw shaft:

$$J_m\ddot{\theta}_m = Ku - D\dot{\theta}_m - F_d R \tag{2}$$

Where J_m- equivalent rotational inertia of feed system; θ_m - Motor shaft-screw rotation angle; K- motor torque constant; u- control quantity; D- equivalent damping coefficient of feed system; F_d- screw load force; R- lead screw radius.

Then:

$$J\ddot{\theta}_m = Ku - D\dot{\theta}_m - M_f - T_r \tag{3}$$

Where $J = J_m + \frac{h^2}{4\pi^2}\cdot m$, $M_f = \frac{h}{2\pi}\cdot F_f$, $T_r = \frac{h}{2\pi}\cdot F_r$, $\frac{F_D}{F_d} = \frac{2\pi R}{h}$, h- led of screw.

LuGre dynamic model can describe all kinds of dynamic and static characteristics of friction accurately, such as creeping, limit cycle shock, pre sliding deformation, friction memory, variable static friction and static Stribeck curve. The friction compensation based on the LuGre friction model can represent the friction torque of the system as:

$$\begin{cases} \dot{z} = \dot{\theta} - \frac{|\dot{\theta}|}{g(\dot{\theta})} z \\ \sigma_0 g(\dot{\theta}) = M_c + (M_s - M_c)\exp\left[-\left(\frac{\dot{\theta}}{\dot{\theta}_s}\right)^2\right] \\ M_f = \sigma_0 z + \sigma_1 \dot{z} + \mu_b \dot{\theta} \end{cases} \tag{4}$$

The LuGre friction model is divided into two parts, one part is related to z, to reflect the changes of z and the relevant parts of the mane deformation, The other part is related to the viscous friction factor, and β is used to reflect the change of viscous friction factor.

Therefore, the friction torque M_f in the LuGre model is modified as follows:

$$M_f = \alpha(\sigma_0 z + \sigma_1 \dot{z}) + \beta\mu_b\dot{\theta}, \quad \alpha > 0, \quad \beta > 0 \tag{5}$$

Then:

$$J\ddot{\theta}_m = Ku - D\dot{\theta}_m - \alpha(\sigma_0 z + \sigma_1 \dot{z}) - k_0\beta\mu_b\dot{\theta}_m - T_r \tag{6}$$

Where $\dot{\theta} = k_0\dot{\theta}_m$, and k_0 is the conversion factor between $\dot{\theta}$ and $\dot{\theta}_m$.

3. Design of friction compensation controller

Because of the poor robustness of the ordinary PID control, the high-precision tracking requirements cannot be achieved, the Backstepping method has unique advantages in dealing with nonlinear control problems. Therefore, for the above dynamic equations, a robust adaptive friction compensation controller is designed by using the Backstepping design method.

In order to observe the z values in the LuGre friction model, a nonlinear observer is used in the simplified manner, and the equation is as follows:

$$\dot{\hat{z}} = \dot{\theta} - \frac{|\dot{\theta}|}{g(\dot{\theta})}\hat{z} + k_z e_2\left(\sigma_0 - \sigma_1\frac{|\dot{\theta}|}{g(\dot{\theta})}\right), k_z > 0 \tag{7}$$

Set $\tilde{z}$ as the error of the state observer:

$$\tilde{z} = z - \bar{z}, \quad \dot{\tilde{z}} = -\frac{|k_0\dot{\theta}_m|}{g(k_0\dot{\theta}_m)}\tilde{z} - k_z e_2\left(\sigma_0 - \sigma_1\frac{|k_0\dot{\theta}_m|}{g(k_0\dot{\theta}_m)}\right) \tag{8}$$

Define position tracking error signal e_1,and set θ_e as expected angular displacement:

$$e_1 = \theta_e - \theta_m, \quad \dot{e}_1 = \dot{\theta}_e - \dot{\theta}_m \tag{9}$$

Using Lyapunov function without solving the system state equation is a general method to judge the stability of the system. The Lyapunov function V1 is selected in order to obtain the expected value of $\dot{\theta}_m$ from the output error e1:

$$V_1 = \frac{1}{2}e_1^2, \quad \dot{V}_1 = e_1\dot{e}_1 = e_1(\dot{\theta}_e - \dot{\theta}_m) \tag{10}$$

Using the Backstepping design method to design the reference speed control signal:

$$\theta_{1d} = \dot{\theta}_e + k_1 e_1 + k_2\chi, \quad \chi = \int_0^t e_1 dt \tag{11}$$

Where θ_{1d} - speed control signals; $k_1>0$; $k_2>0$; χ - integration of position tracking errors.

The error between the reference speed θ_{1d} and the actual speed $\dot{\theta}_m$:

$$e_2 = \theta_{1d} - \dot{\theta}_m = \dot{\theta}_e - \dot{\theta}_m + k_1 e_1 + k_2\chi, \quad \dot{e}_2 = \ddot{\theta}_e - \ddot{\theta}_m + k_1\dot{e}_1 + k_2 e_1 \tag{12}$$

Then:

$$\dot{e}_1 = e_2 - k_1 e_1 - k_2\chi \tag{13}$$

The following control laws are designed for the system:

$$Ku = \bar{J}\left(\ddot{\theta}_e + k_1\dot{e}_1 + k_2 e_1\right) + \frac{e_1}{k_3} + k_c e_2 + \bar{D}\dot{\theta}_m + \bar{\alpha}\bar{z}\left(\sigma_0 - \sigma_1\frac{|k_0\dot{\theta}_m|}{g(k_0\dot{\theta}_m)}\right) + \bar{\alpha}\sigma_1 k_0\dot{\theta}_m + \bar{\beta}\mu_b k_0\dot{\theta}_m + \bar{T}_r \tag{14}$$

Where k_4- is a positive real number, $\bar{J}$ and $\bar{D}$ - measurements of parameter J and D.

Define the Lyapunov function further:

$$V_2 = V_1 + \frac{1}{2}k_2\chi^2 + \frac{1}{2}k_3 J e_2^2 + \frac{1}{2\gamma_1}\tilde{\alpha}^2 + \frac{1}{2\gamma_2}\tilde{\beta}^2 + \frac{1}{2\gamma_3}\tilde{T}_r^2 + \frac{1}{2\gamma_4}\tilde{D}^2 + \frac{1}{2}\alpha\tilde{z}^2 \tag{15}$$

Use $\alpha z - \bar{\alpha}\bar{z} = \alpha\tilde{z} + \tilde{\alpha}\bar{z}$ [8],then:

$$\dot{V}_2 = -k_1 e_1^2 - k_z k_c e_2^2 - \alpha\frac{|k_0\dot{\theta}_m|}{g(k_0\dot{\theta}_m)}\tilde{z}^2 + \tilde{D}\left[k_z e_2\dot{\theta}_m - \frac{1}{\gamma_4}\dot{\bar{D}}\right] + \tilde{\beta}\left(k_z e_2\mu_b k_0\dot{\theta}_m - \frac{1}{\gamma_2}\dot{\bar{\beta}}\right) + \tilde{\alpha}\left[k_z e_2\bar{z}\left(\sigma_0 - \sigma_1\frac{|k_0\dot{\theta}_m|}{g(k_0\dot{\theta}_m)}\right) + k_z e_2\sigma_1 k_0\dot{\theta}_m - \frac{1}{\gamma_1}\dot{\bar{\alpha}}\right] + \tilde{T}_r\left(k_z e_2 - \frac{1}{\gamma_3}\dot{\bar{T}}_r\right) \tag{16}$$

The adaptive law is chosen as:

$$\dot{\bar{\alpha}} = \gamma_1 k_z e_2\left(\bar{z}\sigma_0 - \bar{z}\sigma_1\frac{|k_0\dot{\theta}_m|}{g(k_0\dot{\theta}_m)} + \sigma_1 k_0\dot{\theta}_m\right), \quad \dot{\bar{\beta}} = \gamma_2 k_z e_2\mu_b k_0\dot{\theta}_m, \quad \dot{\bar{T}}_r = \gamma_3 k_z e_2, \quad \dot{\bar{D}} = \gamma_4 k_z e_2\dot{\theta}_m \tag{17}$$

Then:

$$\dot{V}_2 = -k_1 e_1^2 - k_z k_c e_2^2 - \alpha\frac{|k_0\dot{\theta}_m|}{g(k_0\dot{\theta}_m)}\tilde{z}^2 \le 0 \tag{18}$$

So the equation (14) and (17) can guarantee the system to be globally asymptotically stable.

4. Adaptive friction compensation simulation

In order to verify the effectiveness of the integral backstepping adaptive compensation algorithm, the system is simulated and verified by Matlab/Simulink.

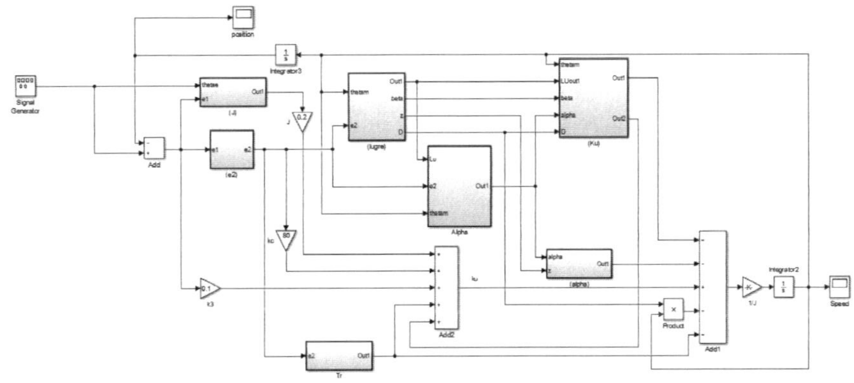

Figure 2. simulation block diagram of adaptive friction compensation system

LuGre simulation parameters are selected as follows: Ms=340N, Mc=280N, μb=0.02Ns/m, σ0=260N/m, σ1=2.5Ns/m, θ_s=0.01m/s, J=0.2kgm2.In simulation, the adaptive parameters are adjusted constantly to achieve good tracking effect. The initial values of the adaptive coefficients are as follows: kc=80, k1=13, k2=1, k3=0.1, k0=0.1, kz=0.1, γ_1=2, γ_2=10, γ_3=10, γ_4=5.

Diagram (a) and (b) in figure 3 is based on the initial values of the above adaptive coefficients. Diagram (a) showing concussion in starting time, to curb this phenomenon need to adjust the parameters.

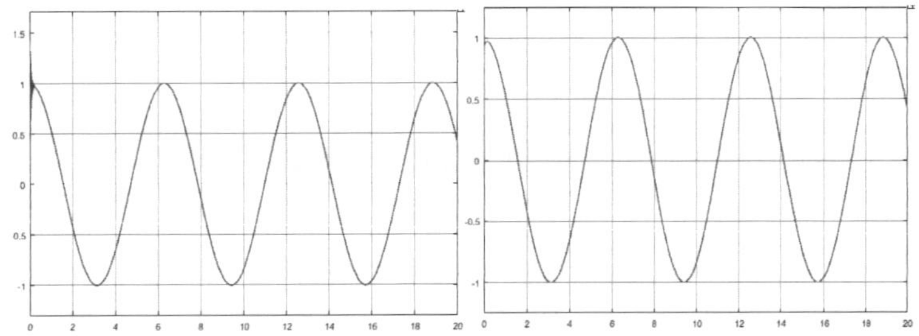

(a) (b)

Figure 3. speed tracking and position tracking effect

When the value of the adaptive coefficients are: kc=80, k1=10, k2=1, k3=13, k0=0.1, kz=0.12, γ_1=2, γ_2=100, γ_3=100, γ_4=75, the speed tracking effect obtained by simulation are shown as in figure 3 (b). When the value of kz is between 0.11-0.13 and gamma 1 is between 1.9-2.1, the tracking effect is shown in figure 3(b). The phenomenon of starting and stopping vibration is obviously eliminated and good tracking result is achieved.

Select sine input signal as figure 4(a) and the simulation time is set to 50s:

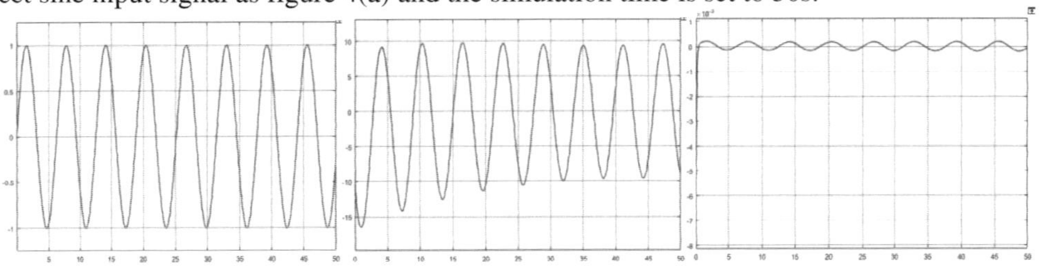

(a) (b) (c)

Figure 4. system input signal and friction torque and tracking error

Figure 4(b) and (c) shows the friction and position error curves. The position tracking error reaches 10-3, and within 0.3 seconds it is stable near 0 values. There is no large chattering in the whole process, and it has higher tracking accuracy. The friction force varies with the speed of motion, and tends to converge.

5. Conclusion

(1) The friction compensation can control the system motion error within 10^{-3} level, and the tracking error is quickly stabilized near 0 values.

(2) The friction compensation method can inhibition the influence of nonlinear factors. By adjusting the adaptive parameters(k_0、k_z and γ_1), the phenomenon of starting and stopping vibration can be avoided effectively. The value of k_z is between 0.11-0.13, and the value of γ_1 is between 1.9-2.1. The compensation method is insensitive to external disturbances and has strong stability.

Reference

[1] Wang Lei, Liu Haitao and Liang Tao. Modeling and dynamic response analysis of feed system under low frequency servo force [J]. *Journal of mechanical engineering*, 2015,**51** (3):80-86.
[2] Huang Lulang, Cha Xian Liang. Modeling and analysis of NC machine tool transmission system [J]. *automation of manufacturing industry*, 2015,**37** (7):92-95.
[3] Gordon D J，Erkorkmaz K. Accurate control of ball screw drives using pole-placement vibration damping and a novel trajectory prefilter[J].*Precision Engineer-ing*,2013,**37**(2):308-22.
[4] Brain A H, Pierre D, Carlos C D W. A survey of models, analysis tools and compensation methods for the control of machines with friction[J]. *Automatica*,1994,**30**(7):1083-138.
[5] Kamalzadeh A，Erkorkmaz K. Compensation of axial vibrations in ball screw drives[J].*CIRP Annals—Manufacturing Technology*,2007,**56**(1):373-8.
[6] Choi J. Development of hysteresis friction model for a precise tracking control system[J].*Proceedings of the Institution of Mechanical Engineers*，Part C:*Journalof Mechanical Engineering Science*,2012,**226**(5):1338-44.
[7] Zhou Jinzhu, Duan Baoyan and Huang Jin.LuGre friction model of servo system and compensation [J]. *control theory and applications*, 2008,**25** (6):990-4.
[8] Zhang Dan. Compensation control of servo system with frictional links [D]. Xi'an: Xi'an Electronic and Science University, 2008:1-50.

IOP Conf. Series: Materials Science and Engineering **339** (2018) 012007 doi:10.1088/1757-899X/339/1/012007

Identification of moving sinusoidal wave loads for sensor structural configuration by finite element inverse method

B Zhang[1]and S.Yu[2]
[1]Zhuji Leimeng Robotic Technology Co., Ltd, Zhuji City, Zhejiang Province, 311800,China.
[2]Department of Mechanical and Industrial Engineering, Ryerson University, Toronto, M5B 2K3,Canada.

E-mail:billbiaobiaozhang@yahoo.ca

Abstract: In this paper, a beam structure of composite materials with elastic foundation supports is established as the sensor model, which propagates moving sinusoidal wave loads. The inverse Finite Element Method (iFEM) is applied for reconstructing moving wave loads which are compared with true wave loads. The conclusion shows that iFEM is accurate and robust in the determination of wave propagation. This helps to seek a suitable new wave sensor method.

1. Introduction

By detecting and localizing sound sources, acoustic sensor surveillance systems are commonly utilized in many fields as a countermeasure for accidents or crimes. These systems collect data from a ranging mechanism using acoustic sensors, by measuring the times-of-flight (TOF) or difference- in-times –of –flight (DTOF) in order to compute the positions. Nevertheless, data collected by the ranging mechanism are often influenced or corrupted by different parameters (e.g., environment noise, delays associated with transducers, responses, and latency of signal processing algorithms), which lead to errors in the estimation of positions [1,2,3]. The goal of this research is to develop a single sensor capable of locating an acoustic source. Unlike acoustic sensors network, this new sensor can give continuous measurement in nature, and it utilizes new tools to reconstruct the wave impinging on the sensor's surface. Zhang et al.[4] first applied the Tikhonov force reconstruction method to identify moving sinusoidal wave loads, and attempted to develop it as a new sensor method. However, due to complexity of moving arbitrary waves, it is difficult to identify them well just by using the conventional Tikhonov combined with the L–curve method. Then Zhang et al.[5] applied the Arnoldi-Tikhonov regularization method coupled with the Generalized Cross-Validation (GCV) for seeking the regularization parameter to identify arbitrary waves of different amplitudes and waveforms. In general, inverse results in their work are in agreement with true values. However, there are still some significant deviations between them. Besides, inverse results obtained from inverse methods above have narrower range of wave frequencies for identification and are easier to be affected by sensor structure geometry. To improve the accuracy of inverse results, we try to apply the inverse finite element method for sensor structural configuration study. Many inverse finite element method (iFEM) applications are available in the literature. For example, Gunawan [6] applied iFEM to identify the parameters of a simple viscoelastic model in form of the Prony series expansion; Gherlone et al [7] used iFEM for shape sensing: reconstruction of structural displacements by using surface-measured strains. However, until now iFEM has not been applied to reconstruct acoustic wave loads as the sensor method. In this paper, iFEM is presented to reconstruct the distribution of moving acoustic wave, then inverse results are compared with true loads. We will find that iFEM is practical in a micro dimension wave sensor model.

In the following sections, a description of iFEM needed for identifying a moving sinusoidal wave load from sensing responses is provided. Its results are evaluated by true wave loads.

2. Moving wave loads identification by using iFEM

To illustrate the practicability and precision of the presented approach in estimating unknown input moving wave loads, a composite beam structure with elastic foundation supports is investigated. As known in Figure 1, a wave load $f(x,t)$ travels at a constant speed c along an Euler-Bernoulli beam with a finite length. The lateral response of the composite beam is governed by the following partial differential equation [5].

$$\mu_{eq}\frac{\partial^2 W(x,t)}{\partial t^2}+2\mu_{eq}\omega_b\frac{\partial W(x,t)}{\partial t}+(EI)_{eq}\frac{\partial W^4(x,t)}{\partial x^4}+C_f W(x,t)=f(x,t), \tag{1}$$

where $L,\mu_{eq},\omega_b,(EI)_{eq},C_f$ are the beam length, equivalent linear mass density, damping circular frequency, equivalent flexural rigidity and foundation elasticity constant respectively. $w(x,t)$ denotes beam lateral displacement. Here we let the mass and bending stiffness of the elastic foundation neglected for simplicity.

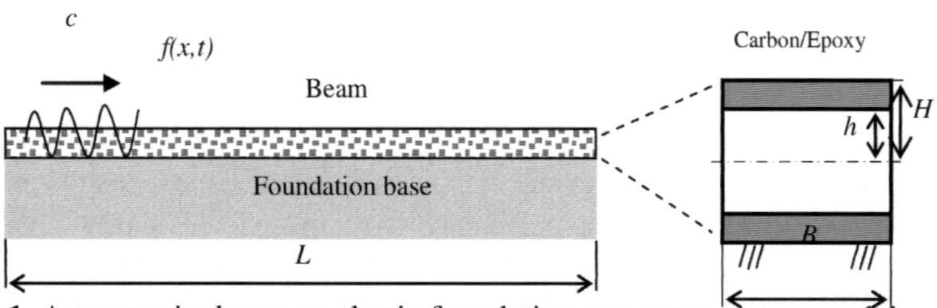

Figure 1. A composite beam on elastic foundation as a wave sensor model

For a composite sandwich structure shown in Figure 1, the equivalent flexural rigidity and the equivalent mass respectively can be expressed as[8]

$$\mu_{eq}=2B\left[\rho_c h+\rho_f (H-h)\right], \tag{2}$$

$$(EI)_{eq}=\frac{2B}{3}\left[E_c h^3+E_f\left(H^3-h^3\right)\right], \tag{3}$$

where B is the sandwich beam width, H and h are terms related with thickness shown in Figure 1, ρ is the material density and E is Young's modulus, Subscripts "c" and "f" represents quantities of the core material and the face material respectively, E_f can be expressed as[8]

$$\frac{1}{E_f}=\frac{1}{E_{11}}\cos^4\theta+\left(\frac{1}{G_{12}}-\frac{2v_{12}}{E_{11}}\right)\cos^2\theta\sin^2\theta+\frac{1}{E_{22}}\sin^4\theta, \tag{4}$$

where E_{11},E_{22},G_{12} and v_{12} are mechanical properties of face lamina along their principal directions; θ is the fiber orientation angle measured from the x-axis in counter-clock wise direction. To solve equation (1), we discretize the entire beam into $\hat{m}$ beam finite elements of length $L_e=L/\hat{m}$. Within a finite element, the beam lateral displacement is

$$W_e(x,t)=Nd_{e1}(x)q_{e1}+Nd_{e2}(x)q_{e2}+Nd_{e3}(x)q_{e3}+Nd_{e4}(x)q_{e4}, \tag{5}$$

where $Nd_{ei}(x)(i=1,2,3,4)$ are interpolation functions described in [9].Suppose that the lateral displacement vector $\{q(t)\}$ is a collection of nodal variables q_{ei} ,where $i=1,2,3,4$.We may obtain the following equations of motion using the Lagrange equations

$$[M]\{\ddot{q}\}+[C]\{\dot{q}\}+[K+K_f]\{q\}=\{F(t)\}, \tag{6}$$

where $\{\ddot{q}\},\{\dot{q}\}$ and $\{q\}$ are the beam element acceleration, velocity and displacement vectors respectively, and $\{F(t)\}$ represents the external force vector which depends on moving wave loads. $[C]$, $[M]$ and $[K+K_f]$ are the beam element damping matrix, mass matrix, beam structural stiffness and foundation base stiffness matrix respectively. Details can be referred to [9]. For the external force vector $\{F(t)\}$, its vector element can be expressed as follows

$$F_i(t)=\int_0^{L_e} f(x,t)Nd_i(x)dx \quad (i=1,2,3,4), \tag{7}$$

the structural system described in equation (6) can be solved by the Newmark β method of direct integration [9].For a typical time step from t to $t+\Delta t$, the acceleration and velocity at $t+\Delta t$ are related to the displacement at $t+\Delta t$ as follows

$$\{\ddot{q}\}_{t+\Delta t}=a_0\left(\{q\}_{t+\Delta t}-\{q\}_t\right)-a_2\{\dot{q}\}_t-a_3\{\ddot{q}\}_t, \tag{8}$$

and

$$\{\dot{q}\}_{t+\Delta t}=\{\dot{q}\}_t+a_6\{\ddot{q}\}_t+a_7\{\ddot{q}\}_{t+\Delta t}, \tag{9}$$

where the integration parameters are given by

$a_0=\frac{1}{\beta\Delta t^2}$, $a_1=\frac{\gamma}{\beta\Delta t}$, $a_2=\frac{1}{\beta\Delta t}$, $a_3=\frac{1}{2\beta}-1$, $a_6=\Delta t(1-\gamma)$ and $a_7=\gamma\Delta t$. In this study, $\beta=0.25$ and $\gamma=0.5$ are selected, which implies a constant average acceleration with unconditional numerical stability. $\{\ddot{q}_{t+\Delta t}\}$ in equation (8) can be substituted into equation (6), and equations for $\{\ddot{q}_{t+\Delta t}\}$ and $\{\dot{q}_{t+\Delta t}\}$ can be expressed in terms of the unknown vector $\{q_{t+\Delta t}\}$. If the response at time t is determined, then the displacement at $t+\Delta t$ may be obtained by solving the following algebraic equations:

$$(a_0[M]+a_0a_7[C]+[K+K_f])\{\ddot{q}\}_{t+\Delta t}=\{F\}+[M]\left(a_0\{q\}_t+a_2\{\dot{q}\}_t+a_3\{\ddot{q}\}_t\right)$$

$$+[C]\left((-1+a_7a_0)\{q\}_t+a_7a_2\{\dot{q}\}_t+a_7a_3\{\ddot{q}\}_t\right), \tag{10}$$

once the displacement vector $\{q_{t+\Delta t}\}$ is found, the velocity and acceleration vectors may be obtained using equations (8) and (9).

The procedure of applying the inverse finite element method to solve our problem is as follows. Assume that displacement velocity and acceleration at a beam element nodes are known, we can use equation (6) and equation (7) to identify moving wave loads from the inverse approach. Using the composite Simpson's rule for numerical integration, the right term in equation (7) can be expressed as

$$\Pi(j)\cong\frac{h_e}{3}\left\{f(0,j)Nd_i(0)+2\sum_{m=1}^{(N/2)-1}f(x_{2m},j)Nd_i(x_{2i})+4\sum_{m=1}^{(N/2)}f(x_{2m-1},j)Nd_i(x_{2m})+f(L,j)Nd_i(x_N)\right\}. \tag{11}$$

For a beam FEM model of 15 elements, if each element is divided into 10 segments, we have in total 151 integration points. Note that $h_e=\frac{L}{15}$ and $x_k=0+k\frac{L}{N}(k=0,1,2,...,N)$, where $N=15\times 10$. The residual error associated with approximation in equation (11) is

$$Error = \left| \frac{Lh_e^4}{180} \left[f(x,t) Nd_i(x) \right]^{(4)} \right| \le 1.05 \times 10^{-5} . \tag{12}$$

Note that

$$\left[a(0,0), a(x_1,0),...,a(x_N,0), a(0,1), a(x_1,1),...,a(x_N,1),...,a(0,N), a(x_1,N),...,a(x_N,N)\right]^T = \sum_{i=1}^{4} \begin{bmatrix} s_0 & 0 & ... & 0 \\ s_0 & s_1 & ... & 0 \\ & ... & ... & \\ s_0 & s_1 & ... & s_N \end{bmatrix} \begin{bmatrix} B_e & 0 & ... & 0 \\ 0 & B_e & .. & 0 \\ ... & & ... & \\ 0 & ... & 0 & B_e \end{bmatrix} \times \tag{13}$$

$\left[f(0,0), f(x_1,0),...., f(x_N,0), f(0,1), f(x_1,1),...., f(x_N,1),..., f(0,N), f(x_1,N),..., f(x_N,N)\right]$, where

$$S_j = Nd_i(x_j) \text{ and } B_e = \begin{bmatrix} \frac{h_e}{3} & \frac{4h_e}{3} & \frac{2h_e}{3} & \frac{4h_e}{3} & \frac{2h_e}{3} & \frac{4h_e}{3} & \frac{2h_e}{3} & ... & ... & \frac{h_e}{3} \end{bmatrix}.$$

Equation (12) can be simplified as

$$V_{[(N+1)\times(N+1)]\times 1} = G_{[(N+1)\times(N+1)]\times[(N+1)\times(N+1)]} \times F_{[(N+1)\times(N+1)]\times 1}, \tag{14}$$

where

$V_{[(N+1)\times(N+1)]\times 1} = \left[a(0,0), a(x_1,0),...,a(x_N,0), a(0,1), a(x_1,1),...,a(x_N,1),...,a(0,N), a(x_1,N),...,a(x_N,N)\right]^T$,

Here $a(x_i, j)$ represents the left hand side term of equation (6). The other two terms in equation (14) are given by

$$G_{[(N+1)\times(N+1)]\times[(N+1)\times(N+1)]} = \sum_{i=1}^{4} \begin{bmatrix} s_0 & 0 & ... & 0 \\ s_0 & s_1 & ... & 0 \\ & ... & ... & \\ s_0 & s_1 & ... & s_N \end{bmatrix} \begin{bmatrix} B_e & 0 & ... & 0 \\ 0 & B_e & .. & 0 \\ ... & & ... & \\ 0 & ... & 0 & B_e \end{bmatrix}$$

and

$F_{[(N+1)\times(N+1)]\times 1} = \left[f(0,0), f(x_1,0),...., f(x_N,0), f(0,1), f(x_1,1),...., f(x_N,1),..., f(0,N), f(x_1,N),..., f(x_N,N)\right]^T$. It should be noted that in this study, we assume that time dependent response such as displacement, velocity and acceleration at distributing $N+1$ fixed points on the beam are known and can be determined beforehand. Through iFEM inverse process, moving wave loads will be reconstructed correctly regardless of arrival time of the wave. Equation (14) is expressed as an inverse problem and $F = G^{-1}V$ is about ill-posed matrix problem. To solve this problem and find a useful and stable solution, the Moore-Penrose inverse method is applied. Since time dependent responses are known and often polluted with random noise disturbance. We develop the random noise filter algorithm to deal with responses before they are used for the inverse process.

3. Moving wave load identification

In this section, we use moving wave loads propagating our sensor model for tests. We suppose that the wave load begins to progressively step on the beam sensor until it is entirely on the beam. It is expressed by

$$f(x,t) = -A(x)\sin\left\{\frac{2\pi}{cT}(x-ct)\right\} \times \left\{1-\hat{H}(x-ct)\right\}, \left(0 \le t \le NT\right) \tag{15a}$$

In this equation, $\hat{N}$ is the number of half-cycles, T is single half-cycle time period and $\hat{H}$ is the Heaviside step function. The speed of wave is denoted by c, which for the cases considered here will be the speed of sound in air. Amplitude is $A(x)$, here x represents the location along beam sensor longitudinal direction. This equation represents a discrete number of half-cycles in a traveling sinusoidal wave. Once the load is completely on the sensor, the force is expressed as

$$f(x,t)=-A(x)\sin\left\{\frac{2\pi}{cT}(x-ct)\right\}\times\left\{\hat{H}\left(x-c\left(t-\hat{N}T\right)\right)-\hat{H}(x-ct)\right\},\left(\hat{N}T\le t\le L/c\right) \tag{15b}$$

until it reaches the other end of the beam. Finally, the load begins to leave the sensor, as

$$f(x,t)=-A(x)\sin\left\{\frac{2\pi}{cT}(x-ct)\right\}\times\left\{\hat{H}\left(x-c\left(t-\hat{N}T\right)\right)-\hat{H}(x-ct)\right\},\left(L/c\le t\le \hat{N}T+L/c\right) \tag{15c}$$

Note that equations (15a~c) represent a sinusoidal wave traveling along the beam sensor in parallel to the beam axial direction. Because the intent here is to demonstrate the effectiveness of the proposed method, waves incident at oblique angles are not considered. Nevertheless, there is no restriction in the approach to parallel traveling waves. The force reconstruction method is effective, no matter what angle the acoustic waves move across the beam sensor.

Suppose that the beam sensor is with sandwich-structured composite as shown in Figure 1, its relative material properties are listed in Table 1.

Table 1. Composite materials

Material	ρ (kg/m^3)	E11(GPa)	E22(GPa)	G12(GPa)	v12
Carbon/Epoxy	1600	177	10.8	76	0.270
Foam	11.2	85.53E-05	85.53E-05	31.40E-05	0.362

Initial values of other parameters are listed in Table 2.

Table 2. Initial parameters

Item	Description	Units	Value
L	Beam length	M	0.5
B	Beam width	M	0.03
H	Half sandwich Beam thickness	m	0.0005
C	Wave speed	ms^{-1}	343
T	Half-cycle duration	S	0.0005
TH	Foundation thickness	M	3
C_f	Foundation Young's modulus	MPa	2.3
K_E	The number of half-cycles		1
H	Thickness	M	0.5*H*

We assume that the wave load amplitude $A(x)$ is constant and equal to 1, and beam structural damping is neglected for simplicity. To simulate noisy measured data, we construct a vector $V=V_{true}+\varepsilon$ where ε is a random noise vector containing pseudorandom entries chosen from a normal distribution, with zero mean and standard deviation equal to one. Figure 2 shows moving wave loads obtained by using the iFEM method. Figure 2(a) and 2(b) represent the comparison of 2-d wave load inverse results with corresponding true values at beam location x=0.2317m and certain time $t=1.0\times10^{-03}$s. We notice that the waveform has good agreement with the true load, but the inverse wave amplitude is slightly larger. Figure 2(c) and 2(d) shows 3-D inverse results and influence of different noise levels on inverse results at the beam mid-span. From Figure 2(d), we notice that inverse results are much less influenced by background noise, even for a noise-to-signal ratio as high as 20%.

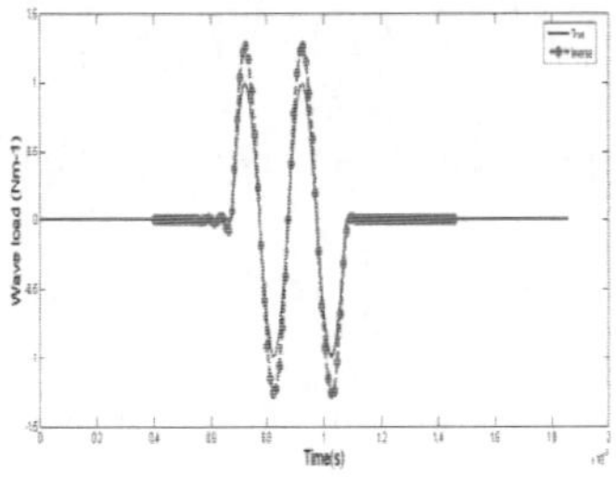

x=0.2317m
(a)

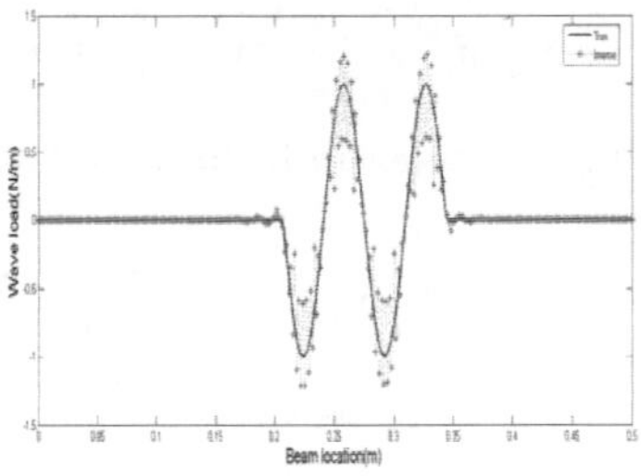

t=1.0e-03s
(b)

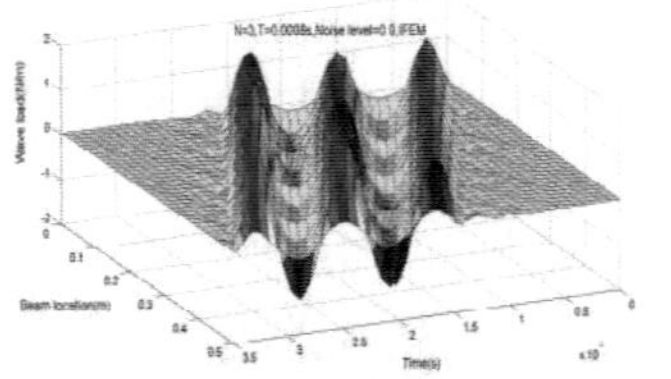

(c) 3-D inverse

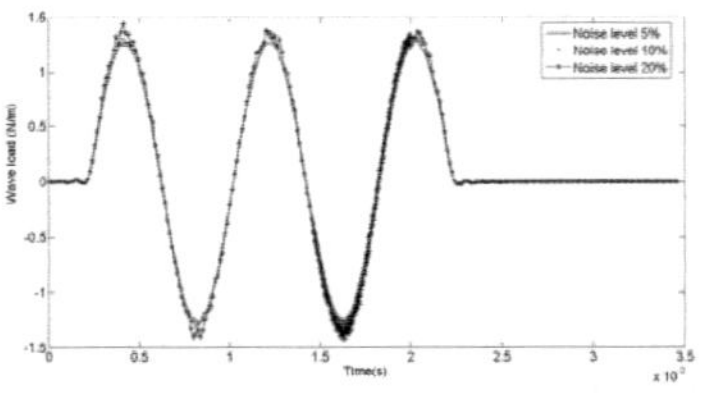

(d) Influence of different noise levels on inverse results at Mid beam location

Figure 2. Moving wave loads inverse results by using IFEM method

In addition, iFEM can always work well even no matter how we vary beam structural dimensions such as thickness.

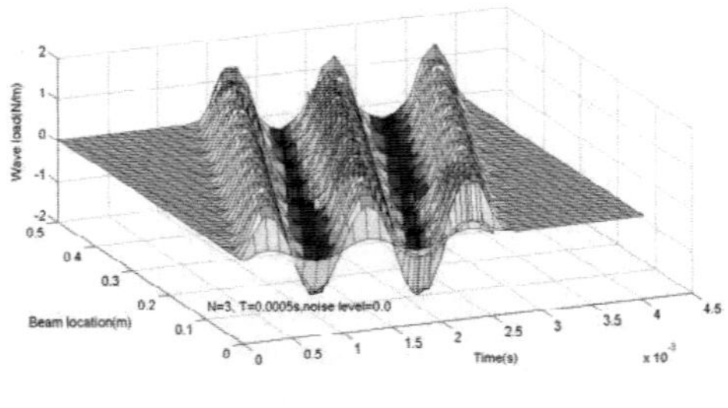

Figure 3. 3-D moving wave inverse results by the iFEM method through varying beam thickness from h=0.001m to h=0.03m.

From Figure 3, we find that even though we reduced the beam thickness from h=0.03m to h=0.001m, inverse results are not affected if the iFEM method is used. Figure 4 shows the minimum beam sensor length for the inverse methods to identify moving acoustic loads. From this figure, we can find that the minimum beam sensor length for iFEM is 0.5mm. From the comparison, we can draw the conclusion that the iFEM method is very practical to be applied for the new micro dimension sensor method for wave prediction.

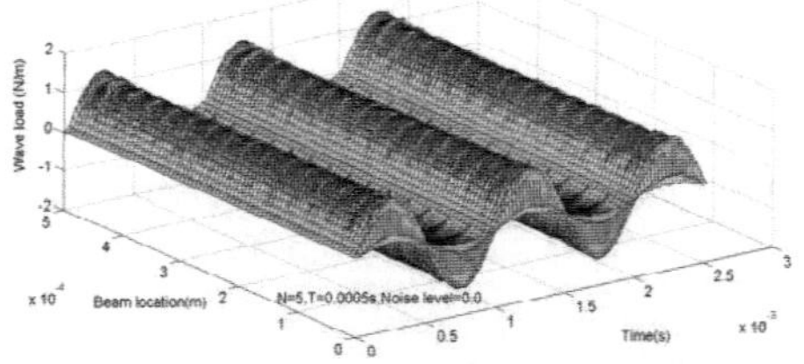

Figure 4. Minimum beam sensor length for the inverse methods to identify moving acoustic loads.

The iFEM can be used to identify wide range frequencies of moving wave loads which propagate through the sensor. Figure 5 shows different frequencies of wave loads identified by the iFEM.

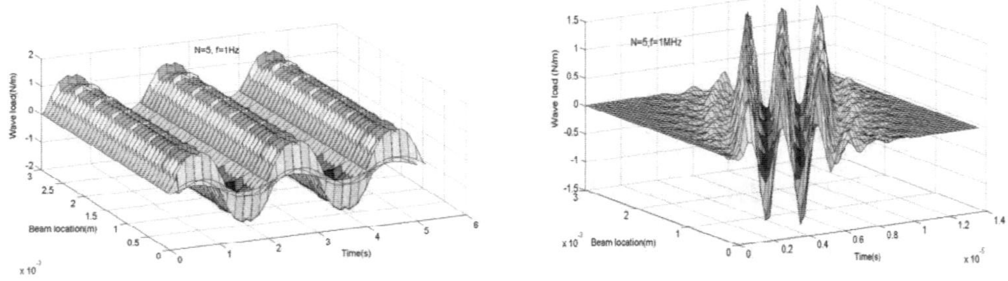

(a) 1Hz *(b)* 1MHz

Figure 5. Minimum and maximum frequencies of sinusoidal wave loads identified by iFEM method

From Figure 6, we know that iFEM method can well identify sinusoidal wave load frequencies between 1 Hz and 1MHz. This demonstrates that iFEM method is capable of predicting the wide range of wave frequencies, and it is very practical to be a new wave sensor method.

4. Conclusions

In this paper, we also apply the iFEM to identify moving loads, iFEM inverse results are compared with corresponding true wave loads. Furthermore, iFEM inverse results are compared with those of the analytical inverse method by changing some sensor structural parameters and noise levels. It is proved that iFEM is suitable in identifying wave loads, especially in the micro dimension sensor wave prediction and therefore it is very practical as a new sensor method.

References

[1] Shang Yu, Meng J, and Shi H, A new algorithm for relative localization in wireless sensor networks. In *18th International Parallel and Distributed Processing Symposium (IPDPS'04)*, Santa Fe, NM, USA, April, 2004, pp..24.

[2] Raykar V, Kozintsev I, and Lienhart R, Position calibration of microphones and loudspeakers in distributed computing platforms. *IEEE Transactions on Speech and Audio Processing*, Vol. 13 (2005), pp. 70–83.

[3] De Marziani C, Urenã J, Hernández Á, Mazo M, Alvarez F, García J J,Villadanos J M, and Jimenez A, Simultaneous measurement of times-of-flight and communications in acoustic sensor networks. In Workshop on Intelligent Signal Processing (WISP'05), Faro, Portugal, September, 2005, pp. 122–127.

[4] Zhang B, Shepard Jr W S. and Jiang Y K, Configuration study of an acoustic wave sensor on a continuous elastic support. *Smart Materials and Structures*, Vol. **21**, No. 2 (2012), pp. 20–36.

[5] Zhang B and Yu S, Research of a new sensor method for predicting arbitrary waves through structural configuration analysis. *Journal of Intelligent Material Systems and Structures*, 2014.pp.120-136

[6] Gunawan F E, Stress wave method for identification of viscoelastic material property based on finite element inverse analysis. *International Journal of Mechanical and Materials Engineering (IJMME)*, Vol. **6**, No.3 (2011), pp. 405–413.

[7] Gherlone M, Cerracchio P, Mattone M, Di Sciuva M and Tessler A, Shape sensing of 3D frame structures using an inverse finite element method. *International Journal of Solids and Structures*, Vol. **49**, No. 22 (2012), pp. 3100–3112.

[8] Kahya V and Mosallam A S, Dynamic analysis of composite sandwich beams under moving mass. *KSU Journal of Engineering Sciences*, Vol. **14**, No. 1 (2011).

[9] Reddy J N, *An Introduction to the Finite Element Method* (Third Edition). McGraw-Hill Higher Education, Boston, 2006.

IOP Conf. Series: Materials Science and Engineering **339** (2018) 012008 doi:10.1088/1757-899X/339/1/012008

A Modified Artificial Bee Colony Algorithm Application for Economic Environmental Dispatch

M Tarafdar Hagh[1,2], Omid Baghban Orandi[1]

[1]Faculty of Electrical and Computer Engineering, University of Tabriz, Tabriz, Iran
[2]Engineering Faculty, Near East University, 99138 Nicosia, North Cyprus, Mersin 10, Turkey

Email: tarafdar@tabrizu.ac.ir

Abstract. In conventional fossil-fuel power systems, the economic environmental dispatch (EED) problem is a major problem that optimally determines the output power of generating units in a way that cost of total production and emission level be minimized simultaneously, and at the same time all the constraints of units and system are satisfied properly. To solve EED problem which is a non-convex optimization problem, a modified artificial bee colony (MABC) algorithm is proposed in this paper. This algorithm by implementing weighted sum method is applied on two test systems, and eventually, obtained results are compared with other reported results. Comparison of results confirms superiority and efficiency of proposed method clearly.

1. Introduction

Traditionally, the main goal of power system operators was only determining of optimal output power of generating units with minimum fuel cost for system and environmental pollution was not considered as an important problem. But in recent years, regulations of governments due to increasing concerns on environmental issues have led to taking this issue into consideration seriously. Therefore, power utilities must plan to reduce emissions of generating units because of mentioned regulations [1]. So, minimizing of fuel cost and emission level of generating units simultaneously is the aim of power system operators nowadays.

In recent years, numerous intelligent optimization methods have been used to solve EED problem, some of which are: the genetic algorithm (GA) [2-6], the evolutionary programming (EP) [7-9], the particle swarm optimization (PSO) [10-11], the interactive honey bee mating optimization (IHBMO) algorithm [12], the artificial bee colony algorithm [14, 16], the multi-objective differential evolution (MODE) [2, 16], the harmony search algorithm [17], the gravitational search algorithm [18], the ant colony optimization [19], the flower pollination algorithm [20] and several hybrid and combined methods.

The ABC algorithm is a relatively new intelligent swarm-based optimization technique which is introduced by Karaboga in 2005 [21]. The ABC algorithm has been widely utilized to handle with various optimization problems due to its advantages including simple concept, easy-to-use, flexibility and powerful exploration capability. Nevertheless, some disadvantages such as poor exploitation capability and low convergence rate can cause major difficulties in more complex problems [22].
In this paper, a modified artificial bee colony algorithm has been used for solution of EED problem. The main advantages and superiority of proposed method are listed below:

1) The original ABC algorithm has some weaknesses such as inappropriate exploitation ability and convergence rate besides its powerful features. To overcome this weakness and improvement of algorithm performance, three modifications have been exerted in EED problem and the ABC algorithm procedure. The purpose of these modifications is to improve the exploitation capability of algorithm and to enhance its convergence rate by preventing of impracticable solutions generation.
2) In comparison with other reported methods, better results are obtained. The superiority of obtained results in terms of minimizing emission and generating cost and computation time, is demonstrated clearly in following sections of paper.
3) The promising performance of proposed method in handling with complex non-convex optimization problems indicates that this method can be considered as an appropriate and powerful technique to solve complicated problems.

To solve EED problem that is a multi-objective problem, the weighted sum technique is employed in solution method of this paper. The weighted sum is a well-known technique for solving of multi-objective problems which converts several objectives to a single objective with specific weights [23]. Eventually, the proposed modified ABC algorithm by employment of weighted sum technique is tested on two different systems and then compared with results of other reported methods in order to assess its efficiency.

The arrangement of this paper is done as follows. Section 2 explains the formulation details of economic environmental dispatch (EED) problem. Section 3 describes the problem solution. Section 4 provides obtained results and comparison with results of other reported methods. Finally, Section 5 concludes the paper.

2. Formulation of the EED Problem

To solve EED problem which is a constrained multi-objective optimization problem, the optimum output power of generating units should be determined in a way that cost and emission level are minimized simultaneously while satisfying all constraints of system and units. Above-mentioned objectives and constraints are explained as follows:

2.1 Fuel cost

By considering valve-point loading effect [6], total fuel cost of generating units is expressed as (1):

$$F = \sum_{i=1}^{N_G} [a_i + b_i P_i + c_i P_i^2 + |d_i \sin\{e_i (P_i^{\min} - P_i)\}|] \qquad (1)$$

where a_i , b_i , c_i, d_i and e_i are cost coefficients of ith generating unit; N_G is total number of the system generating units; P_i^{min} and P_i are lower limit of output power and actual output power of ith unit, respectively.

2.2 Emission

Due to using fossil fuel to generate electric power, some pollutants such as NO_x and SO_x are released into the air. Emission function of these pollutants can be expressed separately [25], but in this paper, emission function is a quadratic and exponential function which includes both mentioned pollutants [26]. Therefore, the emission function is expressed as follows:

$$E = \sum_{i=1}^{N_G} [\alpha_i + \beta_i P_i + \gamma_i P_i^2 + \eta_i \exp(\delta_i P_i)] \qquad (2)$$

where α_i , β_i , γ , η_i and δ_i are emission coefficients of ith generating unit.

2.3 Constraints

Power balance: The constraint of power balance is given as (3)

$$\sum_{i=1}^{N_G} P_i = P_D + P_{LOSS} \qquad (3)$$

where P_D and P_{LOSS} are the demand of system and power loss of transmission lines, respectively. P_{LOSS} is typically expressed by using B-coefficient method [27]. Therefore, the system power loss equation is given as below:

$$P_{LOSS} = \sum_{i=1}^{N_G}\sum_{k=1}^{N_G} P_i B_{ik} P_k + \sum_{i=1}^{N_G} P_i B_{i0} + B_{00} \quad (4)$$

Generation limits: The Output power of each generating unit should be within its lower and upper boundaries. These limits are represented by (5)

$$P_i^{\min} < P_i < P_i^{\max} \qquad (i = 1,\ldots,N_G) \quad (5)$$

where P_i^{min} and P_i^{max} are lower and upper boundaries of power generation for *i*th unit.

3. Problem Solution

3.1 Artificial Bee Colony

The artificial bee colony (ABC) algorithm is an intelligent optimization method introduced by Karaboga in 2005 [21]. The ABC algorithm animated the sagacious behavior of honeybees for food attainment.

The colony of artificial bees consists of three types of bees, which are called as employed, onlookers, and scout bees. The employed bees compose one-half of the colony, and the onlooker bees compose second half of it. For every food source, only one employed bee is allocated.

At first, ABC algorithm produces an initial random population of *P* solutions with *D* elements, where *P* and *D* are the numbers of population members and the number of problem parameters, respectively. In the ABC algorithm, each food source position represents a feasible solution, and its nectar quality shows the quality of the related solution. The initial population is generated as below:

$$x_i^j = x_{\min}^j + rand(0,1)(x_{\max}^j - x_{\min}^j) \qquad i = 1,2,\ldots,P \quad (6)$$

where j=1,2,..,D and x_{min}^j and x_{max}^j Indicate minimum and maximum limits of the parameter *j*, respectively. Furthermore, x_i represents a food source position.

Assessment of solutions is executed subsequently, and then it is applied to repetitive cycles of the algorithm. In each cycle, the employed bees produce a new position (solution) which is called neighboring food source and then assesses the quality of content (fitness value) of the new position (new solution). The mentioned neighboring food source (v) is generated as follows:

$$v_{ij} = x_{ij} + \varphi_{ij}(x_{ij} - x_{kj}) \qquad i = 1,2,\ldots,P \quad (7)$$

where j is a random integer number in the range of [1, D], k is a random integer number in the range of [1, P] which should not be the same as *i* and φ_{ij} is a random number in the range of [-1,1].

The employed bee compares the new and old solution and memorizes the solution with higher quality. When all of the employed bees complete their process, they share the obtained information related to the food sources with the onlooker bees which are waiting in the dancing area of the hive by waggle dancing. After that, the onlooker bee starts to evaluate the information of all employed bees and subsequently selects a food source according to the probability of each source. For defining mentioned probabilities, the fitness value of each solution should be computed. Fitness value of solution i is calculated as follows:

$$Fitness_i = \begin{cases} \dfrac{1}{1+f_i} & f_i \geq 0 \\ 1+abs(f_i) & f_i < 0 \end{cases} \quad (8)$$

where $Fitness_i$ and f_i are the fitness value and the cost value of a solution i, respectively.

For calculating the being chosen probability for each food source, equation (9) is employed:

$$p_i = \frac{Fitness_i}{\sum_{i=1}^{SN} Fitness_i} \quad (9)$$

A food source (solution) with better content clearly will absorb more onlooker bees in comparison with poor one. According to aforementioned probabilities, onlooker bees choose their food sources and then similar to the employed bees, the operations of producing a neighbor food source position, assessment, comparison and choosing the better one are executed.

During the ABC algorithm procedure, every solution has a chance of a predetermined number of cycles to be improved (which is known as a *limit* parameter). When no improvement occurs, the allocated employed bee leaves that food source and after that changes to a scout bee and begins its search process for a new food source. The mentioned procedure is repeated for a certain number of cycles which is called by maximum cycle number (MCN).

3.2 Modifications of problem and ABC algorithm

3.2.1 Specifying slack generating unit and its generation

To satisfy of power balance constraint perennially, one generating unit is determined as a slack unit [28]. In this case, generation of slack unit is specified by total generation of residual $N_G - 1$ units as follows:

$$P_s = P_D + P_L - \sum_{\substack{i=1 \\ i \neq s}}^{N_G} P_i \qquad (10)$$

where s is the chosen unit as slack one.

Therefore, the system power loss equation should be rewritten as (11):

$$P_L = B_{ss} P_s^2 + (2\sum_{\substack{i=1 \\ i \neq s}}^{N_G} B_{si} P_i + B_{0s}) P_s + \sum_{\substack{i=1 \\ i \neq s}}^{N_G} \sum_{\substack{i=1 \\ i \neq s}}^{N_G} P_i B_{ij} P_j + \sum_{\substack{i=1 \\ i \neq s}}^{N_G} B_{0i} P_i + B_{00} \qquad (11)$$

Combining (10) and (11) results in:

$$B_{ss} P_s^2 + (2\sum_{i=1}^{N_G} B_{si} P_i + B_{0s} - 1) P_s + (P_D + \sum_{i=1}^{N_G - 1} \sum_{j=1}^{N_G - 1} P_i B_{ij} P_j + \sum_{i=1}^{N_G - 1} B_{0i} P_i - \sum_{i=1}^{N_G - 1} P_i + B_{00}) = 0 \qquad (12)$$

By solving (12) generation of the slack unit will be determined.

3.2.2 Enhancement of ABC algorithm exploitation ability

To improve the ABC algorithm weakness in exploitation, two changes are implemented in the process of ABC algorithm:

1) In the process of producing new food source, onlooker bees can be propelled towards better positions. For this purpose, influenced by PSO [29], a global best solution which is achieved from prior cycles of the algorithm, can be employed. So the generation of new food source (solution) by onlooker bees is carried out as below:

$$v_{ij} = x_{ij} + \varphi_{ij}(x_{ij} - x_{kj}) + \beta(G_{best} - x_{ij}) \qquad (13)$$

where G_{best} is the best solution of the algorithm in previous cycles and β is a random number in [-1,1].

Scout bees have the analogous procedure. After quitting of a food source by its employed bee, the new scout bee produces new food source (new solution) by taking the global best solution into consideration as follows:

$$x'_{ij} = rand(x_{ij}^{\min}, x_{ij}^{\max}),$$

$$x_{ij}^{new} = x'_{ij} + \alpha(G_{best} - x'_{ij}) \qquad (14)$$

where α is an arbitrary number in the range of [-1,1].

2) Range damping of φ is an effective technique for improvement of poor exploitation capability of ABC algorithm. By range damping of φ in each cycle, the opportunity of solutions to improve will be enhanced and each solution can be exploited adequately to reach better solutions.

3.3 The Weighted Sum Method

The weighted sum is a common method for solving of multi-objective problems that convert some objectives to a single objective with specific weights. The importance of each objective and priority of decision-makers are two basic factors for determining the value of mentioned weights. These values commonly should be chosen in a way that summation of weights is equivalent to one [30]. Therefore, the multi-objective EED problem by using the weighted sum method is formulated as below:

$$\text{minimize} \quad w_1 \times F(P_i) + w_2 \times \alpha \times E(p_i) \qquad (15)$$

$$\text{subject to}: \quad w_1 + w_2 = 1$$

Where w_1 and w_2 are relevant weights of objectives. α is an important factor which solves the probable range difference of objective functions. In this paper, the α factor is calculated by following procedure [31]:

1- Computing the ratio between the maximum fuel cost of each generating unit and its maximum emission:

$$\alpha_i = \frac{F(P_{i,\max})}{E(P_{i,\max})} \qquad (16)$$

2- Sort α_i in ascending order. Start to sum of $P_{i,max}$ from the unit with lowest P_{max} till reaching power demand P_D.

3- The α_i of last generating unit in stage 2 is selected as α factor for the optimization problem.

By implementing the above-mentioned process, for example, the obtained α factor for system 1 is equal to 62.03564.

4. Results

To demonstrate superior performance and efficiency of proposed method, three different systems are studied. The results of studies have been obtained by using MATLAB software on a PC of Pentium-IV, 3.0 GHz and 4 GB RAM. The obtained results have been compared with some reported methods in [2] which are MODE, PDE, NSGA-II and SPEA 2 algorithms.

System 1: This system has 6 generating units with given specifications in Table 1. The system demand is 1200 MW [2]. The loss coefficients of system are given as follows:

$$B = 10^{-6} \times \begin{bmatrix} 140 & 17 & 15 & 19 & 26 & 22 \\ 17 & 60 & 13 & 16 & 15 & 20 \\ 15 & 13 & 65 & 17 & 24 & 19 \\ 19 & 16 & 17 & 71 & 30 & 25 \\ 26 & 15 & 24 & 30 & 69 & 32 \\ 22 & 20 & 19 & 25 & 32 & 85 \end{bmatrix} \qquad (17)$$

Table 1. Specifications of system 1

Unit	P_{min} (MW)	P_{max} (MW)	a ($/h)	b ($/MWh)	c ($/(MW)^2h)	α (lb/h)	β (lb/MWh)	γ (lb/(MW)^2h)
1	10	125	756.7988	38.5390	0.15247	13.8593	0.32767	0.00419
2	10	150	451.3251	46.1591	0.10587	13.8593	0.32767	0.00419
3	35	210	1243.5311	38.3055	0.03546	40.2669	-0.54551	0.00683

4	35	225	1049.9977	40.3965	0.02803	40.2669	-0.54551	0.00683
5	125	315	1356.6592	38.2704	0.01799	42.8955	-0.51116	0.00461
6	130	325	1658.5696	36.3278	0.02111	42.8955	-0.51116	0.00461

For execution of proposed algorithm, its main parameters should be determined. Colony size, limit value and maximum cycle number (MCN) for this system are 20, 30 and 200, respectively.

Besides the aforementioned algorithms, the obtained results of multi-objective MABC algorithm are listed in Table 2.

The best result of proposed method is obtained when $w_1 = 0.7768$ and $w_2 = 0.2232$. From Table 2, it is clear that the MABC has the minimum fuel cost (64842 $) in comparison with others. This method also is the best one in terms of execution time with 2.32 seconds. Although the minimum level of emission is not related to MABC, this algorithm shows the better overall performance compared to its competitors.

Table 2. Obtained results of system 1

	MABC	MODE [2]	PDE [2]	NSGA-II [2]	SPEA 2 [2]
P_1 (MW)	105.9614	108.6284	107.3965	113.1259	104.1573
P_2 (MW)	119.4011	115.9456	122.1418	116.4488	122.9807
P_3 (MW)	206.6863	206.7969	206.7536	217.4191	214.9553
P_4 (MW)	206.1632	210	203.7047	207.9492	203.1387
P_5 (MW)	308.0475	301.8884	308.1045	304.6641	316.0302
P_6 (MW)	305.3453	308.4127	303.3797	291.5969	289.9396
Cost ($)	64842	64843	64920	64962	64884
Emission (lb)	1285.2	1286	1281	1281	1285
CPU time (s)	2.32	3.09	3.52	5.42	7.05

System 2: This system has 10 generating units with given specifications in Table 3 .The system demand is 2000 MW [2]. The loss coefficients of system are given as follows:

$$B = 10^{-6} \times \begin{bmatrix} 49 & 14 & 15 & 15 & 16 & 17 & 17 & 18 & 19 & 20 \\ 14 & 45 & 16 & 16 & 17 & 15 & 15 & 16 & 18 & 18 \\ 15 & 16 & 39 & 10 & 12 & 12 & 14 & 14 & 16 & 16 \\ 15 & 16 & 10 & 40 & 14 & 10 & 11 & 12 & 14 & 15 \\ 16 & 17 & 12 & 14 & 35 & 11 & 13 & 13 & 15 & 16 \\ 17 & 15 & 12 & 10 & 11 & 36 & 12 & 12 & 14 & 15 \\ 17 & 15 & 14 & 11 & 13 & 12 & 38 & 16 & 16 & 18 \\ 18 & 16 & 14 & 12 & 13 & 12 & 16 & 40 & 15 & 16 \\ 19 & 18 & 16 & 14 & 15 & 14 & 16 & 15 & 42 & 19 \\ 20 & 18 & 16 & 15 & 16 & 15 & 18 & 16 & 19 & 44 \end{bmatrix} \quad (18)$$

The main parameters of proposed algorithm which are colony size, limit value and maximum cycle number (MCN) for this system have been determined as 20, 30 and 300, respectively.

Table 4 lists obtained result of MABC algorithm besides results of its competitors.

Table 3. Specifications of system 2

Unit	P_{min} (MW)	P_{max} (MW)	a $(\$/h)$	b $(\$/MWh)$	c $(\$/(MW)^2h)$	d $(\$/h)$	e (rad/MW)	α (lb/h)	β (lb/MWh)	γ $(lb/(MW)^2h)$	η (lb/h)	δ $(1/MW)$
1	10	55	1000.403	40.5407	0.12951	33	0.0174	360.0012	-3.9864	0.04702	0.25475	0.01234
2	20	80	950.606	39.5804	0.10908	25	0.0178	350.0056	-3.9524	0.04652	0.25475	0.01234
3	47	120	900.705	36.5104	0.12511	32	0.0162	330.0056	-3.9023	0.04652	0.25163	0.01215
4	20	130	800.705	39.5104	0.12111	30	0.0168	330.0056	-3.9023	0.04652	0.25163	0.01215
5	50	160	756.799	38.5390	0.15247	30	0.0148	13.8593	0.3277	0.00420	0.24970	0.01200
6	70	240	451.325	46.1592	0.10587	20	0.0163	13.8593	0.3277	0.00420	0.24970	0.01200
7	60	300	1243.531	38.3055	0.03546	20	0.0152	40.2669	-0.5455	0.00680	0.24800	0.01290
8	70	340	1049.998	40.3965	0.02803	30	0.0128	40.2669	-0.5455	0.00680	0.24990	0.01203
9	135	470	1658.569	36.3278	0.02111	60	0.0136	42.8955	-0.5112	0.00460	0.25470	0.01234
10	150	470	1356.659	38.2704	0.01799	40	0.0141	42.8955	-0.5112	0.00460	0.25470	0.01234

Table 4. Obtained results of system 2

	MABC	BSA [32]	MODE [2]	PDE [2]	NSGA-II [2]	SPEA-2 [2]
P_1 (MW)	54.8825	55	54.9487	54.9853	51.9515	52.9761
P_2 (MW)	80	80	74.5821	79.3803	67.2584	72.813
P_3 (MW)	83.9065	86.5308	79.4294	83.9842	73.6879	78.1128
P_4 (MW)	82.5526	86.9844	80.6875	86.5942	91.3554	83.6088
P_5 (MW)	137.3777	129.1542	136.8551	144.4386	134.0522	137.2432
P_6 (MW)	168.3053	146.9258	172.6393	165.7756	174.9504	172.9188
P_7 (MW)	300	300	283.8233	283.2122	289.435	287.2023
P_8 (MW)	308.6352	323.9002	316.3407	312.7709	314.0556	326.4023
P_9 (MW)	429.1395	435.9938	448.5923	440.1135	455.6978	448.8814
P_{10} (MW)	439.6692	440.0149	436.4287	432.6783	431.8054	423.9025
Cost ($)	113450	112807.37	113480	113510	113540	113520
Emission (lb)	4118.3	4188.0926	4124.9	4111.4	4130.2	4109.1
CPU time (s)	3.19	N/A	3.82	4.23	6.02	7.53

The best result of proposed method is obtained when $w_1 = 0.8465$ and $w_2 = 0.1535$. From Table 4, it can be observed that the MABC algorithm has minimum fuel cost (113450 $) and minimum competition time (3.19 seconds) in comparison with mentioned methods. The minimum emission amount is obtained by SPEA-2 method but this result cannot be an appropriate solution because of its higher generation cost. Overall, it can be seen that the proposed method has the better performance compared to its competitors.

5. Conclusion

This paper has applied multi-objective MABC algorithm to solve the environmental economic dispatch (EED) problem. The weighted sum technique has been used to handle with this multi-objective optimization problem.

The presented algorithm has been tested on two systems with different specifications for assessment of its efficiency and subsequently, its results have been compared with other reported methods. Obtained results clearly show that proposed method has competitive and suitable performance. The shorter computation time and the better solutions confirm this matter that the multi-objective MABC algorithm has a great capability for solving more complex optimization problems.

References

[1] J. Resek, K. Le, J. Golden, C. Stansberry, R. Vice, J. Wood, *et al.*, "Potential impacts of clean air regulations on system operations. Discussion," *IEEE transactions on power systems,* vol. 10, pp. 647-656, 1995.

[2] M. Basu, "Economic environmental dispatch using multi-objective differential evolution," *Applied Soft Computing,* vol. 11, pp. 2845-2853, 2011.

[3] R. T. King, H. Rughooputh, and K. Deb, "Solving the multiobjective environmental/economic dispatch problem with prohibited operating zones using NSGA-II," in *Communications, Computers and Signal Processing (PacRim), 2011 IEEE Pacific Rim Conference On*, 2011, pp. 298-303.

[4] Y. Song, G. Wang, P. Wang, and A. Johns, "Environmental/economic dispatch using fuzzy logic controlled genetic algorithms," in *Generation, Transmission and Distribution, IEE Proceedings-*, 1997, pp. 377-382.

[5] M. Basu, "Dynamic economic emission dispatch using nondominated sorting genetic algorithm-II," *International Journal of Electrical Power & Energy Systems,* vol. 30, pp. 140-149, 2008.

[6] M. Rajkumar, K. Mahadevan, S. Kannan, and S. Baskar, "Combined Economic and Emission Dispatch with Valve-point loading of Thermal Generators using Modified NSGA-II," *Journal of Electrical Engineering and Technology,* vol. 8, pp. 490-498, 2013.

[7] M. A. Abido, "Environmental/economic power dispatch using multiobjective evolutionary algorithms," *Power Systems, IEEE Transactions on,* vol. 18, pp. 1529-1537, 2003.

[8] M. A. Abido, "Multiobjective evolutionary algorithms for electric power dispatch problem," *Evolutionary Computation, IEEE Transactions on,* vol. 10, pp. 315-329, 2006.

[9] J. S. Dhillon and D. Kothari, "Economic-emission load dispatch using binary successive approximation-based evolutionary search," *Generation, Transmission & Distribution, IET,* vol. 3, pp. 1-16, 2009.

[10] M. Abido, "Multiobjective particle swarm optimization for environmental/economic dispatch problem," *Electric Power Systems Research,* vol. 79, pp. 1105-1113, 2009.

[11] K. Mandal, S. Mandal, B. Bhattacharya, and N. Chakraborty, "Non-convex emission constrained economic dispatch using a new self-adaptive particle swarm optimization technique," *Applied Soft Computing,* vol. 28, pp. 188-195, 2015.

[12] A. Ghasemi, "A fuzzified multi objective interactive honey bee mating optimization for environmental/economic power dispatch with valve point effect," *International Journal of Electrical Power & Energy Systems,* vol. 49, pp. 308-321, 2013.

[13] Sh. Sharifi, et. al. "Environmental economic dispatch using improved artificial bee colony algorithm, "*Evolving Systems, Springer*, Vol. 8, Issue 3, pp 233–242, 2017.

[14] D. C. Secui, "A new modified artificial bee colony algorithm for the economic dispatch problem," *Energy Conversion and Management,* vol. 89, pp. 43-62, 2015.

[15] S. Bhongade and S. Agarwal, "An optimal solution for Combined Economic and Emission Dispatch problem using Artificial Bee Colony Algorithm," in *Power and Energy Systems: Towards Sustainable Energy (PESTSE), 2016 Biennial International Conference on*, 2016, pp. 1-7.

[16] L. Wu, Y. Wang, X. Yuan, and S. Zhou, "Environmental/economic power dispatch problem using multi-objective differential evolution algorithm," *Electric Power Systems Research,* vol. 80, pp. 1171-1181, 2010.

[17] S. Sivasubramani and K. Swarup, "Environmental/economic dispatch using multi-objective harmony search algorithm," *Electric power systems research,* vol. 81, pp. 1778-1785, 2011.

[18] U. Güvenç, Y. Sönmez, S. Duman, and N. Yörükeren, "Combined economic and emission dispatch solution using gravitational search algorithm," *Scientia Iranica,* vol. 19, pp. 1754-1762, 2012.

[19] I. Karakonstantis and A. Vlachos, "Ant Colony Optimization for Continuous Domains applied to Emission and Economic Dispatch Problems," *Journal of Information and Optimization Sciences,* vol. 36, pp. 23-42, 2015.

[20] A. Abdelaziz, E. Ali, and S. A. Elazim, "Implementation of flower pollination algorithm for solving economic load dispatch and combined economic emission dispatch problems in power systems," *Energy,* vol. 101, pp. 506-518, 2016.

[21] D. Karaboga, "An idea based on honey bee swarm for numerical optimization," Technical report-tr06, Erciyes university, engineering faculty, computer engineering department2005.

[22] B. Akay and D. Karaboga, "A modified artificial bee colony algorithm for real-parameter optimization," *Information Sciences,* vol. 192, pp. 120-142, 2012.

[23] K. Deb, *Multi-objective optimization using evolutionary algorithms* vol. 16: John Wiley & Sons, 2001.

[24] D. C. Walters and G. B. Sheble, "Genetic algorithm solution of economic dispatch with valve point loading," *Power Systems, IEEE Transactions on,* vol. 8, pp. 1325-1332, 1993.

[25] J. Talaq, F. El-Hawary, and M. El-Hawary, "A summary of environmental/economic dispatch algorithms," *Power Systems, IEEE Transactions on,* vol. 9, pp. 1508-1516, 1994.

[26] M. Gent and J. Lamont, "Minimum-emission dispatch," *IEEE Transactions on power apparatus and systems,* vol. 6, pp. 2650-2660, 1971.

[27] A. Bhattacharya and P. K. Chattopadhyay, "Solving economic emission load dispatch problems using hybrid differential evolution," *Applied Soft Computing,* vol. 11, pp. 2526-2537, 2011.

[28] V. N. Dieu, P. Schegner, and W. Ongsakul, "A newly improved particle swarm optimization for economic dispatch with valve point loading effects," in *Power and Energy Society General Meeting, 2011 IEEE*, 2011, pp. 1-8.

[29] A. Mahor, V. Prasad, and S. Rangnekar, "Economic dispatch using particle swarm optimization: a review," *Renewable and sustainable energy reviews,* vol. 13, pp. 2134-2141, 2009.

[30] B. Gjorgiev and M. Čepin, "A multi-objective optimization based solution for the combined economic-environmental power dispatch problem," *Engineering Applications of Artificial Intelligence,* vol. 26, pp. 417-429, 2013.

IOP Conf. Series: Materials Science and Engineering **339** (2018) 012008 doi:10.1088/1757-899X/339/1/012008

[31] S. K. Kotagiri, N. R. Narra, and P. No, "Combined Economic And Emission Dispatch Using Artificial Neural Network," *IOSR Journal of Engineering (IOSRJEN),* vol. 3, pp. 01-09, 2013.

[32] Modiri-Delshad, Mostafa, and Nasrudin Abd Rahim. "Multi-objective backtracking search algorithm for economic emission dispatch problem." Applied Soft Computing 40, 479-494, 2016.

IOP Conf. Series: Materials Science and Engineering **339** (2018) 012009 doi:10.1088/1757-899X/339/1/012009

The Influence Analysis of the Rainfall Meteorological Conditions on the Operation of the Balloon Borne Radar in Plateau

Qiong Li[1], **Fangzhi Geng**[1,]

[1]Air Force Early Warning Academy, 288 HUANGPU Road, Wuhan 430019, China

Email: gfz0691@163.com

Abstract. Based on the characteristics of complex terrain and different seasons' weather in Qinghai Tibet Plateau, through statistic the daily rainfall that from 2002 to 2012, nearly 11 years, by Bomi meteorological station, Bomi area rainfall forecast model is established, and which can provide the basis forecasting for dangerous weather warning system on the balloon borne radar in the next step, to protect the balloon borne radar equipment's safety work and combat effectiveness.

1. Introduction

Over the past decade, balloon borne radar system are more and more popular in the armed forces around the world, with its strong ability of low altitude warning and leaving in the sky for a long time, high cost-effective, good mobility and maintenance ability. Because of the balloon borne radar works highly in the sky, its range of detection to the low altitude target is large, and it is mainly used to warning on some important area in military. With conservative calculation [1], the balloon borne radar's coverage capability of low altitude defense is equivalent to a few or even more than a dozen conventional ground-based radar.

In practical, the survival ability of the balloon borne radar system is closely related to rain and other weather conditions in the work area. In addition, the balloon borne radar detection performance is equally affected by meteorological clutter which is mainly the rain and cloud, especially, the radar worked in the S and C band are more serious. Therefore, it is necessary to research for the rain and cloud meteorological characteristics, which can better grasp the influence to the survivability and operas ability and other aspects of the balloon borne radar, so as to take effective measures to maximize play equipment operational effectiveness.

2. Research status

In foreign countries, many military departments have a perfect weather forecasting model, which can make a more accurate forecasting of long-term, medium-term, short-term and near dangerous weather forecasting, the forecasting factors are rich, and the timeliness is obvious. Such as a total of 350 civil site and 150 military site provide a set of 6 to 60 hours of meteorological element forecast guidance in United States, publish the following weather elements forecast guidance two times every day: the highest and lowest temperature, humidity, dew point, precipitation probability, types of precipitation, precipitation, thunderstorm probability, snowfall, cloud cover and ceiling, visibility, solar energy, solar radiation, surface wind and so on. It provide guidance for both civil disaster prevention and safety production, but also for military carrying the normal work of the balloon borne radar.

IOP Conf. Series: Materials Science and Engineering **339** (2018) 012009 doi:10.1088/1757-899X/339/1/012009

In our country, there are comparatively perfect weather forecast system, which is relatively rich in civilian business product, but it is very few in the field of military balloon borne radar. At present, we have basically mastered the influence of meteorological conditions and atmospheric environment to the balloon borne radar equipment in the low altitude plain area, but have not adequately studied in some specific regions of plateau all the time, lack of an auxiliary decision-making measure for development and safe use of equipment in complex weather conditions, which can help the development organization and combat troops. According to the operation situation of the two type of balloon borne equipment have been developed [2], it occurs frequently that solders have to wait until the weather conditions reach at the safe operation requirements, and caused many accidents due to the non-standard weather, causing significant losses.

Due to the geographical features and the climate of the plateau is different with the low altitude area, it results in that there are some distinctions on design, production and use of the balloon borne radar equipment. Therefore, it is urgent and realistic to carry out the analysis and research on the meteorological environment in the specific region and the accurate forecasting of the weather phenomena which affect the safety of the equipment, in the early stage of the development of the balloon borne radar equipment in plateau.

2.1. The analysis method

Time series analysis is a kind of processing method which uses the parameter model to analyze the observed ordered data. The dynamic characteristics and the state of the system can be analyzed and predicted by the time series, which can provide the basis for the system state monitoring. Matlab toolbox contains a lot of functions, with these functions, you can easily carry out time series analysis.

Meteorological data is characterized by time series data, Matlab time series function can find the regularity in the data which meet the characteristic, and could find the value which contained in the data. First, save the Meteorological data in the text file format after arranging, which obtained by radar. The file name can be X.txt, X refers to the year. Second, enter the command! Type X.txt, then the data of the text file will store in the Matlab software cache, and then, the X.txt data will be red in the time series format by ascii2fts function. The function is: ak = ascii2fts ('X.txt', 2,1), which is named akin Malab. By the ascii2fts function, the data all are stored in Matlab with the form of objects, the method of calling the data as follows: ak.dates read out the time field; ak.all read outall the weather field; ak.Rainball read out the rainfall field. Finally, draw the bar graph and enter bar (ak) to get Figure 1 to Figure 3.

2.2. The summary of geographical environment and meteorological condition on Bomi

Qinghai Tibet Plateau is 3000 to 5000 meters above sea level, with an average altitude of 4500 meters. Bomi area is located in the south eastern part of Tibet, is located in the middle reaches of the Yarlung Zangbo River north shore, shows a trend of North higher and south lower trend, has a wide disparity between the highest and the lowest terrain elevation, with a large fluctuation. At the same time, the meteorological characteristics of this region also has a significantly difference with the plain area, for example, dramatic and quickly changes in the weather, large distinction of temperature between day and night , lower air temperature in winter and spring, gusty wind, lightning, blizzard, long time for sunshine, strong ultraviolet. Bomi formed a special variety of climate zones which impacted by Indian Ocean of the southwest monsoon, coexisting in tropical, subtropical, temperate, frigid, humid and semi humid zone.

The characteristics of the plateau geographic and the meteorological environment determine that the design and use of the balloon borne radar equipment is very different within the Plain.

2.3. Analysis of the varied characteristics about rainfall in the last eleven years on Bomi

The weather station is located at 29.86 degrees north latitude, 95.76 degrees east longitude, altitude 2737 meters. This paper draws and analysis's the rainfall varied characteristic diagram in the past eleven years from 2002 to 2012 on Bomi region, carried through a large number of statistical data which comes from Bomi meteorological station daily rainfall data, used matlab development environment for computing platform.

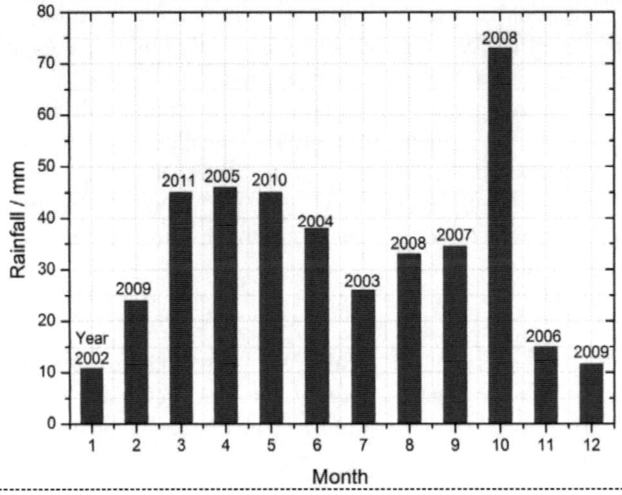

Figure 1. Daily maximum rainfall per month.

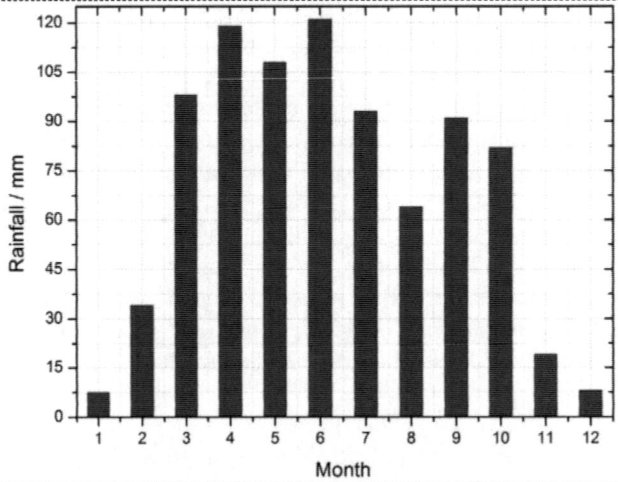

Figure 2. Annual average rainfall per month.

Figure 1 show, daily maximum rainfall per month all most exceeds 30mm from March to October in the past eleven years, especially, reached 86mm on October 2008. So the commander should avoid the tethered balloon flying, to guarantee the security of the radar equipment. From Figure 2 it can be seen that the annual average maximum rainfall in the past 11 years is June, reached 120.4mm. And the annual average rainfall is more than 65mm from March to October, reduced suddenly from November to February of the following year. From Figure 3 we can see that annual rainfall all most exceeds 120mm from March to October in the paste eleven years. Especially, the annual maximum rainfall reached 241mm on May 2010.

Generally, Bomi has a heavy rain more frequently than usual in summer. Due to the special terrain features of the plateau, Bomi Country appears cumulonimbus which can produce lightning throughout the entire year, and annual average rainfall is 977mm by statistic.

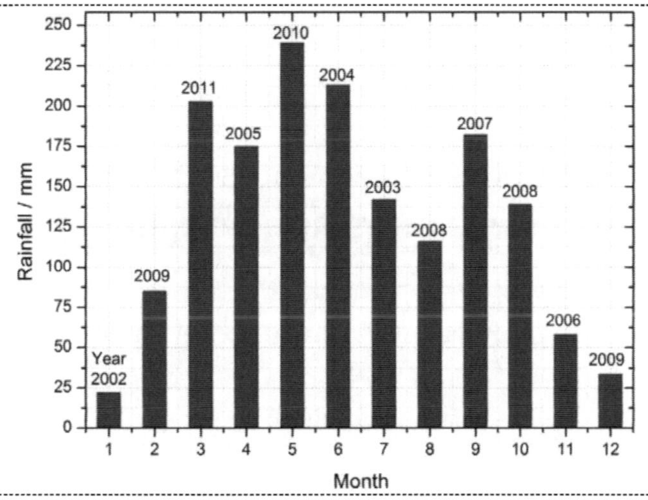

Figure 3. Annual maximum rainfall per month.

3. Concluding remarks

Plateau type of balloon borne radar [3] equipment is one of the key equipment is being developed by the air force; it will be deployed in the region of the Qinghai Tibet Plateau. From the performance of balloon borne radar's debugging and use in recent years, rainfall is one of the important factors affecting equipment safety, even severe weather disasters will directly endanger the sphere, resulting in sphere plunging to the ground or damaged [4].

Rainfall over the Tibetan Plateau is more concentrated. Monthly average rainfall exceeds 65mm from March to October each year, affected by warm moist air of the Indian Ocean and the bay of Bengal. During the time when the weather cannot reach safe requirement, the meteorological station should keep warning advanced and the balloon borne radar should avoid operation. From November to February, there is heavy wind and little rain because of the ground controlled by cold high atmospheric pressure. The rainfall is about 977mm each year from 2002 to 2012. the daily maximum rainfall reached 86mm. the monthly maximum on May 2010 reached 241mm. In the rainy season, the rainfall proportion is more than 80% on total year. In the alternately season of dry and wet, the rainfall characteristic of swelled and plummeted suddenly is obvious.

References

[1] Lixiao ZHAN, Ziyue TANG, etc. Modeling and Simulation of Cloud and Rain Clutter of Balloon Borne radar, Radar Science and Technology, 2015(1), pp 15-19.

[2] Shan GAO, Balloon Borne Radar Structure Overall Design, *Radar Science and Technology*, 2016 (6), pp 487-490.

[3] Li Jie, Wang Yang, Tethered Aerostat Balloon Borne Radar System Load Configuration Research, *Modern radar,* 2016 (9), pp 1-3.

[4] Yulin MA, The Use of the Balloon Borne Radar System in the Blockade of the Island, *The ship electronic countermeasure*, 2017 (1), pp 71-74.

Vibration characteristics of two-stage planetary transmission system with thin-walled ring gear on elastic supports

JianYing Li [1], QingChun Hu [2*], ChangFu Zong [1], TianJun Zhu [1] and ZeXing Zhang [1]

[1] School of Mechanical and Automotive Engineering, Zhaoqing University, Zhaoqing 526061, China
[2] School of Mechanical and Automotive Engineering, South China University of Technology, Guangzhou 510640, China

School of Natural Science Fund Project (201728); Practical Teaching Reform Project (sjjx201627); Guangdong Province Natural Science Fund (2014A030311045)

Email: huqingchunscut@163.com

Abstract. A dual-clutch and dual-speed planetary gears mechanism of a hybrid car coupled-system is taken as research subject, in which the ring gear of planet set II is a thin-walled structure and the clutch friction plates of planet set II are used as its elastic supports. Based on the lumped parameter-rigid elastic coupled dynamic model of two-stage planetary transmission system with thin-walled ring gear on elastic supports, the motion differential equations are established and the dynamic responses are solved by the Runge-Kutta method considering each stage internal and external time-varying mesh stiffness. The vibration displacements of each stage ring gear have been affected differently in time-domain, the translational vibration displacement of the ring gear of planet set I are obviously more than the torsional vibration displacement, but it is opposite for the ring gear of planet set II; The translational and torsional vibration responses of each stage ring gear arrive the peak in low-frequency. The analysis results of this paper can enrich the theoretical research of multistage planetary transmission and provide guidance for dynamic design.

1. Introduction

Planetary gears mechanism is a key part of the transmission system of a hybrid car, which is often solved multi-power sources coupled problem [1,2]. Although many studies on the planetary gears are carried out, its vibration and noise is still an important research subject, such as dynamic modeling, natural frequency analysis and dynamic response solving, parameters stability in Refs. [3-11], in which the basic components and planet gears are seen as rigid bodies, and the support stiffness and time-variant meshing stiffness are considered. When the ring gear is a thin-walled structure, its elasticity should be taken into account in certain situations [12-15].

The dynamic characteristics of transmission system have been affected by the elastic deformations of thin-walled ring gear, so the ring gear is no longer seen as a rigid body by some researchers, the modal properties of planetary gears having elastic continuum ring gear are studied by the perturbation and a candidate mode method, in which the planet are spaced equally and unequally, and its parametric instability are analyzed in Refs. [16-18]. Zhang Jun et al [19,20] pointed out the elasticity of the ring gear has a certain influence on the dynamic characteristics of planetary gears. Tao Qing et al[21] proposed a novel modeling method of rigid-flexible coupled planetary gear transmission systems that consist of rigid gear subsystems and elastic ring gear subsystems, its model is established by using the finite element method. Bao Heyun et al[22] established a translational-rotational dynamics model with meshing beyond pitch point in consideration of planetary gear sliding friction, time-varying mesh stiffness and damp, integrated mesh errors and flexible ring. Li Jianying and Hu Qingchun et al in Refs. [23,24] established the same lumped parameter-rigid elastic coupled dynamic model of two-stage planetary transmission system for a hybrid car through inter-stage coupled method, after including the elasticity of the ring gear of planet set II and simplifying its elastic supports. In Ref. [23], the natural frequencies are solved and the associated vibration modes are discussed, the rules are revealed which are the influences of the elastic supports stiffness and rim thickness of the ring gear of planet set II on natural frequencies of two-stage planetary transmission system, and the vibration modes are classified into seven types: I / II stage coupled rotational mode, I stage translational mode, I stage planet mode, II stage translational mode, II stage degenerate planet mode, II stage distinct planet mode and purely ring gear mode, its properties are summarized. In Ref. [24], the dynamic characteristics of two-stage planetary transmission system with thin-walled ring gear on elastic supports under different working conditions are discussed by the theoretical analysis and experimental testing.

In order to know the vibration characteristics of two-stage planetary transmission system with thin-walled ring gear on elastic supports, a dual-clutch and dual-speed planetary gears mechanism of a hybrid car coupled-system is taken as research subject in this paper, and its composition is described in detail in Refs. [23, 24], in which the ring gear of planet set II is a thin-walled structure and the clutch friction plates of planet set II are used as its elastic supports. The translational and torsional directions time-domain and frequency-domain vibration displacement of planet set I and II of two-stage planetary transmission system with thin-walled ring gear on elastic supports are solved by numerical method considering each stage internal and external time-varying mesh stiffness. The results of this paper can enrich the theoretical research of multistage planetary transmission and provide guidance for dynamic design.

2. The motion differential equations of planetary transmission with thin-walled ring gear on elastic supports [23, 24]

Two-stage planetary gears mechanism for a hybrid car is composed of planet set I and II, clutches c_1, c_2 and shell, there are three assumptions when modeling in this paper (Seen Refs.[23,24]). The carrier of planet set I is the input, the sun of planet set I and II is the output. The sun of planet set I and II connect together, the ring gear of planet set I and the carrier of planet set II connect together. Clutch c_1 simultaneously control the ring gear of planet set I and the carrier of planet set II .Clutch c_2 only control the ring gear of planet set II . Figure 1 is dynamic model of two-stage planetary system with thin-walled ring gear on elastic supports, its establishing method in detail can be seen Refs.[23,24].

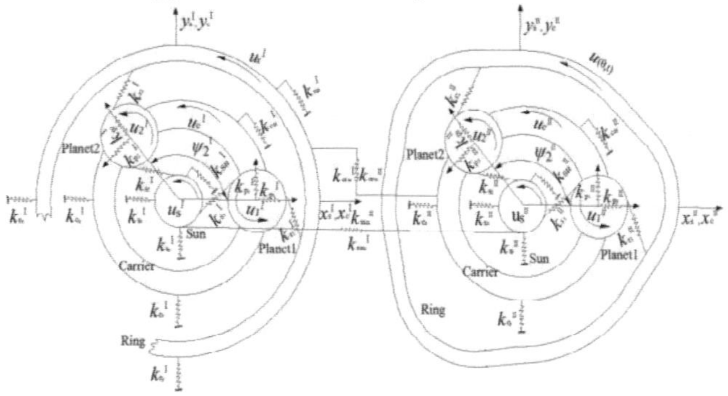

Figure 1. Dynamic model of two-stage planetary transmission system with thin-walled ring gear on elastic supports [23,24].

Establishing the motion differential equations of planet set Ⅰ and Ⅱ according to Newton's second law

$$M_{e}\ddot{u}+k_{bend}L_{1}u+k_{rn}^{\mathrm{II}}L_{2}u+k_{rn}^{\mathrm{II}}\sum_{n=1}^{N_{\mathrm{II}}}L_{3}^{n}(x_{n}\sin\alpha_{r}-y_{n}\cos\alpha_{r}-u_{n})=0 \tag{1}$$

$$\begin{cases} m_{r}^{i}\ddot{x}_{r}^{i}-\sum_{n=1}^{N_{i}}k_{rn}^{i}\delta_{rn}^{i}\sin\psi_{rn}^{i}+\Delta_{rx}^{i}+\nabla_{rx}^{i}=0 \\ m_{r}^{i}\ddot{y}_{r}^{i}+\sum_{n=1}^{N_{i}}k_{rn}^{i}\delta_{rn}^{i}\cos\psi_{rn}^{i}+\Delta_{rx}^{i}+\nabla_{ry}^{i}=0 \\ \left[I_{r}^{i}/\left(r_{r}^{i}\right)^{2}\right]\ddot{u}_{r}^{i}+\sum_{n=1}^{N_{i}}k_{rn}^{i}\delta_{rn}^{i}+\Delta_{ru}^{i}+\nabla_{ru}^{i}=0 \end{cases} \tag{2}$$

$$\begin{cases} m_{c}^{i}\ddot{x}_{c}^{i}+\sum_{n=1}^{N_{i}}k_{pn}^{i}(\delta_{nt}^{i}\sin\psi_{n}^{i}-\delta_{nr}^{i}\cos\psi_{n}^{i})+k_{cx}^{i}x_{c}^{i}=0 \\ m_{c}^{i}\ddot{y}_{c}^{i}-\sum_{n=1}^{N_{i}}k_{pn}^{i}(\delta_{nt}^{i}\cos\psi_{n}^{i}+\delta_{nr}^{i}\sin\psi_{n}^{i})+k_{cy}^{i}y_{c}^{i}=0 \\ \left[I_{c}^{i}/\left(r_{c}^{i}\right)^{2}\right]\ddot{u}_{c}^{i}-\sum_{n=1}^{N_{i}}k_{pn}^{i}\delta_{nt}^{i}+\Delta_{cu}^{i}=\frac{T_{c}^{i}}{r_{c}^{i}} \end{cases} \tag{3}$$

$$\begin{cases} m_{s}^{i}\ddot{x}_{s}^{i}-\sum_{n=1}^{N_{i}}k_{sn}^{i}\delta_{sn}^{i}\sin\psi_{sn}^{i}+k_{sx}^{i}x_{s}^{i}=0 \\ m_{s}^{i}\ddot{y}_{s}^{i}+\sum_{n=1}^{N_{i}}k_{sn}^{i}\delta_{sn}^{i}\cos\psi_{sn}^{i}+k_{sy}^{i}y_{s}^{i}=0 \\ \left[I_{s}^{i}/\left(r_{s}^{i}\right)^{2}\right]\ddot{u}_{s}^{i}+\sum_{n=1}^{N_{i}}k_{sn}^{i}\delta_{sn}^{i}+\Delta_{su}^{i}=-\frac{T_{s}^{i}}{r_{s}^{i}} \end{cases} \tag{4}$$

$$\begin{cases} m_{n}^{i}\ddot{x}_{n}^{i}+k_{rn}^{i}\delta_{rn}^{i}\sin\alpha_{r}^{i}-k_{sn}^{i}\delta_{sn}^{i}\sin\alpha_{s}^{i}+k_{pn}^{i}\delta_{nr}^{i}+\nabla_{nx}^{i}=0 \\ m_{n}^{i}\ddot{y}_{n}^{i}-k_{rn}^{i}\delta_{rn}^{i}\cos\alpha_{r}^{i}-k_{sn}^{i}\delta_{sn}^{i}\cos\alpha_{s}^{i}+k_{pn}^{i}\delta_{nt}^{i}+\nabla_{ny}^{i}=0 \\ \left[I_{n}^{i}/\left(r_{n}^{i}\right)^{2}\right]\ddot{u}_{n}^{i}-k_{rn}^{i}\delta_{rn}^{i}+k_{sn}^{i}\delta_{sn}^{i}+\nabla_{nu}^{i}=0 \end{cases} \tag{5}$$

where M_{e} represents equivalent mass of the ring gear of planet set Ⅱ; k_{bend} is bending stiffness of the ring gear of planet set Ⅱ, $k_{bend}=EI/R^{3}\left(1-v^{2}\right)$, where E is Yong's modulus of the ring gear of planet set Ⅱ, $E=2.07\mathrm{e}10^{11}$, v is Poisson's ratio of the ring gear of planet set Ⅱ, $v=0.3$, R is neutral radius of the ring gear of planet set Ⅱ, I is area moment of inertia of the ring gear of planet set Ⅱ; L_{1}, L_{2} and L_{3} are dimensionless operators(see Appendix of Refs.[23,24]). m_{h}^{i}, I_{h}^{i} are mass and moment of inertia of component h of planet set i; T_{h}^{i} is external torque of component h of planet set i; ∇_{rj}^{i}, ∇_{nj}^{i} represent elastic coupled term of the ring gear, the nth planet in the j direction, respectively; Δ_{rj}^{i} represent bearing coupled term of the ring gear in the j direction; Δ_{hu}^{i} represent torsional coupled term of component h of planet set i. All coupled terms are

$$\nabla_{rx}^{\mathrm{I}}=\nabla_{ry}^{\mathrm{I}}=\nabla_{ru}^{\mathrm{I}}=0,\nabla_{rx}^{\mathrm{II}}=\nabla_{ry}^{\mathrm{II}}=\nabla_{ru}^{\mathrm{II}}=(\sum_{n=1}^{N_{\mathrm{II}}}b_{r}\chi\Big|_{\theta=\psi_{n}^{\mathrm{II}}})$$

$$\nabla_{nx}^{\mathrm{I}}=\nabla_{ny}^{\mathrm{I}}=\nabla_{nu}^{\mathrm{I}}=0,\nabla_{nx}^{\mathrm{II}}=\nabla_{ny}^{\mathrm{II}}=\nabla_{nu}^{\mathrm{II}}=b_{p}\chi\Big|_{\theta=\psi_{n}^{\mathrm{II}}}$$

$$\Delta_{rx}^{i}=\begin{cases}k_{rx}^{\mathrm{I}}x_{r}^{\mathrm{I}},i=\mathrm{I}\\ \pi\left(k_{rbs}+k_{rus}\right)x_{r}^{\mathrm{II}},i=\mathrm{II}\end{cases},\Delta_{ry}^{i}=\begin{cases}k_{ry}^{\mathrm{I}}y_{r}^{\mathrm{I}},i=\mathrm{I}\\ \pi k_{rbs}y_{r}^{\mathrm{II}},i=\mathrm{II}\end{cases},\Delta_{ru}^{i}=\begin{cases}k_{ru}^{\mathrm{I}}u_{r}^{\mathrm{I}}+k_{cru}^{\mathrm{I}}\frac{r_{e}}{r_{r}^{\mathrm{I}}}\left(\frac{r_{e}}{r_{r}^{\mathrm{I}}}u_{r}^{\mathrm{I}}-\frac{r_{e}}{r_{c}^{\mathrm{II}}}u_{c}^{\mathrm{II}}\right),i=\mathrm{I}\\ \frac{2\pi k_{rus}}{\cos^{2}\alpha_{r}^{\mathrm{II}}u_{r}^{\mathrm{II}}},i=\mathrm{II}\end{cases}$$

$$\Delta_{su}^{i}=\begin{cases}k_{su}^{\mathrm{I}}u_{s}^{\mathrm{I}}+k_{ssu}^{\mathrm{I}}\frac{r_{b}}{r_{s}^{\mathrm{I}}}\left(\frac{r_{b}}{r_{s}^{\mathrm{I}}}u_{s}^{\mathrm{I}}-\frac{r_{b}}{r_{s}^{\mathrm{II}}}u_{s}^{\mathrm{II}}\right),i=\mathrm{I}\\ k_{su}^{\mathrm{II}}u_{s}^{\mathrm{II}}+k_{ssu}^{\mathrm{II}}\frac{r_{b}}{r_{s}^{\mathrm{II}}}\left(\frac{r_{b}}{r_{s}^{\mathrm{II}}}u_{s}^{\mathrm{II}}-\frac{r_{b}}{r_{s}^{\mathrm{I}}}u_{s}^{\mathrm{I}}\right),i=\mathrm{II}\end{cases},\Delta_{cu}^{i}=\begin{cases}k_{cu}^{\mathrm{I}}u_{c}^{\mathrm{I}},i=\mathrm{I}\\ k_{cu}^{\mathrm{II}}u_{c}^{\mathrm{II}}+k_{cru}^{\mathrm{II}}\frac{r_{e}}{r_{c}^{\mathrm{II}}}\left(\frac{r_{e}}{r_{c}^{\mathrm{II}}}u_{c}^{\mathrm{II}}-\frac{r_{e}}{r_{r}^{\mathrm{I}}}u_{r}^{\mathrm{I}}\right),i=\mathrm{II}\end{cases}$$

where $\chi=\frac{\partial}{\partial\theta}\sin\alpha_{r}^{\mathrm{II}}+\cos\alpha_{r}^{\mathrm{II}}$, $b_{r}=\left(-\sin\psi_{rn}^{\mathrm{II}},\cos\psi_{rn}^{\mathrm{II}},1\right)^{T}$, $b_{p}=\left(\sin\alpha_{r}^{\mathrm{II}},-\cos\alpha_{r}^{\mathrm{II}},1\right)^{T}$. r_{e} is connecting shaft radius of the ring gear of planet set I with the planet of planet II, r_{b} is connecting shaft of the sun gear of planet set I with the sun gear of planet set II, all other symbol of equations (1) -(5) can be referred in Refs. [23,24].

The internal and external time-varying mesh stiffness with the nth planet of planet set i $k_{sn}^{i}(t)$, $k_{rn}^{i}(t)$

$$k_{sn}^{i}(t)=\bar{k}_{sn}^{i}+2\mu^{i}\sum_{l=1}^{\infty}(a_{sn}^{i(l)}\sin l\omega_{\mathrm{m}}^{i}t+b_{sn}^{i(l)}\cos l\omega_{\mathrm{m}}^{i}t),k_{rn}^{i}(t)=\bar{k}_{rn}^{i}+2\varepsilon^{i}\sum_{l=1}^{\infty}(a_{rn}^{i(l)}\sin l\omega_{\mathrm{m}}^{i}t+b_{rn}^{i(l)}\cos l\omega_{\mathrm{m}}^{i}t) \quad (6)$$

where $\bar{k}_{sn}^{i}$, $\bar{k}_{rn}^{i}$ are internal and external average mesh stiffness; $2\mu^{i}$, $2\varepsilon^{i}$ are peak to peak amplitudes of internal and external mesh stiffness variations; $a_{h'n}^{i(l)},b_{h'n}^{i(l)}$ $(h'=s,r)$ are Fourier series expansion coefficient; l is Fourier series expansion number; ω_{m}^{i} is mesh frequency [24].

3. Dynamic responses of planetary transmission with thin-walled ring gear on elastic supports

The Runge-Kutta method is used for numerical solutions of equations (1) -(6) in this paper and numbers of each stage planet are 4. The input shaft rotation speed $n_{c}^{\mathrm{I}}=100r/\min$ and the input load torque $T_{c}^{\mathrm{I}}=35.2\mathrm{Nm}$, the ring gear of planet set I and the carrier of planet set II are fixed components, the numerical simulation parameters are shown in Table1.

Table 1. Parameters of two-stage planetary transmission system with thin-walled ring gear on elastic supports [23,24].

	Parameters	Ring gear	Carrier	Sun gear	Planet
Planet set I	Teeth	83	—	33	25
	Pitch diameter/mm	103.75	—	41.25	31.25
	Base diameter/mm	97.49	72.5	38.76	29.37
	Modulus/mm	1.25	—	1.25	1.25
	Pressure angle/(°)	20°	20°	20°	20°
	Mass/kg	0.547	0.573	0.106	0.064
	Moment of inertia/kg • mm^2	933.060	552.328	18.443	6.963
	Bearing stiffness/(N/μm)	170	109	126	108
	Torsional stiffness/(N/μm)		$k_{su}^{\mathrm{I}}=k_{cu}^{\mathrm{I}}=0$ $k_{ru}^{\mathrm{I}}=15750$		
	Mesh stiffness/(N/μm)		$\bar{k}_{rn}^{\mathrm{I}}=373.5$ $\bar{k}_{sn}^{\mathrm{I}}=322.8$		
	Tooth width/mm	17	—	17	17
Planet set II	Teeth	83	—	33	25
	Pitch diameter/mm	103.75	—	41.25	31.25
	Base diameter/mm	97.49	72.5	38.76	29.37
	Modulus/mm	1.25	—	1.25	1.25
	Pressure angle/(°)	20°	20°	20°	20°
	Mass/kg	0.741	0.613	0.143	0.086
	Moment of inertia/ kg • mm^2	1277.185	730.328	27.812	11.145

Bearing stiffness/(N/μm)	0.31	109	126	108
Torsional stiffness/(N/μm)		$k_{su}^{\text{II}}=k_{ru}^{\text{II}}=0$ $k_{cu}^{\text{II}}=6850$		
Mesh stiffness/(N/μm)		$\bar{k}_{rn}^{\text{II}}$=505.3 $\bar{k}_{sn}^{\text{II}}$=433.8		
Tooth width/mm	23	—	23	23
Coupling stiffness/(N/μm)		$k_{ssu}^{\text{I}}=k_{ssu}^{\text{II}}=151.0$ $k_{cru}^{\text{I}}=k_{cru}^{\text{II}}=5630$		
Contact ratios		$c_s^{\text{I}}=c_s^{\text{II}}=1.66$ $c_r^{\text{I}}=c_r^{\text{II}}=1.79$		

The vibration displacement time-domain curves of the ring gear of planet set Ⅰ and Ⅱ are shown in figure 2 and figure 3, where the negative value indicates the vibration displacement direction is opposite to the specified positive direction in the dynamic model. It can be known that the vibration displacements of each stage ring gear show fluctuation trends, in generally. The vibration displacement amplitudes of the ring gear of planet set Ⅰ in the x, y translational directions are clearly more than in the u torsional direction, which in the y translational direction are more than in the x translational direction, it is illustrated that the dynamic load acting on the support bearing of planet with the ring gear of planet set Ⅰ meshing in the y translational direction is more than in the x translational direction. The vibration displacement amplitudes of the ring gear of planet set Ⅱ in the x, y translational directions are clearly less than in the u torsional direction, which illustrated that the vibration displacement of the ring gear of planet set Ⅱ is significantly affected by torque.

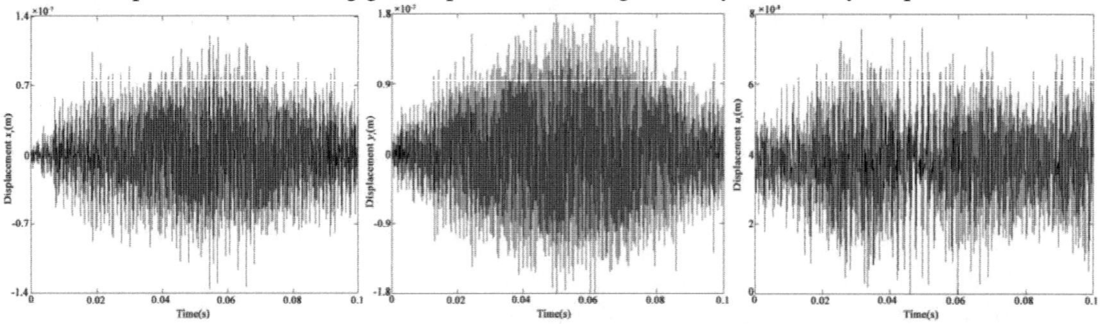

(a) x translational direction (b) y translational direction (c) u torsional direction

Figure. 2 Vibration displacement time-domain curves of the ring gear of planet set Ⅰ.

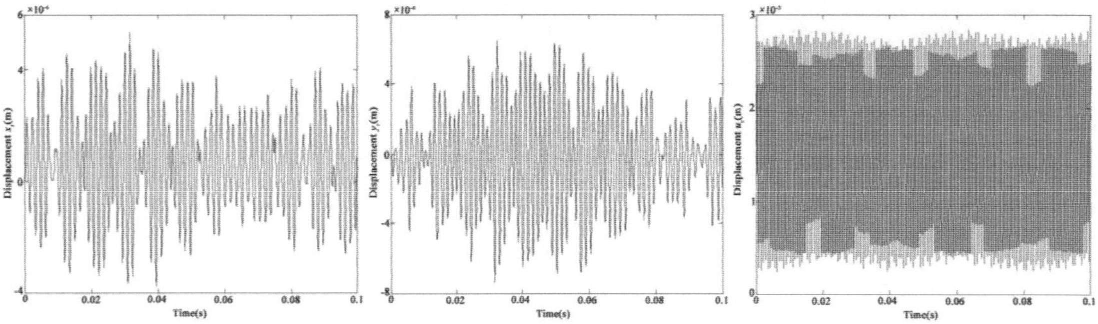

(a) x translational direction (b) y translational direction (c) u torsional direction

Figure. 3 Vibration displacement time-domain curves of the ring gear of planet set Ⅱ.

The vibration displacement frequency-domain curves of the ring gear of planet set Ⅰ and Ⅱ are shown in figure 4 and figure 5, in which f_m, f_h ($h=\text{r,c,s,p}$) are meshing fundamental frequency and rotating frequency of each stage component h', respectively. As shown in figure 4(a), it can be known that the combination frequencies $f_m/2-f_s-2f_c$, $2f_m-f_p-2f_s$, $5f_m\pm3f_p$ and $3f_m$, $5f_m$ appear in low-frequency, the vibration response peak at $3f_m$; The odd multiple meshing fundamental frequencies are main components in middle-frequency, such as $7f_m$, $9f_m$ and $11f_m$, where the vibration response peak at $7f_m$; Except $13f_m$ and the combination frequency $13f_m+3f_p-3f_c$ appearing in high-frequency, the combination frequency $12f_m+f_p+f_s$ is main component. As shown in figure 4(b), except $5f_m$ and the combination frequency $5f_m \pm 3f_p$, $2f_m+2f_p-2f_s$, f_p-2f_s appear in low-frequency, the meshing fundamental frequency f_m also appears, but the vibration response peak at $3f_m$; Except the odd multiple meshing fundamental frequencies appearing in middle-frequency, the combination frequency $8f_m-f_p+f_s$ also appears, the vibration response peak at $7f_m$; The combination frequencies $13f_m+f_p-f_s$ and $14f_m-f_p-f_s$ appear in high-frequency; As shown in figure 4(c), the combination frequencies f_s-3f_c, $5f_m+f_p+2f_s$ and $13f_m$ appear in low, middle and high-frequency of torsional direction, but the vibration displacement amplitudes are more.

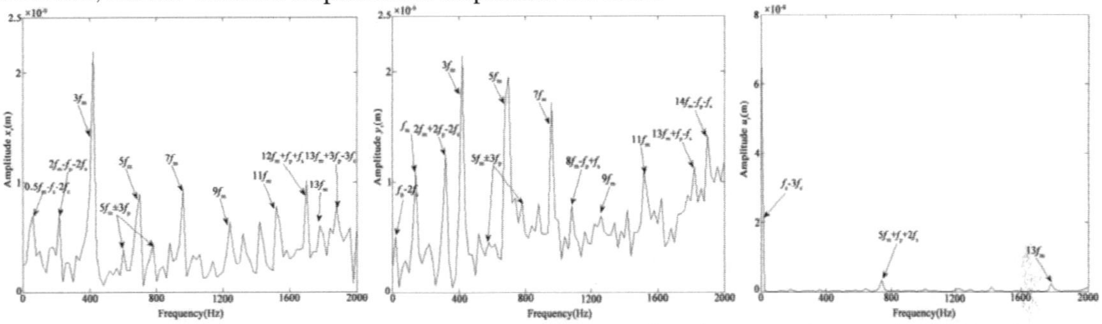

(a) *x* translational direction (b) *y* translational direction (c) *u* torsional direction

Figure. 4 Vibration displacement frequency-domain curves of the ring gear of planet set Ⅰ.

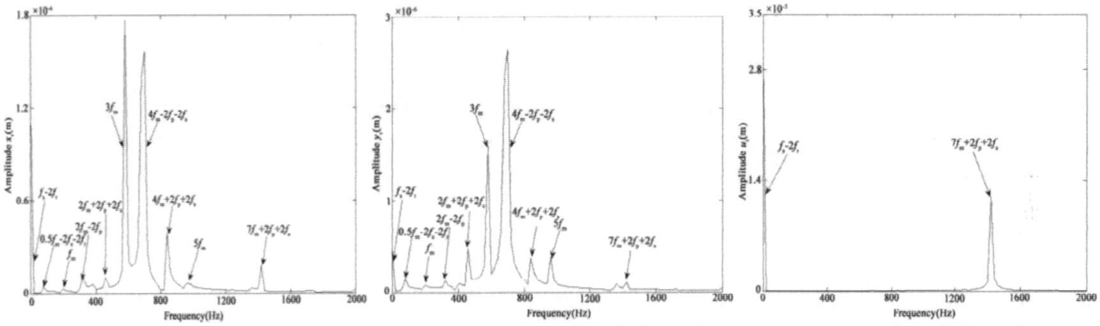

(a) *x* translational direction (b) *y* translational direction (c) *u* torsional direction

Figure. 5 Vibration displacement frequency-domain curves of the ring gear of planet set Ⅱ.

As shown in figure 5(a) and 5(b), it can be found that the spectra characteristic of the vibration displacements of the ring gear of planet set Ⅱ are similar in the x and y translational direction. Except the combination frequencies f_s-2f_r, $f_m/2-2f_s-2f_r$, $2f_m-2f_p$ and $2f_m+2f_p+2f_s$ appearing in low-frequency, the meshing fundamental frequency f_m also appear, the vibration response peak at $3f_m$ in the x translational direction, but in the y translational direction the vibration response peak at $4f_m-2f_p-2f_s$; Except the combination frequencies $4f_m+2f_p+2f_s$ and $7f_m+2f_p+2f_s$ appearing, meanwhile $5f_m$ also appears in middle-frequency; The multiple meshing fundamental frequencies and rotating frequency of each stage component don't appear in high-frequency. As shown in figure 5(c), the combination frequencies f_s-2f_c and $7f_m+2f_p+2f_s$ appear in the low, middle-frequency, the meshing fundamental frequency and rotating frequency of each stage component don't appear obviously in high-frequency.

4. Conclusions

The dynamic responses of two-stage planetary transmission are solved by the Runge-Kutta method considering each stage internal and external time-varying mesh stiffness, which base on the lumped parameter-rigid elastic coupled dynamic model of two-stage planetary transmission for a hybrid car, the translational and torsional vibration displacements of each stage ring gear are obtained and the time-frequency responses are analyzed. The vibration displacements of each stage ring gear are affected differently in time-domain, the translational vibration displacement of the ring gear of planet set Ⅰ are obviously more than the torsional vibration displacement, but it is opposite for the ring gear of planet set Ⅱ; The translational and torsional vibration responses of each stage ring gear arrive the peak in low-frequency.

References

[1] Y T Sun and H T Zhang 2011 *J. Heilongjiang Institute of Technol.* **25** 13–16 (in Chinese)
[2] Y T Luo and Y S Chen 2012 *J. Mech. Engineering.* **48** 70–75 (in Chinese)
[3] A Kahraman 1994 *J. Sound Vib.* **173** 125–130
[4] Y M Song, W D Xu, C Zhang and S Y Wang 2006 *J. Mech. Engineering.* **42** 16–21 (in Chinese)
[5] F H Duan, Q C Hu and C X Xie 2010 *J. Mech. Engineering.* **46** 62–67 (in Chinese)
[6] J Lin and R G Parker 1999 *ASME J. Vib. Acoust.* **121** 316–321
[7] T M Ericson and R G Parker 2013 *J. Vib. Acoust.* **135** 061002–1–13
[8] D R Kiracofe and R G Parker 2007 *J. Vib. Acoust.* **129** 1–16
[9] C G Cooley and R G Parker 2012 *J. Vib. Acous.* **134** 061014–1–11
[10] J Lin and R G Parker 2002 *J. Sound and Vib.* **249** 129–145
[11] C G Cooley and R G Parker 2013 *Int. J. Mech. Sci.***69** 59–71
[12] T Hidaka, Y Terauchi and K Nagamura 1979 *Bulletin of the JSME.* **22** 1142–49
[13] J P Vaujany and H C Kim 1996 *ASME Design Engineering Division.* **88** 73–80
[14] A Kahraman and S Vijayakar 2001 *J. of Mech. Design.* **123** 408–415
[15] A Kahraman, A A Kharazi and M Umrani 2003 *J. Sound Vib.* **262** 752–768
[16] X H Wu and R G Parker 2008 *J. Applied Mech.* **75** 0310141–12
[17] R G Parker and X H Wu 2010 *J. Sound Vib.* **329** 2265–75
[18] R G Parker and X H Wu 2012 *J. Vib. Acoust.***134** 0410111–11
[19] J Zhang, Y M Song and J J Wang 2009 *J. Mech. Engineering.* **45** 29–36 (in Chinese)
[20] J Zhang, X Z Liu, Y Jiao and Y M Song 2014 *J. Mech. Engineering.* **50** 104–112 (in Chinese)
[21] T Qing, W L Sun and J X Zhou 2015 *J. Xi'an Jiaotong University.* **49** 113–120 (in Chinese)
[22] H Y Bao, X J Zhou, R P Zhu and F X Lu 2016 *J. Central South University* (*Sci. Technol.*). **47** 3005–10 (in Chinese)
[23] J Y Li, Q C Hu, M Chai and X F Peng 2017 *J. Donghua University*(*Eng. Ed*). **34** 102–110
[24] J Y Li and Q C Hu 2016 *Proc. Int. Conf. ASIAN MMS 2016& CCMMS 2016* (*Guangzhou*) vol 408 pp 1023–1042

IOP Conf. Series: Materials Science and Engineering **339** (2018) 012011 doi:10.1088/1757-899X/339/1/012011

Analysis of Nonlinear Dynamics in Linear Compressors Driven by Linear Motors

Liangyuan Chen

Guangxi Power Grid Company Limited Electric Power Research Institute, Nanning, 530023, China

Email: xajdcly@163.com

Abstract. The analysis of dynamic characteristics of the mechatronics system is of great significance for the linear motor design and control. Steady-state nonlinear response characteristics of a linear compressor are investigated theoretically based on the linearized and nonlinear models. First, the influence factors considering the nonlinear gas force load were analyzed. Then, a simple linearized model was set up to analyze the influence on the stroke and resonance frequency. Finally, the nonlinear model was set up to analyze the effects of piston mass, spring stiffness, driving force as an example of design parameter variation. The simulating results show that the stroke can be obtained by adjusting the excitation amplitude, frequency and other adjustments, the equilibrium position can be adjusted by adjusting the DC input, and to make the more efficient operation, the operating frequency must always equal to the resonance frequency.

1. Introduction

Domestic and industrial applications, such as compressors, valves, pumps, vibrators, and so on, involve linear reciprocating motion. Usually, such motion is achieved by using rotary motors and rotary-to-linear mechanical converters. However, the overall efficiency of the system is relatively low, due to the inherent low efficiency of induction motors and the mechanical friction which is associated with the converters. A reciprocating compressor driven by a linear motor[1] eliminates the side force on the cylinder wall caused by the crank shaft and therefore not only significantly reduces the frictional loss but also provides a simple means for modulating the refrigerator load according to the demand and results in additional energy saving. Linear oscillating motors are inherently more potential in energy efficiency than other types for light duty compressors. This is why research has focused on this topic for the past 30 years. The major concern in previous studies[2] was that system performance under stable operations and, system instability problems, which might arise in practice, were not well understood probably because of complexities in the dynamic modeling of the system.

Configuration of a linear compressor under study is simple: a circular cylinder fixed onto a stator is supported by ground springs and a piston fixed onto a mover is suspended by piston springs. Piston movement, however, is determined in a rather complicated way by interaction of mechanical parts with electromagnetic and thermodynamic subsystems. The gas force in compression chamber possesses inherently nonlinear characteristics with respect to the piston motion and, hence, its dynamic components are coupled to static components. Furthermore, motion of the piston is not mechanically constrained. Therefore, piston motion becomes uncontrollable under some operating conditions. Chen et al.(2007) [3] and Park et al.(2002)[4] studied the static and dynamics characteristics and the effect of operating parameters through the linearized gas force load, respectively. In the paper, steady-state nonlinear response characteristics and the effect of operating parameters of a linear compressor were investigated considering the linearized and nonlinear gas force load.

2. Dynamic Modeling of Linear Compressor System

The mechanical system can be modeled as a mass-spring-damper[5-6], as is shown in Fig. 1. The moving mass consists of the mass of the piston, the carrier, the link, the mover, and part of the leaf springs. The governing equation of motion is

$$m\frac{d^2x}{dt^2}+c\frac{dx}{dt}+K_s(x-x_0)=F_g-F_e \tag{1}$$

where x is the mover position, m is the total moving mass, c is the viscous damping coefficient that represents the frictional effect between the cylinder and piston, K_s is the stiffness of the leaf springs, x_0 is the initial position where the spring force is zero, F_e is the motor force, F_D is the damping force, and F_g is the gas force.

The gas force is given by the following equation:

$$F_g(t)=[p(t)-P_s]A_p \tag{2}$$

$$s_0=x_0+\frac{(P_d-P_s)A_p}{2K_s} \tag{3}$$

$$p(t)=\begin{cases} P_s & p<P_s \quad \text{Suction} \\ \left[\dfrac{x_{max}-X_{top}}{x(t)-X_{top}}\right]^q P_s & P_s\le p\le P_d \quad \text{Compression} \\ P_d & p>P_d \quad \text{Exhaust} \\ \left[\dfrac{x_{min}-X_{top}}{x(t)-X_{top}}\right]^q P_d & P_s\le p\le P_d \quad \text{Expansion} \end{cases} \tag{4}$$

where A_p, $P(t)$, and P_s denote the cross sectional area of the piston, compression chamber pressure and suction pressure respectively, S_{piston} is the cross-sectional area of the piston, q is the polytropic coefficient, X_{Top} is the position of the cylinder top head. It is assumed that the gas behaves as an ideal gas in the compression cycle, that is, compression and expansion processes are described by polytropic processes as follows and shown in Fig. 2 and given as Eq. (4).

At the moment the mover's velocity is zero, the mover is at the maximum (or minimum) position, denoted x_{max} (or x_{min}). The actual stroke is given by $S = x_{max} - x_{min}$. The equilibrium position s_0 is given by $(x_{max} + x_{min}) / 2$ and is given as Eq. (3), which is required to be controlled to zero.

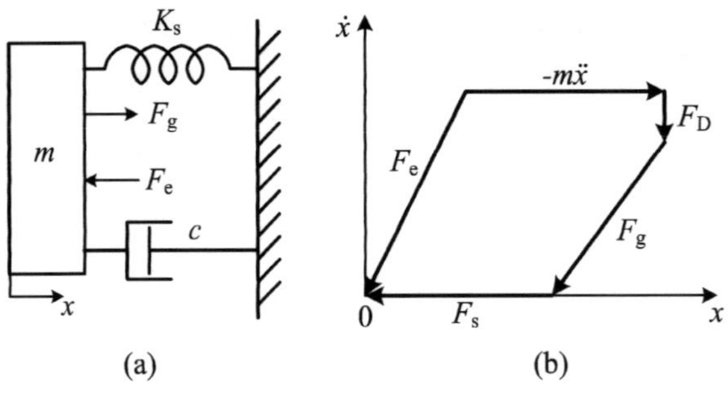

Figure 1. Dynamic model and force vector diagram. (a) Dynamic model. (b) Force vector diagram.

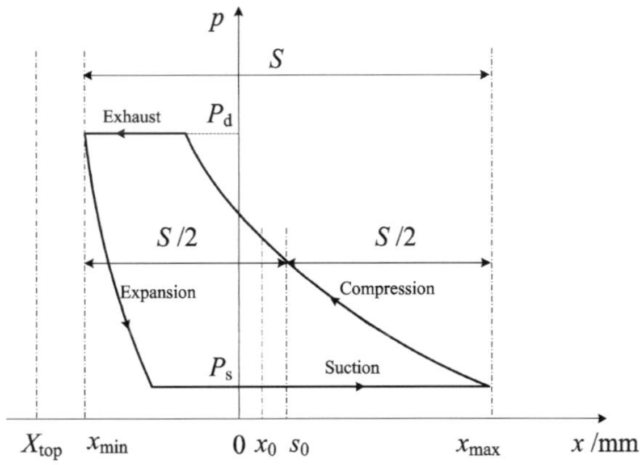

Figure 2. The polytropic processes

3. Dynamics Analysis Based on the Linearized Model

As is known above, K_g is a nonlinear function of the piston stroke and is subjected to the operating conditions. Kornhauser studied the gas spring[7], and regarded the mean spring stiffness as the equivalent stiffness, which is given by the following equation:

$$K_g = \frac{\beta q (P_d - P_s) A_p}{S} \tag{5}$$

where β is the stiffness coefficient of gas spring.

As long as the motion is harmonic, the energy loss of the mechanical system can be simulated by equivalent damping.The work on the gas, the heat exchange of the main cylinder and the loss of friction are considered in the form of energy dissipation. The first two items reflect the equivalent damping of the gas spring and can be estimated by the adiabatic work instead of the variable work. The indicated power of the compressor is given by the following equations:

$$W_g + W_h = \frac{q}{q-1} P_s [V_1 - (\frac{P_d}{P_s})^{\frac{1}{q}} V_s][(\frac{P_d}{P_s})^{\frac{q-1}{q}} - 1] \tag{6}$$

$$c_g = \frac{W_g + W_h}{\pi \omega A^2} \tag{7}$$

where W_g and W_h is the work on the gas and the heat exchange of the main cylinder respectively, V_1 is the maximum intake volume of cylinder, V_s is clearance volume at the top of a cylinder, A is the displacement amplitude of reciprocating piston, ω is the angular frequency of reciprocating motion of piston, c_g is the equivalent damping coefficient.

Let $c_{eq}=c+c_g$, $K_{eq}=K_s+K_g$, Eq. (1) is linearized as follows:

$$m\frac{d^2x}{dt^2} + c_{eq}\frac{dx}{dt} + K_{eq}(x(t) - x_0) = -F_e(t) \tag{8}$$

Let $x_0=0$, $\dot{x}=0$, the harmonic excitation force $F_e(t) = F_{em}\sin(t)$, the steady-state solution of Eq.(8) is given by the following equations:

$$x(t) = \frac{F_{em}}{K_{eq}} \frac{1}{\sqrt{(1-\frac{\omega^2}{\omega_n^2})^2 + (\frac{2n_D}{\omega_n}\frac{\omega}{\omega_n})^2}} \sin(\omega t - \varphi) \tag{9}$$

$$f_d = \frac{1}{2\pi}\sqrt{\frac{K_s}{m} + \frac{K_g}{m} - (\frac{c+c_g}{2m})^2} \tag{10}$$

where φ is the phase difference between displacement and exciting force, and f_d is the resonance frequency.

IOP Conf. Series: Materials Science and Engineering **339** (2018) 012011 doi:10.1088/1757-899X/339/1/012011

Resonant frequency is one of the important operating parameters[8], for the function of piston diameter, compressor stroke, mass of the moving part, spring stiffness and exhaust pressure, and many other factors.Fig.3 shows the result that the resonance frequency varies with the exhaust pressure and stroke. As can be seen from the diagram, the resonance frequency increases with the increase of the exhaust pressure due to the damping of the gas. When the exhaust pressure is constant, the resonance frequency decreases with the increase of the stroke. Fig.4 shows the gas force with the same stroke and different equilibrium positions.The gas pressure at the initial equilibrium position, both sides of the piston and the mechanical spring stiffness changes will alter the equilibrium position. The greater the exhaust pressure and the smaller spring stiffness, then the equilibrium position deviation is bigger, and the clearance is greater. All that is resulting in decrease in emissions even without exhaust. The equilibrium position in the operation can be adjusted by adjusting the initial balance position of the installation.

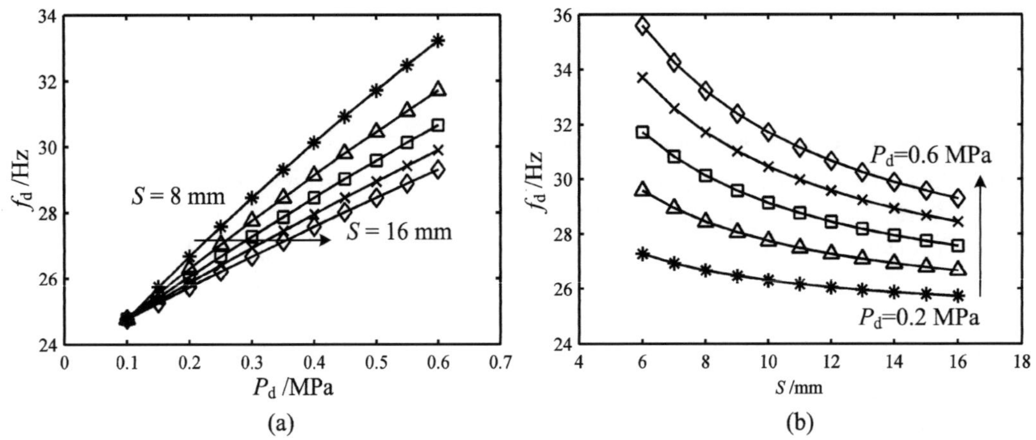

Figure 3. The resonance frequency varies with the exhaust pressure and stroke. (a) The resonance frequency varies with the exhaust pressure. (b)The resonance frequency varies with stroke.

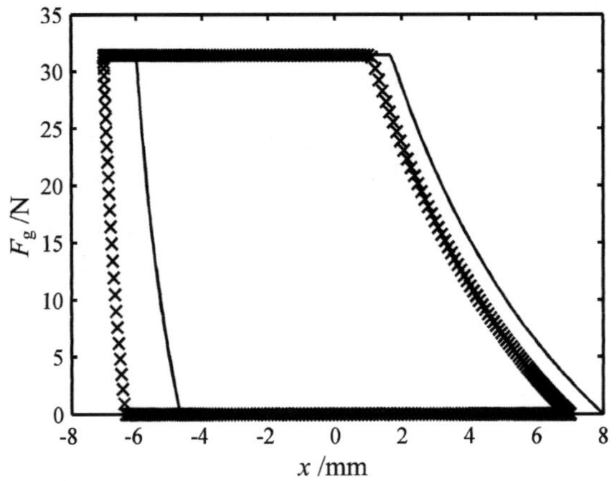

Figure 4. The gas force with the same stroke and different equilibrium positions

4. Dynamics Analysis Based on the Nonlinear Model

For the resonance frequency, it is important to reasonably select the ratio of m and K_s. When K_s keeps unchanged, m changes the resonance frequency of the mechanical system to affect the overall performance of the linear compressor. The compressor with light weight moving mass has higher dynamic response and system bandwidth, thus can work at high frequency, output much power, and the motor efficiency will be higher. So m should be as small as possible. And the smaller the moving mass, the smaller the stiffness of the spring, and the less material to make spring. But spring is in high frequency, the structure and material of the spring, the dynamic structure and material support are put forward very high requirements. And the moving mass m is not as small as possible. While the inertia of motion is small, anti disturbance ability is weak, on the other hand, the matched mechanical spring stiffness is small, energy storage capacity weak equilibrium position also affected.

Based on the nolinear model, under the same excitation force, the mechanical position of a small mechanical system with a small mass movement is offset greatly, and the displacement waveform is non sinusoidal and asymmetrical, as is shown in Fig.5 and Fig.6.

The power and the stroke vary with the frequency, and near the resonance frequency, the stroke and the input power of the exciting force is the biggest, but the balance position of the motion is shifted, as is shown in Fig.7.

When the frequency is unchanged, the stroke increases with the increase of the exciting force amplitude. When the exciting force amplitude is unchanged, the stroke at the resonant frequency is the largest. The resonance frequency decreases with the increase of the stroke, as is shown in Fig.8.

The model of an actual linear compressor is set up based on the nonlinear dynamics system, and the simulation result of displacement compared with the actual experimental result is shown in Fig. 9. The results are consistent, and the modeling method is proved to be correct.

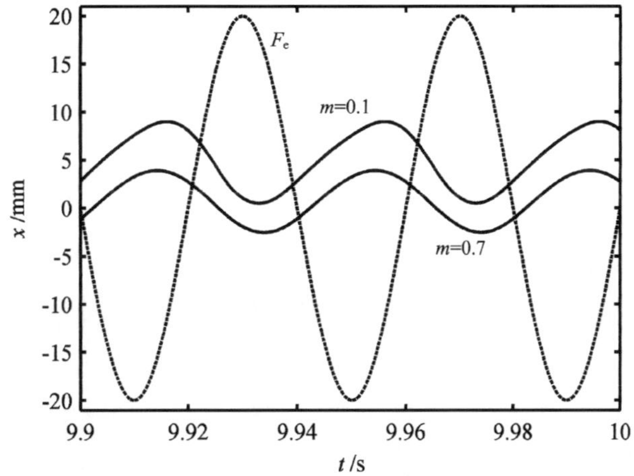

Figure 5. Displacement response at P_d=0.2 MPa, *m*=0.1kg and *m*=0.7 kg

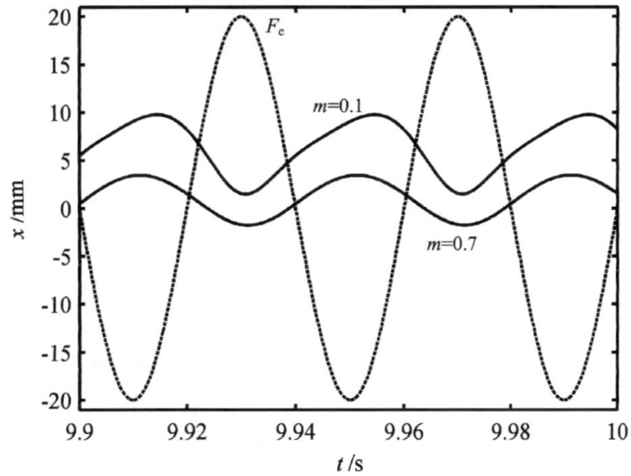

Figure 6. Displacement response at P_d=0.4 MPa, *m*=0.1 kg and *m*=0.7 kg

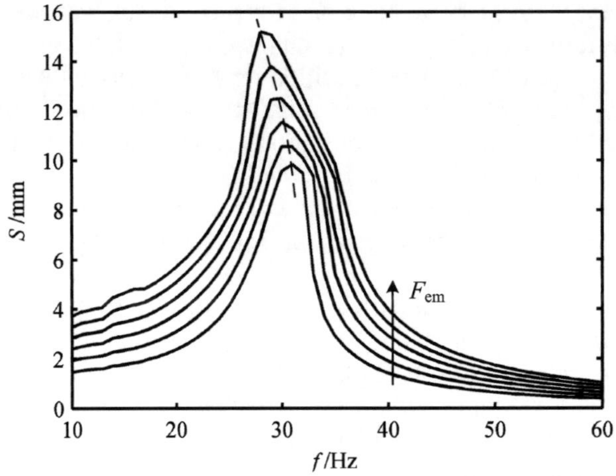

Figure 7. The relationship of stroke and phase with various frequency (F_{em}=25N)

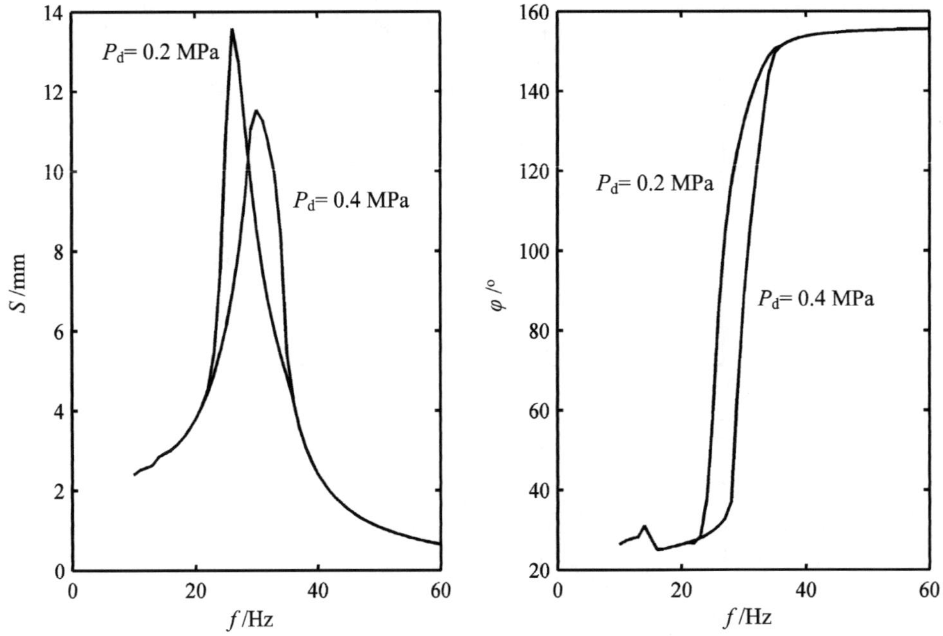

(a) (b)

Figure 8. The relationship between stroke and frequency under different excitation amplitudes (P_d= 0.4 MPa)

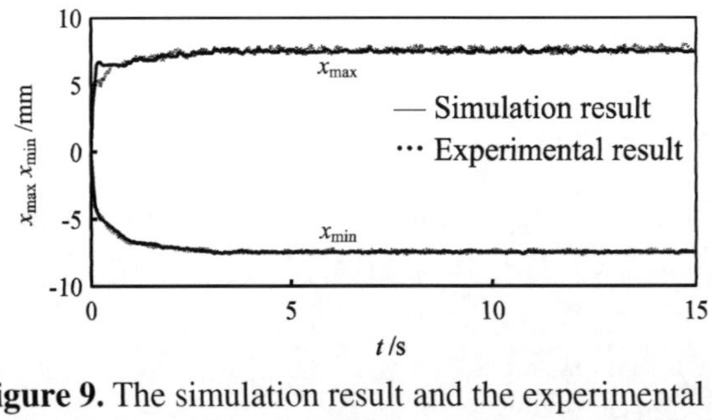

Figure 9. The simulation result and the experimental result

5. Summary

Steady-state nonlinear response characteristics of a linear compressor are investigated based on the linear and nonlinear model. The gas force can be expressed by the equivalent stiffness and equivalent damping, thus the nonlinear system is simplified as a linear model, which can be very convenient for the parameter analysis. Dynamics analysis based on the nonlinear model indicates that the moving mass is not the smaller the better; the mass-spring system can only absorb the excitation energy and have the maximum stroke when it works at the resonance frequency; the stroke increases with the increase of the excitation amplitude; the resonance frequency decreases with the increase of the stroke however increases with the increase of exhaust pressure.

References

[1] K. Park, E. P. Hong and K. H. Lee, Development of a linear motor for compressors of household refrigerators, Proceeding of 2001 Linear Drives and Industry Applications (LDIA '2001), Nagano, Japan, pp.283-286, 2001.

[2] C. Minas, Nonlinear Dynamics of an Oilless Linear Drive Reciprocating Compressor, ASME J. Vibr. Acoust., vol. 116 , pp. 79–84, 1994.

[3] N. Chen, Y. J. Tang, Y. N. Wu, Study on static and dynamic characteristics of moving magnet linear compressors, Cryogenics, vol. 47, pp.457–467, 2007.

[4] S. J. Park, Y. J. Hong, H. B. Kim, The effect of operating parameters in the Stirling cryocooler. Cryogenics, vol. 42, pp.419–425, 2002.

[5] G. S. Choe and K. J. Kim, Theoretical and experimental analysis of nonlinear dynamics in a linear compressor, Trans. ASME J. Vibr. Acoust. , vol. 124, pp.152–154, Jan. 2002.

[6] Yan et al. Memoryless variable structure control for affine nonlinear systems using only output information, International Journal of Robust and Nonlinear Control, Vol. 25, No.17, pp. 3316-3329, 2015.

[7] M. Xia and X. Chen, Analysis of resonant frequency of moving magnet linear compressor of Stirling cryocooler, Int. J. Refrig., vol.33, pp. 739–744, Jun. 2010.

[8] A. A. Kornhauser, Dynamic modelling of gas springs, Journal of Dynamic Systems, Measurement, and Control, 116: 414-418, 1994.

IOP Conf. Series: Materials Science and Engineering **339** (2018) 012012 doi:10.1088/1757-899X/339/1/012012

A New Fault Location Algorithm for T-type Transmission Lines Based on Phase Recognition

Zongfeng Zhu[1], Xuefeng Chen[1], Yichao Zhang[2,3], Cong Guo[1] and Dongwei Qiao[1]

[1] State Grid Taian Power Supply Company, Taian 271000, China;

[2] Shandong Hising Power Tech Co. LTD. Jinan 250101, China

Email: 352655727@qq.com

Abstract. In view of the traditional method for fault location of T-type transmission lines is not able to realize fault location correctly when the fault occurs near the T-type node, A new algorithm based on phase recognition is proposed about fault location of T-type transmission lines in this article. The algorithm constructs a range function by using the phase property of hyperbolic tangent function, regard the fault position parameter as a given condition, and the phase of function equals to zero when the reference point matches the fault position. Finally, the PSCAD/EMTDC simulation results verify the effectiveness of the algorithm.

1. Introduction

High voltage transmission line is an important component of power system. It not only undertakes the important task of power transmission, but also undertakes the interconnection of regional power grid. Its security and stability are really important. However, due to the long transmission distance and adverse external environment, it is easy to be affected by natural disasters such as wind, icing, lightning and so on, thus causing the line fault. For the long distance power transmission, the fault patrol has many difficulties, so searching a fast and accurate algorithm of fault location has very important significance to security, stability and economic operation of the power system[1].

With the rapid development of power system and national economy, T-type transmission lines are widely used in power systems because of their high transmission capacity and low cost of transmission construction. However, due to its large transmission power, the fast and accurate fault location of it is more and more important. Because of the complex structure of the T-type transmission lines, the conventional method of two-terminal fault location[2-3] can not be well applied to it. At present, T-type transmission lines are mainly based on traveling wave [4-5] and fault analysis[6-7]. The traveling wave method needs to be put into special equipment, the technology is more complex, the application cost is higher, and the difficult problem of wave-head identification exists. The method of fault analysis is of low requirement for equipment and small investment, so as to obtain wider application. However, the traditional method for fault location of T-type transmission lines is not able to realize fault location correctly when a high impedance fault occurs near the T-type node.

A new algorithm based on phase recognition is proposed about fault location of T-type transmission lines in this article. The algorithm regards the fault position parameter as a given condition, and the phase of function equals to zero when the reference point matches the fault position. The method does not need to judge fault types in advance, and has good applicability to nonlinear resistance faults and various fault types, and solves the problem of measuring dead zone near the T-type node in traditional method.

IOP Conf. Series: Materials Science and Engineering **339** (2018) 012012 doi:10.1088/1757-899X/339/1/012012

2. Fault analysis of double terminal line based on phase identification

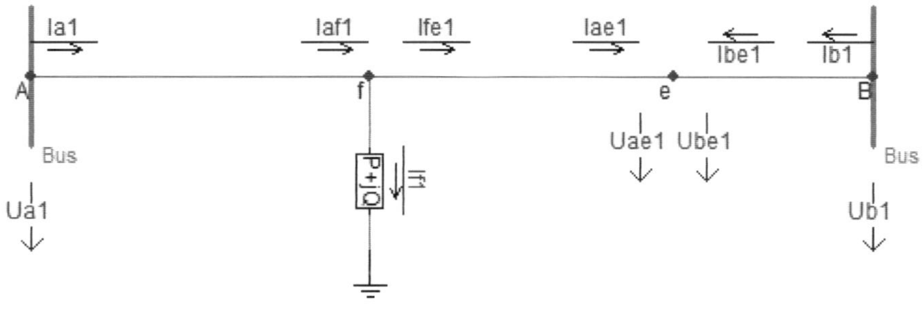

Figure 1. Positive sequence circuit diagram of two-terminal transmission lines

According to figure 1, when there is an unbalanced fault at the f point on the line, there is:

$$\dot{U}_{f1} = \dot{U}_{a1} cosh\gamma l_{af} - \dot{I}_{a1} Z_{c1} sinh\gamma l_{af} (1\text{-}1)$$

$$\dot{I}_{af1} = \dot{I}_{a1} cosh\gamma l_{af} - \frac{\dot{U}_{a1}}{Z_{c1}} sinh\gamma l_{af} (1\text{-}2)$$

$$\begin{cases} \dot{I}_{fe1} = \dot{I}_{af1} - \dot{I}_{f1} \\ \dot{U}_e = \dot{U}_{be1} \quad (1\text{-}3) \\ \dot{I}_e = -\dot{I}_{be1} \end{cases}$$

$$\dot{U}_e = \dot{U}_{f1} cosh\gamma l_{fe} - \dot{I}_{fe1} Z_{c1}\, sinh\gamma l_{fe} (1\text{-}4)$$

$$\dot{I}_e = \dot{I}_{fe1} \cosh\gamma l_{fe} - \frac{\dot{U}_{f1}}{Z_{c1}} sinh\gamma l_{fe} (1\text{-}5)$$

Taking（1-1）、（1-2）、（1-3）into（1-4）,we can get:

$$\dot{U}_e = \dot{U}_{a1} cosh\gamma\left(l_{af} + l_{fe}\right) - \dot{I}_{a1} Z_{c1} sinh\gamma\left(l_{af} + l_{fe}\right) + \dot{I}_{f1} Z_{c1} sinh\gamma l_{fe}$$
$$= \dot{U}_{ae1} + \dot{I}_{f1} Z_{c1} sinh\gamma l_{fe} (1\text{-}6)$$

$$\dot{U}_{ae1} - \dot{U}_{be1} = \dot{U}_e - \dot{I}_{f1} Z_{c1} sinh\gamma l_{fe} - \dot{U}_{be1} = \dot{I}_{f1} Z_{c1} sinh\gamma (l_f - l_e) (1\text{-}7)$$

Taking（1-1）、（1-2）、（1-3）into（1-5）,we can get:

$$\dot{I}_e = \frac{\dot{U}_{a1} cosh\gamma l_{af} - \dot{I}_{a1} Z_{c1} sinh\gamma l_{af}}{Z_{c1}} sinh\gamma l_{fk} + \left(\dot{I}_{a1}\, cosh\gamma l_{af} - \frac{\dot{U}_{a1}}{Z_{c1}}\, sinh\gamma l_{af} - \dot{I}_{f1}\right) cosh\gamma l_{fe}$$
$$= \dot{I}_{ae1} - \dot{I}_{f1} cosh\gamma l_{fe} (1\text{-}8)$$

$$\dot{I}_{ae1} + \dot{I}_{be1} = \dot{I}_e + \dot{I}_{f1} cosh\gamma l_{fe} + \dot{I}_{be1} = \dot{I}_{f1} cosh\gamma (l_f - l_e) (1\text{-}9)$$

where$\dot{U}_{a1}$、$\dot{U}_{b1}$ are positive sequence voltage of point A and point B; $\dot{I}_{a1}$、$\dot{I}_{b1}$ are positive sequence current of A and B;$\dot{U}_{f1}$、$\dot{I}_{f1}$ is positive sequence voltage and current of fault point f; $\dot{U}_{ae1}$、$\dot{U}_{be1}$ is the positive sequence voltage of reference point e from A and B by calculation; $\dot{U}_e$、$\dot{I}_e$ is positive sequence voltage and current of reference point e; γ is positive sequence propagation coefficient of transmission line; Z_{c1} is wave impedance.

According to（1-7）and（1-9）distance function can be constructed:

$$\mathrm{f}(l_e) = \frac{\dot{U}_{ae1} - \dot{U}_{be1}}{\dot{I}_{ae1} + \dot{I}_{be1}} = Z_{c1} tanh\gamma (l_f - l_e) \quad (1\text{-}10)$$

The hyperbolic tangent function is a singular function, so $\mathrm{f}(l_e)$ phase properties is similar to the odd functions, when $l_e < l_f$, $\arg \mathrm{f}(l_e)$ located near 90 degrees; when $l_e > l_f$, $\arg \mathrm{f}(l_e)$ located near -90 degrees, the phase is zero degree only when the reference point coincides with the fault point, so it has better phase recognition characteristics.

According to the phase characteristic, using the dichotomy or secant method to make the reference point E approaching the fault point F, Only when the reference point E coincides with the fault point F, the phase of the ranging function is zero, so we get the fault location.

3. Fault analysis of T-type transmission line based on phase identification

IOP Conf. Series: Materials Science and Engineering **339** (2018) 012012 doi:10.1088/1757-899X/339/1/012012

According to the second chapter, referencing the double terminal fault location, a new algorithm for fault location of T-type transmission lines is proposed.

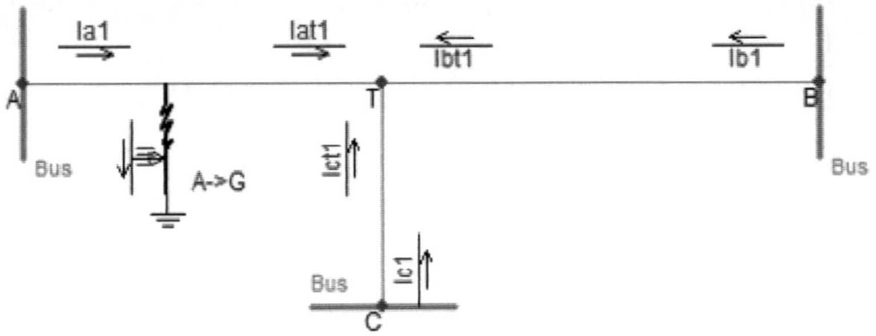

Figure 2. Positive sequence circuit diagram of T-type transmission lines

Figure 2 is positive sequence circuit diagram of T-type transmission lines when taking the AT line fault as an example. $\dot{U}_{a1}$、$\dot{U}_{b1}$、$\dot{U}_{c1}$ are positive sequence voltage of point A , point B and point C; $\dot{I}_{a1}$、$\dot{I}_{b1}$、$\dot{I}_{c1}$ are positive sequence current of point A , point B and point C; $\dot{U}_{t1}$ is positive sequence voltage of T node; $\dot{I}_{at1}$、$\dot{I}_{bt1}$、$\dot{I}_{ct1}$ are positive sequence current form each branch.

$\dot{U}_{at1}$、$\dot{U}_{bt1}$、$\dot{U}_{ct1}$ are voltage of T node by using parameter of point A, point B and point C:

$$\begin{cases} \dot{U}_{at1} = \dot{U}_{a1} cosh\gamma l_{at} - \dot{I}_{a1} Z_{c1} sinh\gamma l_{at} \\ \dot{U}_{bt1} = \dot{U}_{b1} cosh\gamma l_{bt} - \dot{I}_{b1} Z_{c1} sinh\gamma l_{bt} \\ \dot{U}_{ct1} = \dot{U}_{c1} cosh\gamma l_{ct} - \dot{I}_{c1} Z_{c1} sinh\gamma l_{ct} \end{cases} \quad (2\text{-}1)$$

$\dot{I}_{at1}$、$\dot{I}_{bt1}$、$\dot{I}_{ct1}$ are current of T node by using parameter of point A, point B and point C:

$$\begin{cases} \dot{I}_{at1} = \dot{I}_{a1} cosh\gamma l_{at} - \frac{\dot{U}_{a1}}{Z_{c1}} sinh\gamma l_{at} \\ \dot{I}_{bt1} = \dot{I}_{b1} cosh\gamma l_{bt} - \frac{\dot{U}_{b1}}{Z_{c1}} sinh\gamma l_{bt} \\ \dot{I}_{ct1} = \dot{I}_{c1} cosh\gamma l_{ct} - \frac{\dot{U}_{c1}}{Z_{c1}} sinh\gamma l_{ct} \end{cases} \quad (2\text{-}2)$$

When located at branch AT, we can get the voltage and current of T node by parameter of point B and point C, $\dot{U}_{t1} = \frac{\dot{U}_{bt1}+\dot{U}_{ct1}}{2}$, $\dot{I}_{t1} = \dot{I}_{bt1} + \dot{I}_{ct1}$; according to (1-7) and (1-9) ,we can get:

$$\dot{U}_{ak1} - \dot{U}_{tk1} = \dot{I}_{f1} Z_{c1} sinh\gamma(l_f - l_k) \quad (2\text{-}3)$$

$$\dot{I}_{ak1} + \dot{I}_{tk1} = \dot{I}_{f1} cosh\gamma(l_f - l_k) \quad (2\text{-}4)$$

$$\mathrm{f}(l_{ak}) = \frac{\dot{U}_{ak1}-\dot{U}_{tk1}}{\dot{I}_{ak1}+\dot{I}_{tk1}} = Z_{c1} tanh\gamma(l_f - l_k) \quad (2\text{-}5)$$

Similarly, the ranging function of branch BT and branch CT is:

$$\mathrm{f}(l_{bk}) = \frac{\dot{U}_{bk1}-\dot{U}_{tk1}}{\dot{I}_{bk1}+\dot{I}_{tk1}} = Z_{c1} tanh\gamma(l_f - l_k) \quad (2\text{-}6)$$

$$\mathrm{f}(l_{xk}) = \frac{\dot{U}_{ck1}-\dot{U}_{tk1}}{\dot{I}_{ck1}+\dot{I}_{tk1}} = Z_{c1} tanh\gamma(l_f - l_k) \quad (2\text{-}7)$$

Based on the above analysis and the conclusion of the second chapter, a new T-type fault analysis algorithm based on phase identification is proposed:

1)determine the fault branch

Choosing the reference point K at the node T, according to the formula (2-5) 、 (2-6) 、 (2-7) separately calculate $\mathrm{argf}(l_{ak})$、$\mathrm{argf}(l_{bk})$、$\mathrm{argf}(l_{ck})$. If there is only one branch, the phase is near -90 degrees, the phase of other branches are near 90 degree, it can be judged that this branch is the fault branch; if more than one branch judge as fault branch, the fault occurred at the T node.

2)fault distance measurement

Assuming that the branch AT fails, we divide branch AT into n equal to the right length, and calculate $\arg f(l_{ak})$ at each equal point. There must be a point of h let $\arg[f(\frac{h}{n}l_{ak})] > 0$ and $\arg[f(\frac{h+1}{n}l_{ak})] < 0$. Dichotomy is used within the interval $[\frac{h}{n}l_{ak}, \frac{h+1}{n}l_{ak}]$, the point whose $\arg f(l_{ak}) = 0$ is the fault location.

3. Simulation and results

Simulation model of T-type transmission line as shown in the following figure, using PSCAD/EMTDC for simulation calculation.

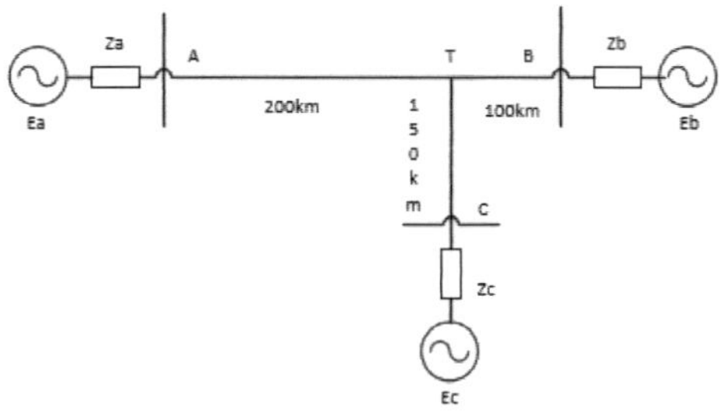

Figure 3. simulation model of T-type transmission line

system voltage: 220kV

initial phase: $\arg\dot{E}_a = \angle 0°$, $\arg\dot{E}_b = \angle 20°$, $\arg\dot{E}_c = \angle 40°$

Parameter of line: $R_1 = 0.02083\Omega/km$, $L_1 = 0.8948mH/km$, $C_1 = 0.0129\mu F/km$, $R_0 = 0.1148\Omega/km$, $L_0 = 2.2886mH/km$, $C_0 = 0.00523\mu F/km$

system parameters of A: $R_{a1} = 1.0515\Omega$, $L_{a1} = 80.154mH$, $R_{a0} = 0.6\Omega$, $L_{a0} = 63.4mH$

system parameters of B: $R_{b1} = 8.76\Omega$, $L_{b1} = 102.54mH$, $R_{b0} = 2.53\Omega$, $L_{b0} = 72.823mH$

system parameters of C: $R_{c1} = 3.672\Omega$, $L_{c1} = 138.46mH$, $R_{c0} = 5.7\Omega$, $L_{c0} = 90.8mH$

The sampling rate of the three terminal data is 6kHz, and the fundamental wave vector is extracted by full wave Fourier algorithm.

Table 1 takes the branch AT fault as an example, and gives the influence of different transition resistances and different fault types on the accuracy of fault location, From the table we can see that the algorithm is not affected by the fault type, and has good applicability to nonlinear impedance faults.

Table 1. influence of different fault types and transition resistance on distance measurement **results.**

Fault Type	Transition resistance	Actual Fault Distance /km			Maximum Deviation
		50	100	150	
Single phase grounding	50+25i	49.865	99.905	149.945	0.068%
	100+50i	49.845	99.875	149.905	0.078%
	150+100i	49.805	99.865	149.915	0.098%
Two-phase grounding	50+25i	50.075	100.115	150.075	0.058%
	100+50i	49.123	99.865	149.905	0.068%
	150+100i	49.785	99.895	149.945	0.108%
Two-phase short circuit	50+25i	49.665	99.755	149.885	0.168%
	100+50i	50.025	99.975	150.025	0.013%
	150+100i	50.025	100.075	150.015	0.038%
Three-phase short circuit	50+25i	49.965	149.965	149.965	0.018%
	100+50i	50.025	149.985	149.985	0.013%
	150+100i	50.095	150.085	150.085	0.048%

Table 2 gives the result when the single phase grounding occurs near the T nodes, we can see that the fault location can still be judged well through the algorithm, and solves the problem of measuring dead zone near the T-type node in traditional method.

Table 2. identification of single phase grounding where short circuit fault near T node.

Fault branch	Fault distance/km	$\arg(f_{at})$ /(°)	$\arg(f_{bt})$ /(°)	$\arg(f_{ct})$ /(°)	Fault branch judgment	Result
AT	199	-85.726	81.365	94.534	AT	+
AT	199.5	-85.175	82.152	95.372	AT	+
AT	199.8	-84.332	90.564	89.879	AT	+
AT	199.9	-84.233	86.665	92.862	AT	+
BT	99	90.766	-103.31	89.751	BT	+
BT	99.5	90.717	-103.26	89.754	BT	+
BT	99.8	90.655	-103.38	89.816	BT	+
BT	99.9	90.614	-103.39	89.687	BT	+
CT	149	96.226	82.125	-84.299	CT	+
CT	149.5	97.158	83.142	-83.387	CT	+
CT	149.8	90.767	90.877	-82.642	CT	+
CT	149.9	90.859	91.065	-82.231	CT	+

4.Conclusion

This paper presents a new fault location method of T-type fault location based on phase identification. The algorithm constructs a range function by using the phase property of hyperbolic tangent function, using the phase difference to judge the fault branch, and using the phase of function equals to zero when the reference point matches the fault position to realize fault location.

The method firstly judges the fault branch line, and then makes precise location of the fault, which reduces the scale of operation and improves the efficiency of fault location. Compared with the traditional T-type transmission line fault location algorithm, the algorithm is not affected by the load current, no need to know fault types in advance, has good applicability for nonlinear resistance fault. At the same time, it solves the problem of measuring dead zone near the T-type node in traditional method.

Reference

[1] WANG Zhaolei, LIU Mingguang. Traveling Wave Fault Location for Teed Transmission Line Based on Solutions of Linear Equations [J]. Power System Technology,2014,38(04):1046-1050. [2017-08-27]. DOI: 10.13335/j.1000-3673.pst.2014.04.034

[2] JIN Xianing, WANG Fuping, WANG Zanji. Research on Fault Location Based on Dynamic Synchronous Phasor Measurement by PMU [J]. Power System Technology,2013,37(10):2932-2937. [2017-08-27]. DOI: 10.13335/j.1000-3673.pst.2013.10.024

[3] WANG Bo, JIANG Quanyuan, GU Wei, CHEN Donghai. A Novel Fault Location Method Suitable for Transmission Line with Multi Branches[J]. Power System Technology,2013,37(04):1152-1158. [2017-08-27]. DOI: 10.13335/j.1000-3673.pst.2013.04.045

[4] Li botong, li Yongli, Hu Yu. A New Fault Location Algorithm for Extra-high Voltage Mixed Lines Based on Fault Zone Identification [J/OL]. Automation of Electric power system, 2013,37(05):93-98.

[5] Zhang Yongjian，Xu Jie，Sun Jia．Travelling wave fault location for three-terminal transmission line based on static wavelet transform[J]．Power System Technology，2012，36(6)：84-88(in Chinese)．

[6] LIN Fuhong, WANG Zengping, ZENG Huimin. A Novel Fault Location Algorithm Based on Phase Characteristics of Fault Location Function for Three-terminal Transmission Lines,2011,31(13):107-113. [2017-08-27]. DOI: 10.13334/j.0258-8013.pcsee.2011.13.015

[7] Shi Shihong，He Benteng，Zhang Wujun. Fault location for HV three-terminal transmission lines[J]. Proceedings of the CSEE，2011，31(13)：105-110(in Chinese).

The statistical analysis of the Geomagnetically Induced Current events occurred in Guangdong, China during the declining phase of solar cycle 23 (2003–2006)

Y Y Ni[1]

[1] School of Mathematics and Physics, North China Electric Power University, 2 Beinong Road, Changping District, Beijing 102206, China

Email: niyiyi058@163.com

Abstract. We study the interplanetary causes of intense geomagnetic storms (Dst ≤ -100 nT) and the corresponding Geomagnetically Induced Current (GIC) events occurred in Ling'ao nuclear power station, Guangdong during the declining phase of solar cycle 23 (2003–2006). The result shows that sMC (a magnetic cloud with a shock), SH (sheath) and SH+MC (a sheath followed by a magnetic cloud) are the three most common interplanetary structures responsible for the storms which will cause GIC events in this period. As an interplanetary structure, CIR (corotating interaction regions) also plays an important role, however, the CIR-driven storms have a relatively minor effect to the GIC. Among the interplanetary parameters, the solar wind velocity and the southward component of the IMF (interplanetary magnetic field) are more important than solar wind density and the temperature to a geomagnetic storm and GIC.

1. Introduction

The solar plasma emited by the Sun interact with the Earth's magnetic field, affecting the physical processes in the magnetosphere and ionosphere. Their net result is recorded at the Earth's surface as variations of geomagnetic and geoelectric fields. The electric fields induced at the Earth's surface will drive geomagnetically induced currents (GIC) in the circuit of power transmission lines and neutral-grounded transformers[1]. Intense variations of the geomagnetic field caused by solar activity are called geomagnetic storms, which means GIC is one of the manifestations of geomagnetic storms. Previous studies have found that GIC poses a potential threat to the technological systems such as power lines, pipelines and railway systems with severely distorting exciting currents, which will damage the transformers and even threat the safe operation of the whole power system[2].

The GIC events are used to be thought that only occurred in high latitudes areas such as North America and northern Europe. For example, the most well-known GIC events are the province wide blackout in Quebec, Canada, in March 1989 and the blackout in Malmö, a city of southern Sweden, during the Halloween storm of October 2003[3]. But with the study proceeding, people have found that the mid and low latitudes regions also can be influenced by GIC. In China, it has been found several times of GIC events in Guangdong and Jiangsu Province. Although GIC has not caused severe damage, it is still significant and urgent to estimate the GIC levels in Chinese power grids and to determine the effects of GIC on Chinese present and planned power systems.

When investigating the drivers of space weather at Earth, in particular the GIC, it is important to distinguish between the different interplanetary structures, which have been previously reported to produce significantly varying responses in the Earth magnetosphere and it is therefore reasonable to expect different responses for GICs[4]. Gonzalez have found that the most common interplanetary structures leading to the development of an intense geomagnetic storm (Dst ≤ -100 nT) are magnetic cloud (MC), sheath field (SH), sheath fields followed by a magnetic cloud (SH+MC) and corotating interaction regions (CIR)[5]. However, the relative importance of each driving structures has been found to vary with the solar cycle phase.

The solar cycle 23 (1996–2006) can be divided into three phases: the rising phase (1996–1998), the maximum phase (1999–2002) and the declining phase (2003–2006). In this paper, the cases we chosen are all in the declining phase of solar cycle 23. The Dst index is a measure of the strength of the ring current and widely used for measuring the intensity of geomagnetic storms. According to the peak Dst index, the storms can be divided into three types: intense storms (Dst ≤ -100 nT), moderate storms (-100 nT < Dst ≤ -50 nT) and small storms (Dst > -50 nT). The intense storms usually happen in the maximum and declining phases of the solar cycle.

The purpose of this paper is to study the intense geomagnetic storms (Dst ≤ -100 nT) occurred in the declining phase of solar cycle 23 (2003–2006) and the GIC events caused by these storms. We will study the interplanetary sources of the storms, finding what kind of interplanetary structures may cause GIC events and the difference between the storms drived by different interplanetary structures. We use the solar wind and interplanetary magnetic field (IMF) data observed by the ACE (Advanced Composition Explorer) satellite, combining with the Dst data, to analyze the influence of each interplanetary parameter to the geomagnetic storms and GICs. The GIC data we used is from Ling'ao nuclear power station (N22E114) in Guangdong Province, so we choose the Dst data of Zhaoqing geomagnetic station (N23E112).

2. Statistical analysis

Echer et al. have identified 90 intense geomagnetic storms (Dst ≤ -100 nT) during solar cycle 23[6] and we choose 29 of them listed in table 1 which occurred during 2003–2006. The columns in the table are the date of peak Dst, the value of peak Dst, the interplanetary structure identified as the cause of the storm main phase and the peak value of the GIC intensity if it has.

Table 1. The 29 major geomagnetic storms and 13 GIC events during 2003–2006.

Date	Dst_p (nT)	IP Structure	I_{GIC} (A)	Date	Dst_p (nT)	IP Structure	I_{GIC} (A)
2003/05/29*	-144	SH	11.89	2004/11/10*	-289	SH+MC	42.82
2003/06/18	-141	SH		2004/11/12	-109	SH	
2003/07/12	-105	CIR		2005/01/18*	-121	CIR	6.98
2003/08/18	-148	sMC		2005/01/22*	-105	SH	27.44
2003/10/29*	-353	SH+MC	57.05	2005/05/08	-127	CIR	
2003/10/30*	-383	SH	48.57	2005/05/15	-263	sMC	
2003/11/20*	-422	sMC	23.86	2005/05/20	-103	nsMC	
2004/01/22*	-149	nonMC	5.71	2005/05/30	-138	nonMC	
2004/02/11	-109	CIR		2005/06/13	-106	sMC	
2004/04/04	-112	SH		2005/08/24*	-216	sMC	13.20
2004/07/23	-101	nonMC		2005/08/31*	-131	CIR	5.34
2004/07/25	-148	sMC		2005/09/11	-123	SH	
2004/07/27*	-197	SH+MC	18.26	2006/04/14	-111	sMC	
2004/08/30	-126	sMC		2006/12/15*	-147	sMC	14.52
2004/11/08*	-373	sMC	35.36				

* The date of GIC events recorded by Ling'ao nuclear power station, Guangdong.

The interplanetary remnants of coronal mass ejections (CME) are called as ICME. MC (magnetic cloud) is a subset of the ICME. When the B_z or B_y component of IMF display a 180° rotation, we can consider an ICME to be a MC. If the MC is preceded by an interplanetary fast shock but only the MC field causes the storm, we call this type of structure as a MC with a shock (sMC). Otherwise, we call it as a MC with no shock (nsMC). NonMC means any ICME that does not follow the MC criteria. SH (sheath field) is the fields present in the region between the interplanetary shock and the driver ICME. SH+MC stands for a sheath followed by a magnetic cloud. CIR (corotating interaction regions) corresponds to the interaction regions between high and low speed solar wind streams[6].

One can see in table 1 that the three most common interplanetary structures of intense storms during 2003-2006 are sMC, SH and CIR. These three types of structures account for 76% of the whole interplanetary structures causing intense storms during the declining phase of solar cycle 23, with CIRs causing 17%, sMCs 35% and SHs 24% of the storms. The sMC, SH+MC and SH are the three most dominant interplanetary structures responsible for the GIC events, with a number of 4, 3 and 3 cases respectively. The sMC, SH+MC, SH, nsMC and nonMC all can be classified as ICME[7]. We can compute that the GIC events caused by ICMEs account for 85% with 11 cases and CIRs account for 15% with 2 cases. Borovsky and Denton have found that the biggest GIC events are from CME drivers, whereas CIR-driven storms had a relatively minor effect[8]. As the table 1 shown, among the 13 events there are only three cases that the strength of GIC less then 10A, and two of them are drived by CIR. This result tallies with Borovsky's conclusion.

3. Analysis of typical cases

For the purpose to find the differences between the storms drived by different interplanetary sources, we choose two cases drived by CIR and sMC, respectively.

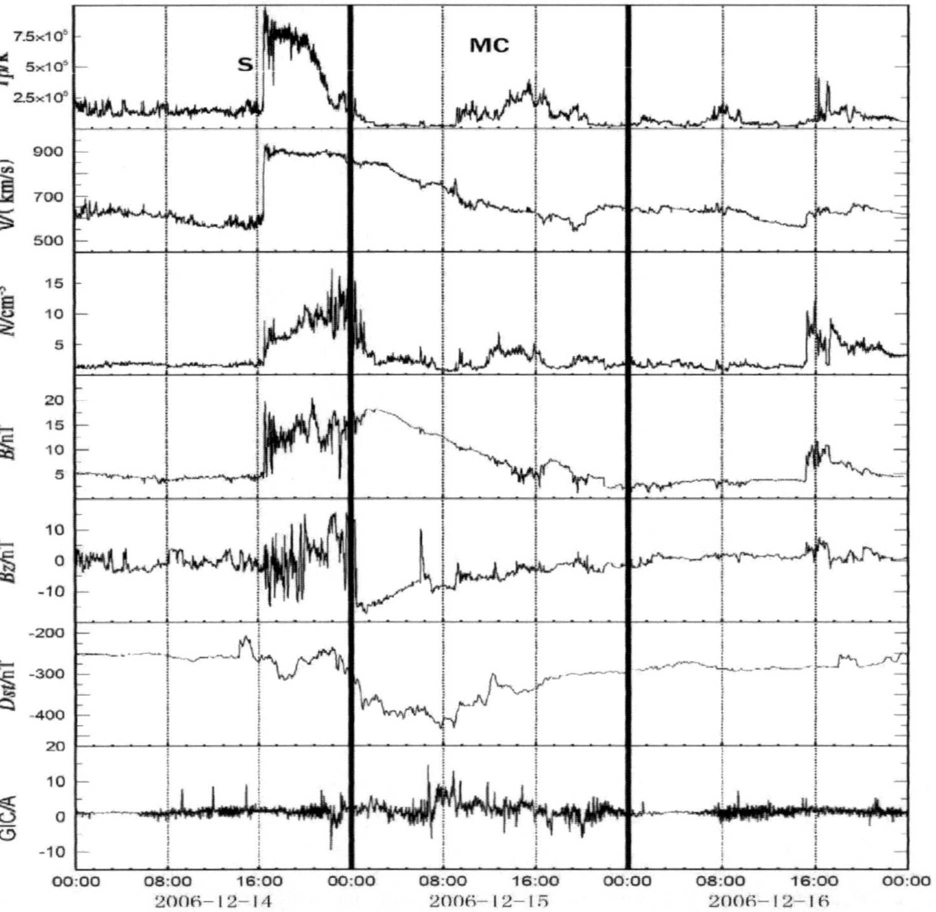

Figure 1. The geomagnetic storm and the GIC event occurred from 14 to 16 December 2006.

Figure 1 shows a sMC-driven storm during 14–16 December 2006. The parameter V stands for the solar wind velocity, N for the number density of solar wind plasma, T_p for the proton temperature, B for the strength of IMF and B_z for the north-south component of IMF. B_s is the southward component of IMF (B_s = -B_z for southward field, B_s = 0 for northward). As the China National Space Weather Center reported, a X 3.4 solar flare erupted at 10:40 UT 13 December 2006, accompanied by a CME. The CME arrived at the earth and caused the intense geomagnetic storm in the early morning of day 15. We can see an interplanetary shock occurred at ~16:00 on day 14 December, then the parameter N, T_p, V and B all start to increase quickly. The component of the IMF B_z began to fluctuate in the north-south direction and declined suddenly at ~00:00 on day 15. In the main phase of the storm, the N and T_p were back to the normal value basically, while the speed of the solar wind and the strength of IMF began to decline from the peak value. The component B_z had a long time continuous southward turning and reached a minimum value at ~08:00 on day 15. The peak of the storm occurred at ~08:00 UT on 15 December, with a peak Dst of -433 nT and a peak GIC of 14.52A.

Figure 2 shows an example of a CIR-driven geomagnetic storm caused by the southward components of oscillating B_z fields within the CIR between 31 August and 1 September 2005. The high speed stream (HSS) and low speed stream (LSS), as well the region of compressed magnetic field and plasma, are indicated in the figure. The storm peak occurred at ~15:30 UT on 31 August, with a peak Dst of -378 nT. The GIC event caused by this storm occurred from 14:30 UT 31 August to 02:30 UT 2 September. The peak value of this GIC is 5.34A. According to statistics, the CME-driven storms have a greater value of B_z, a higher speed of the solar wind and a smaller peak Dst, while the CIR-driven storms have a bigger numerical interval of solar wind proton density[9].

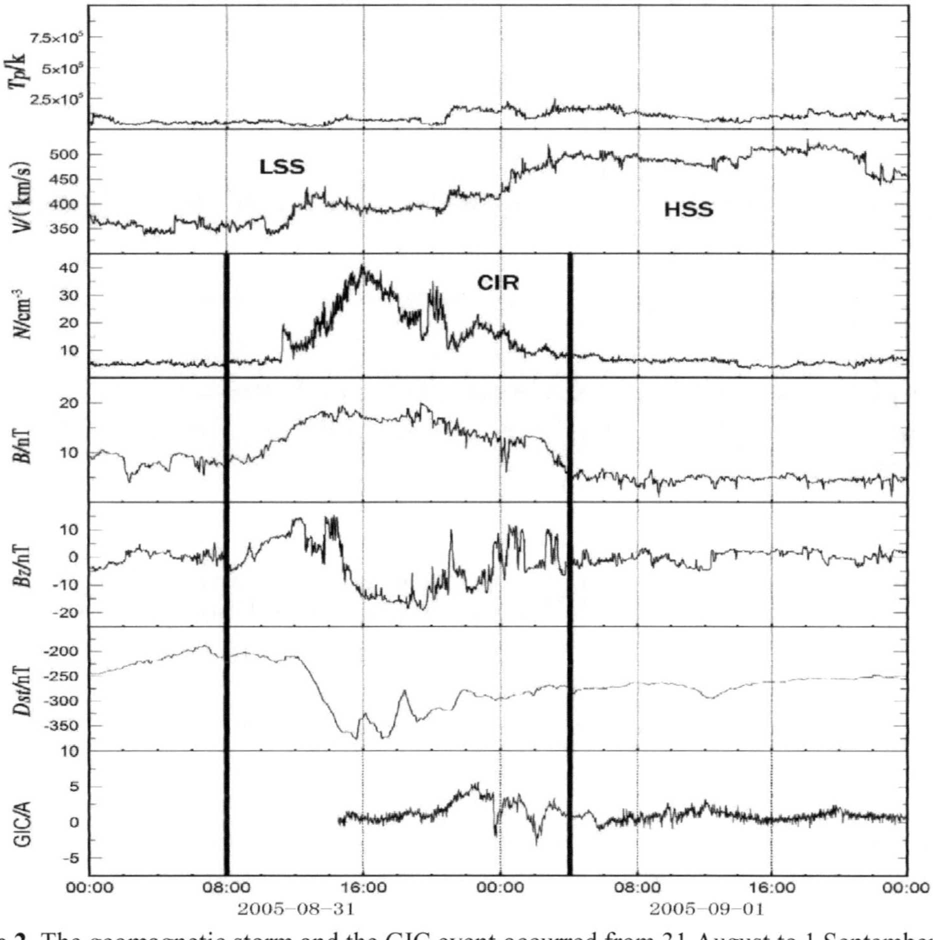

Figure 2. The geomagnetic storm and the GIC event occurred from 31 August to 1 September 2005.

4. Discussion and Conclusion

For the declining phase of solar cycle 23, the three most common interplanetary structures driving intense storms are sMCs (35%), SHs (24%) and CIRs (17%). These three driving structures are responsible for almost 76% of intense storms. However, the geo-effective of CIRs is less, leading to the storms with peak Dst values only between -100 nT and -150 nT. The sMC, SH+MC and SH are the three dominant structures responsible for the GIC events, with a percentage of 31%, 23% and 23% respectively. Considering sMC, SH+MC, SH, nsMC and nonMC as ICMEs, about 85% of GIC events are caused by ICMEs and 15% by CIRs.

From the typical cases, we can see that the storms caused by different interplanetary structures have different change process. The ICME-driven storms are well matched with the corresponding GIC events with bigger peak values, whereas CIR-driven storms are not matched so well. During a ICME-driven geomagnetic storm such as case 1, the Dst often rises in a short duration because of the arrival of the CME shock. The decrease of the Dst during the main phase is closely related to the southward component of the IMF. Comparing with other structures, the CIR-driven storms have a relatively minor effect to the GIC. In the two cases caused by CIRs, the strength of GIC are both less than 10A.

Among the interplanetary parameters, the solar wind velocity V, the southward component of the IMF B_s and the duration time T, are the most dominant factors to geomagnetic storms. According to statistics, when $B_s > 10$ nT and $T \geq 3$h, it is more likely to cause an intense storm[10]. The roles of the solar wind density N and the temperature T_p in space weather studies are complex. At present, the study shows that they have a weak correlation with geomagnetic indices such as Dst index. We will delve more deeply into the problem to find a possible linear relation between the interplanetary parameters and the GIC values for the prediction of the potential GIC events in the future.

References

[1] Liu C M, Liu L G, Pirjola R and Wang Z Z 2009 *Space Weather* **7** S04005
[2] Boteler D H, Pirjola R J and Nevanlinna H 1998 *Adv. Space Res.* **22** 17–27
[3] Viljanen A 2011 *Space Weather* **9** S07007
[4] Savani N P, Vourlidas A, Pulkkinen A, Nieves-Chinchilla T, Lavraud B and Owens M J 2013 *Space Weather* **11** 245–61
[5] Gonzalez W D, Echer E, Gonzalez A L and Tsurutani B T 2007 *Geophys. Res. Lett.* **34** L06101
[6] Echer E, Gonzalez W D, Tsurutani B T and Gonzalez A L 2008 *J. Geophys. Res.* **113** A05221
[7] Chen C, Zhao Z W, Sun S J, Ban P P and Wang B J 2016 *Chinese Journal of Radio Science* **31** 670–5
[8] Borovsky J E and Denton M H 2006 *J. Geophys. Res.* **111** A07S08
[9] Shen X F, Ni B B, Gu X D, Zhou C, Liu Y, Xiang Z and Zhao Z Y 2015 *Chinese Journal of Geophysics* **58** 362–70
[10] Tong Y N, Liu S Q and Gong J C 2008 *Chinese Journal of Space Science* **28** 513–21

Acknowledgments

We would like to thank to the Ling'ao nuclear power station for the GIC data, the Zhaoqing geomagnetic station for Dst data and the ACE satellite for the solar wind data.

IOP Conf. Series: Materials Science and Engineering **339** (2018) 012014 doi:10.1088/1757-899X/339/1/012014

Application of affinity propagation algorithm based on manifold distance for transformer PD pattern recognition

B G Wei[1], K X Huo[2], Z F Yao[1], J Lou[2] and X Y Li[2]

[1] Electric Power Research Institute, Shanghai Electric Power Company, No.171 Handan Road, Shanghai, China.

[2] School of Electrical Engineering, Shandong University, No.17923 Road, Jinan, Shandong Province, China.

E-mail: wbgsj@126.com

Abstract. It is one of the difficult problems encountered in the research of condition maintenance technology of transformers to recognize partial discharge (PD) pattern. According to the main physical characteristics of PD, three models of oil-paper insulation defects were set up in laboratory to study the PD of transformers, and phase resolved partial discharge (PRPD) was constructed. By using least square method, the grey-scale images of PRPD were constructed and features of each grey-scale image were 28 box dimensions and 28 information dimensions. Affinity propagation algorithm based on manifold distance (AP-MD) for transformers PD pattern recognition was established, and the data of box dimension and information dimension were clustered based on AP-MD. Study shows that clustering result of AP-MD is better than the results of affinity propagation (AP), *k*-means and fuzzy *c*-means algorithm (FCM). By choosing different *k* values of *k*-nearest neighbor, we find clustering accuracy of AP-MD falls when *k* value is larger or smaller, and the optimal *k* value depends on sample size.

1. Introduction

Power transformer is one of the more important equipment in the power grid. If a fault occurs, power transformer can cause that partial or even large area blackouts and great economic losses. A large number of fault statistics show that most of the accidents of transformers are caused by aging and damage of insulation, and PD is one of the important reasons for the aging and damage of insulation. It has a great effect on judging the damage condition of insulation and finding the insulation defects of transformers that the types of PD are effectively identified.

A great deal of research work has been done on the diagnosis and identification of PD defects in power transformer at home and abroad [1-3]. Fingerprint recognition technology based on PD statistical spectrum has received extensive attention, and scholars have proposed PRPD, phase resolved pulse sequence analysis (PRPSA) and time-resolved partial discharge (TRPD) model based on different measurement methods and statistical features of PD. With the development of computer and mathematical methods, a variety of pattern recognition methods have emerged, such as pattern recognition method based on clustering analysis[4], pattern recognition method based on artificial neural network [5], pattern recognition method based on fractal theory [6].

As an unsupervised analysis method, clustering analysis has been widely used in pattern recognition. Krivda [7], of Delft University of Technology in Holland, used cluster analysis and artificial neural networks to discuss the automatic recognition of local scenes. Chen [8] identified the developing stages of air-gap discharge in oil-paper insulation based on cluster-wavelet neural network. Since Frey and Dueck put forward Affinity Propagation (AP) in Science magazine in 2007 [9], AP has been widely used in image recognition, image retrieval, data mining and other fields. In this paper, we propose AP-MD and recognize three kinds of PD models by AP-MD.

2. AP algorithm
AP algorithm is a kind of clustering algorithm based on passing messages between adjacent points. In contrast to *k*-means and FCM, AP uses the similarities between data points as the input. First, all data points are regarded as potential category center with equal status. Then, the nearest central points are the optimal representative points of all categories when the sum of the similarities between data points and the nearest central point is the largest [10].

The data set $\boldsymbol{X}=\{x_1, x_2, \dots, x_N\}$ contains the category set $\boldsymbol{C}=\{C_1, C_2, \dots, C_K\}$, $K \le N$, Each data point only belongs to one category. The central point of category which data x_i belongs to is $x_{C(i)}$, $i=1,2, \dots, K$. The error function of clustering is Equation (1).

$$J(\boldsymbol{C}) = \sum_{i=1}^{N} d^2(x_i, x_{C(i)}) \tag{1}$$

The objective function of the AP algorithm is to minimize the error function:

$$C^* = \arg\min[J(\boldsymbol{C})] \tag{2}$$

The similarity matrix $\boldsymbol{S}_{N\times N}$ is calculated by Equation (3), the non-diagonal element $s\,(i, j)$ of the matrix is the similarity between the points x_i and x_j, the diagonal element $s\,(i, i)$ is the preferences $p\,(i)$ which is negative. The initial values of $p\,(i)$ generally are the same value that is the minimum or mean of all non-diagonal elements in the similarity matrix.

$$s(i, j) = \begin{cases} -\left\| x_i - x_j \right\|^2 & i \ne j \\ p & i = j \end{cases} \tag{3}$$

The core of AP algorithm is passing messages between data points. The messages of the AP algorithm are 'Responsibility' and 'Availability'. Responsibility $r\,(i, j)$ is sent from data point i to candidate exemplar point j and it reflects the accumulated evidence for how well-suited point j is to serve as the exemplar for point i. Availability $a\,(i, j)$ is sent from candidate exemplar points j to point i and it reflects the accumulated evidence for how appropriate it would be for point i to choose point j as its exemplar. The iterative process of the AP algorithm is the process of alternately updating the messages. At the beginning, the responsibilities and the availabilities are initialized to zero: $r\,(i, j)=0$, $a\,(i, j)=0$. The iterative equations of $r\,(i, j)$ and $a\,(i, j)$ are written as Equation (4) and Equation (5).

$$r(i, j) \leftarrow s(i, j) - \max_{j' \ne j}\left[a(i, j') + s(i, j')\right] \tag{4}$$

$$a(i, j) \leftarrow \begin{cases} \min_{i \ne j}\{0, r(j, j) + \sum_{i' \ne i, i' \ne j} \max[0, r(i', j)]\}, i \ne j \\ \sum_{i' \ne j} \max[0, r(i', j)], i = j \end{cases} \tag{5}$$

Finally, after a large number of iterations, find the optimal central points of categories and the subordinate relationship between the central points and the data points by the objective function Equation (2).

3. AP-MD algorithm
As we can see from Equation (3), the similarity matrix $\boldsymbol{S}_{N\times N}$ of AP is calculated based on Euclidean distance between two data points. It makes $\boldsymbol{S}_{N\times N}$ can only reflect the local consistency of data, and can't reflect the global consistency and the potential complex structure of data. As shown in Figure 1, the Euclidean distance between point *a* and point *b* is longer than the Euclidean distance between *a* and *b*, which can cause the similarity between *b* and *a* is less than the similarity between *c* and *a*. point *b* and point *a* can be wrongly classified as one category.

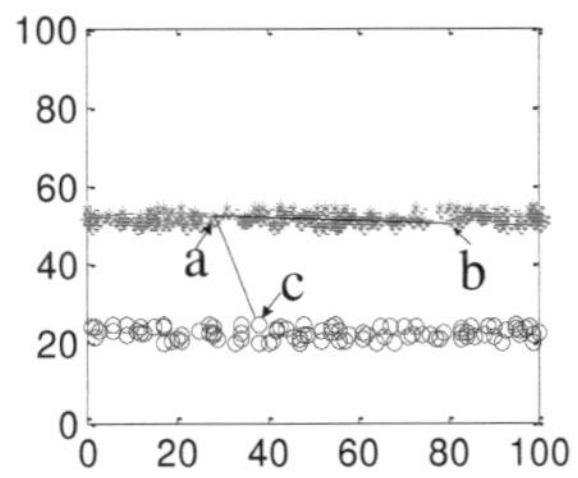

Figure 1. Two sets of data points in linear.

Because the similarity calculated by Euclidean distance is defective, this paper uses the manifold distance which can reflect the global consistency of data points to calculate similarity matrix. Manifold distance violates the theorem that a straight line is the shortest distance between two points in Euclidean space. It can solve the problem mentioned above.

3.1. Manifold distance

Before defining the manifold distance, the length of the line between two data points in manifold is defined. The definition is as follows, search the k-nearest neighbor of each sample point, and then construct the weighted graph $G(V, E)$ by the k-nearest neighbor. V is the set of vertices that correspond to sample points, E is the set of edges. $L(x_i, x_j)$ represents the distance between point x_i and point x_j. The calculation equation is Equation (6).

$$L(x_i,x_j)=\begin{cases}1-\exp[-\beta d(x_i,x_j)^2] & x_i \text{ and } x_j \text{ are } k\text{-nearest neighbor of each other}\\ \infty & \text{else}\end{cases} \tag{6}$$

In Equation (6), $d(x_i, x_j)$ is the Euclidean distance between two points; β is a regulatory factor to prevent the rapid growth of $L(x_i, x_j)$. The value of β is related to the density of the data and it's usually the reciprocal of the mean Euclidean distance of all sample points.

The data points are considered as the vertexes of the weighted graph $G(V, E)$, $\boldsymbol{p}=\{p_1, p_2, \dots, p_l\}\in V$ represents a path between points x_i and x_j and the edges $(p_k, p_{k+1})\in E$, $1\le k\le l-1$. $\boldsymbol{P}_{i,j}$ represents the set of all paths between the points x_i and x_j. The manifold distance between point x_i and point x_j is defined as Equation (7) [11].

$$D(x_i,x_j)=\min_{p\in \boldsymbol{P}_{i,j}}\sum_{k=1}^{|p|-1}L(p_k,p_{k+1}) \tag{7}$$

3.2. The steps of AP-MD algorithm

The steps of AP-MD algorithm are as follows:

- Input: k of k-nearest neighbor and preference p.
- Initialization: $r(i,j)=0$, $a(i,j)=0$ for all i, j.
- $\forall x_i, x_j \in \boldsymbol{X}$, calculate the Euclidean distance between point x_i and point x_j by Equation (8).

$$d(x_i,x_j)=\left\|x_i-x_j\right\| \tag{8}$$

- Search the k-nearest neighbor of each sample point based on Euclidean distance, and then construct the weighted graph $G(V, E)$.
- $\forall x_i, x_j \in \boldsymbol{X}$, calculate the manifold distance between point x_i and point x_j by Equation (7).
- Construct similarity matrix $\boldsymbol{S}_{N\times N}$ based on manifold distance. The non-diagonal element $s(i,j)=-|D(x_i,x_j)|^2$, the diagonal element $s(i,i)$ is the mean of all non-diagonal elements in the similarity matrix $\boldsymbol{S}_{N\times N}$.
- Use AP algorithm to cluster data. Take Equation (2) as the objective function, and perform iterative computations by Equation (4) and Equation (5).
- Determine whether the number of categories satisfies the requirement, If not satisfied, change the value of p and repeat the whole iterative process until the number of categories meets the requirement, and output the final clustering results.

3.3. Comparison of AP-MD with other algorithms

Besides the AP algorithm, the *k*-means algorithm and FCM are two commonly used clustering algorithms, and they both have good results in clustering. We use AP-MD, AP, *k*-mean and FCM to cluster the data in the Figure 1. The results are shown in Figure 2.

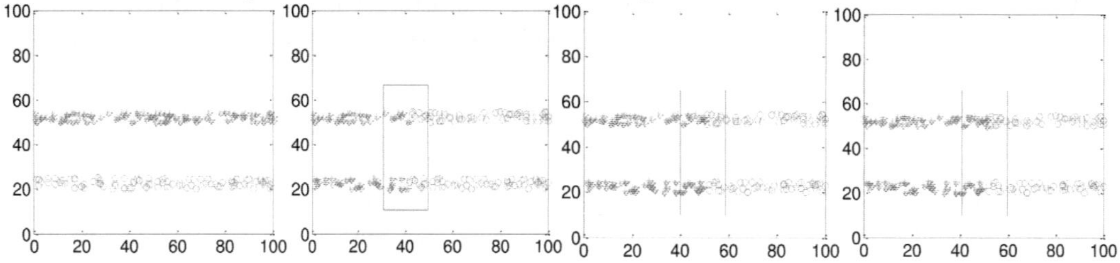

(a) AP-MD (b) AP (c) *k*-means (d) FCM

Figure 2. Clustering results of four algorithms.

Figure 2 shows clearly that the clustering result of AP-MD is better than the results of AP, *k*-means and FCM and the accuracy of the recognition reaches 100%. Clustering results of AP, *k*-means and FCM are only different in the black square. AP, *k*-means and FCM use Euclidean distance to compute the similarity matrix $S_{N\times N}$. So their clustering results are not only very similar, but also have the same problem which the clustering results are completely wrong. The clustering result of AP-MD is better than results of other algorithms for data which are distributed regularly.

3.4. The effects of k value on AP-MD

According to Equation (6) and Equation (7), *k* value of *k*-nearest neighbor is the key for computing manifold distance. In order to study effects of *k* value on clustering results of AP-MD, values of *k* are all integers between 1 and 50 and we cluster the data of Figure 1 by AP-MD. Clustering results are shown in Figure 3 when *k* values are 4, 10, 28 and 30. When $1 \le k \le 4$ and $27 \le k \le 29$, the accuracy of clustering is between 50% and 100%. When $5 \le k \le 26$, the accuracy of clustering can reach up to 100%. Clustering results of AP-MD are similar with clustering results of AP when $k \ge 30$ and they are wrong.

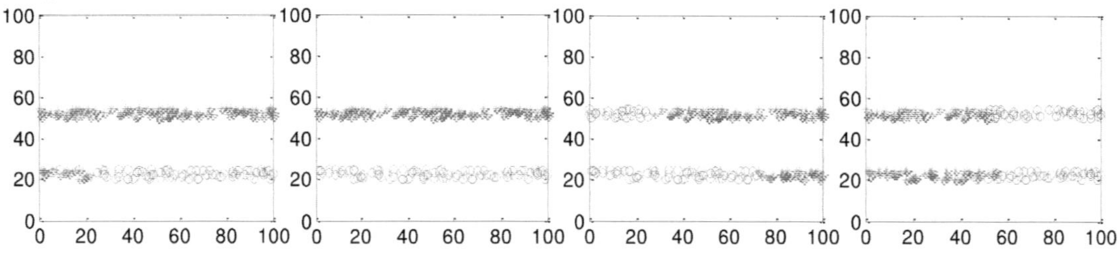

(a) $k = 4$ (b) $k = 10$ (c) $k = 28$ (d) $k = 30$

Figure 3. Clustering results of k =4, 10, 28 and 30.

According to the steps of AP-MD algorithm, it needs iterations and adjusting preference *p* to make number of categories meets the requirement. As shown in Table 1, with the increase of *k* value from 5 to 26, the number of iterations is reduced and *p* value is increased. The number of iterations reduces rapidly when $5 \le k \le 15$, and it reduces slowly as the *k* value continues to increase. Considering all effects of *k* value on AP-MD, The optimal *k* value is 10% ~ 15% of the amount of data contained in one category and doesn't be less than 4.

Table 1. The effects of *k* value on iterations and preference *p*.

k	The number of iterations	preference *p*
5	55	$-2.26\times e^{4}$
10	48	$-6.32\times e^{3}$
15	42	$-2.11\times e^{3}$

20	41	-1.76xe^3
25	39	-1.22xe^3

4. Application of AP-MD for PD Pattern recognition

4.1. Model experiment of PD in transformers
According to the main physical characteristics of PDs in the transformers, three oil-paper insulation defects were designed which were used to simulate three typical PDs in the transformers. As shown in Figure 4. All models were immersed in oil during the experiment. The diameter of circular plate electrodes is 80mm, and the thickness is 10mm. The thickness of all paperboard is 0.5mm. Figure 4 (a) is the cavity discharge model inside solid insulation. The cavity consists of three layers of paperboard with a diameter of 60mm, and the center of the middle paperboard has a circular hole with a diameter of 20mm. Figure 4 (b) is surface discharge in oils model. The sample is a paperboard of 100mm in diameter. Figure 4 (d) is corona discharge in oils mode. The radius of curvature of the tip is less than 0.1mm. Paperboard is placed between the tip and the plate electrode, and the *d* is 1mm [12].

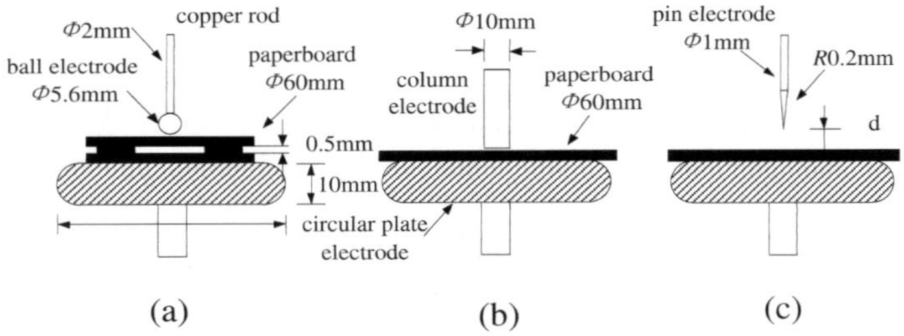

Figure 4. Three oil-paper insulation defects: **(a)** cavity discharge, **(b)** surface discharge, **(c)** corona discharge.

4.2. Constructing the grey-scale images
Use sampled data from the experiment to contract grey-scale images. Steps are as follows. First, extract the PD pulse, and then construct PRPD and space curved face of *φ-q-n* (*φ* is power frequency phase angle of discharge, *q* is discharge amplitudes, *n* is discharge times). Finally, construct grey-scale images of space curved face of *φ-q-n* according to the principles that the maximum value of space curved face corresponds to the maximum grey level and minimum value of space curved face corresponds to the minimum grey level.

4.3. Feature extraction of grey-scale images
We extract box dimension and information dimension as the features of the grey-scale images. Fractal dimension is a measure for describing the complexity of fractal sets. Box dimension is one of the most commonly used methods of fractal dimension, and it is defined as follows.

$$D_B = \lim_{r\to 0}\left\{\left[\ln N_r(\boldsymbol{F})\right]/\ln(r^{-1})\right\} \tag{9}$$

In Equation (9), $\boldsymbol{F}$ is a nonempty bounded subset belonging to d-dimensional Euclidean space $\boldsymbol{R}^d$. $N_r(\boldsymbol{F})$ is the minimum number of sets which can contain subset $\boldsymbol{F}$.

Information dimension is defined as Equation (10). $\{X_i\}_{i=1}^N$ is the number of points in nonempty bounded subset $\boldsymbol{F}$ belonging to d-dimensional Euclidean space $\boldsymbol{R}^d$. The value of N is great. $M(r)$ is the number of d-dimensional cubes which can contain points $\{X_i\}_{i=1}^N$. N_i is the number of points contained in the ith cube, $P_i = N_i/N$.

$$D_I = -\lim_{r\to 0}\lim_{N\to\infty}\left\{\left[\sum_{i=1}^{M(r)} p_i \ln(p_i)\right]\Big/\ln(r)\right\} \tag{10}$$

According to Equation (9) and Equation (10), the box dimension and information dimension of grey-scale images are obtained by the least square method and each defects model obtains 28 groups

IOP Conf. Series: Materials Science and Engineering **339** (2018) 012014 doi:10.1088/1757-899X/339/1/012014

data. Because box dimension and information dimension are on the different order of magnitude, they need to be standardized by Equation (11).Sample data of box dimension information dimension are shown in Figure 5.

$$x^* = (x - X_{\min})/(X_{\max} - X_{\min}) \tag{11}$$

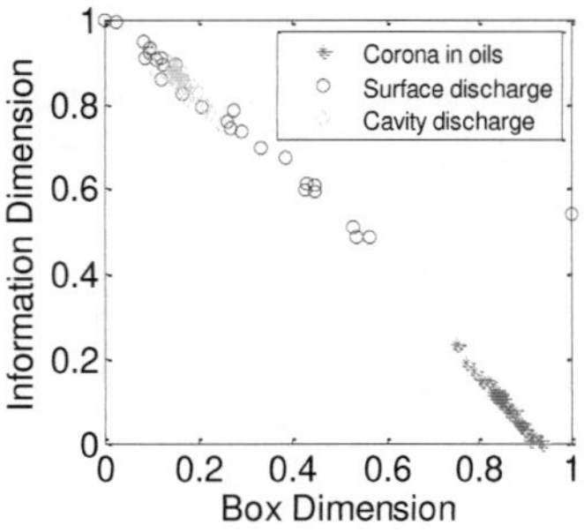

Figure 5. Box Dimension-Information Dimension.

4.4. PD pattern recognition by AP-MD

Cluster the sample data of box dimension and information dimension by AP-MD, AP, *k*-means and FCM. The results are shown in Figure 6. The most obvious difference of Figure 6 (a), Figure 6 (b). Figure 6 (c) and Figure 6 (d) is that point x_a is assigned to the category P_2 in Figure 6 (a). Clustering results of AP, *k*-means and FCM are very similar. The recognition rates of AP-MD for corona discharge, surface discharge and cavity discharge are 100%, 50%, 65.9%. The recognition rates of other algorithms are 96.4%, 32.1%, 60.8%.

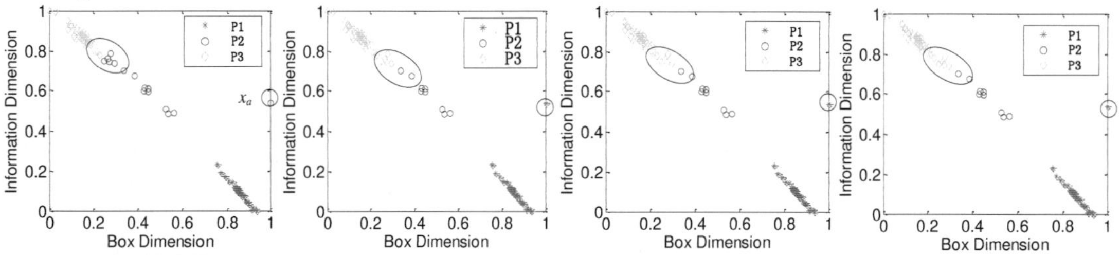

(a) AP-MD (b) AP (c) *k*-means (d) FCM

Figure 6. Clustering results of AP-MD, AP, *k*-means and FCM.

5. Conclusion

The *k* value has effects on clustering result of AP-MD and the number of iterations. When $k \leq 4$ or $k \geq 27$, the accuracy of clustering is reduced. The optimal *k* value is 10% ~ 15% of the amount of data contained in one category and doesn't be less than 4.

The clustering result of AP-MD is better than results of AP, *k*-means and FCM for data which are distributed regularly. The clustering results of AP, *k*-means and FCM for those data are wrong. But it is very applicable to clustering these data.

The recognition rates of AP-MD for corona discharge, surface discharge and cavity discharge are 100%, 50%, 65.9%. Compared with AP, *k*-means and FCM, AP-MD improves the accuracy of partial discharge pattern recognition.

References

[1] Wang M H 2005 Partial discharge pattern recognition of current transformers using an ENN *IEEE Transactions on Power Delivery* **20** (3) pp 1984–90.

[2] Wang K, Yang L J, Liao R J, *et al* 2011 Partial discharge pattern recognition based on discrete hidden Markov models *Transactions of China Electrotechnical Society* **26** (8) pp 205–12.

[3] Liu S B, Lv C, Yu J L, *et al* 2008 Application of Hilbert-Huang transform in pattern recognition for partial discharge of transformers *Proceedings of the CSEE* **28** (31) pp 114–9.

[4] Rodrigo Mor A, Castro Heredia L C and Muñoz F A 2017 New clustering techniques based on current peak value, charge and energy calculations for separation of partial discharge sources *IEEE Transactions on Dielectrics and Electrical Insulation* **24** (1) pp 340–8.
[5] Tan K X and Li F Q 1996 Some types of artificial neural networks used for recognition of partialdischarge *High Voltage Engineering* **22** (4) pp 3–7.
[6] Sun C X, Xu G F, Tang J, *et al* 2005 PD pattern recognition method using box dimension and information dimension as discriminating features in GIS *Proceedings of the CSEE* **25** (3) pp 100–4.
[7] Karivda A 1995 *Recognition of discharges, Discrimination and classification* (Delf: Delft University Press).
[8] Chen W G, Ling Y, *et al* 2012 Method to identify developing stages of air-gap discharge in oil-paper insulation based on cluster-wavelet neural network *Power System Technology* **36** (7) pp 126–32.
[9] Frey B J and Dueck D 2007 Clustering by passing messages between data points *Science* **315** (5814) pp 972–6.
[10] Sun L L, Guo C H, Liu C R, *et al* 2017 Fast affinity propagation clustering based on incomplete similarity matrix *Knowl Inf Syst* **51** (3) pp 941–63.
[11] Wang R P, Shen S G, Chen X L, et al 2012 Manifold-manifold distance and its application to face recognition with image sets *IEEE Transactions on Image Processing* **21** (10) pp 4466–79.
[12] Hussein R, Shanban K B and EI-Hag A H 2017 Robust feature extraction and classification of acoustic partial discharge signals corrupted with noise *IEEE Transactions on Instrumentation and measurement* **66** (3) pp 405–13.

Grid Transmission Expansion Planning Model Based on Grid Vulnerability

Quan Tang[1], Xi Wang[1], Ting Li[1], Quanming Zhang[1], Hongli Zhang[2], Huaqiang Li[2]

[1] Power Economic Research Institute of Sichuan Electric Power Corporation ,Chengdu 61000, China
[2] School of Electrical Engineering and Information, Sichuan University, Chengdu 61000, China

Email: 1123738014@qq.com

Abstract. Based on grid vulnerability and uniformity theory, proposed global network structure and state vulnerability factor model used to measure different grid models. established a multi-objective power grid planning model which considering the global power network vulnerability, economy and grid security constraint. Using improved chaos crossover and mutation genetic algorithm to optimize the optimal plan. For the problem of multi-objective optimization, dimension is not uniform, the weight is not easy given. Using principal component analysis (PCA) method to comprehensive assessment of the population every generation, make the results more objective and credible assessment. the feasibility and effectiveness of the proposed model are validated by simulation results of Garver-6 bus system and Garver-18 bus.

1. Introduction

With the large-scale access of new energy and load, power system will face with more uncertainties. The traditional power grid planning system which is dominated by economy and security, can no longer meet the needs of the current power grid construction, and needs to be supplemented and improved[1]. As an extension of the security and stability of power grid, the power system vulnerability will be introduced into the power grid planning, which will have guiding and referential significance for planning and building a strong grid.

Most of the traditional researches on vulnerability just from the viewpoint of grid components, and only consider the relative vulnerability of single component of power grid, which can't measure the degree of vulnerability between different power grids simply and effectively[2]. Therefore, this paper proposes the concept of global structure and state vulnerability of power grid, and applies it to power grid planning effectively.

Existing researches have pointed out that the non-uniformity of power grid is the important reason that the grid tends to self-organized criticality[3-4]. Therefore, it is necessary to consider the uniformity of power grid structure and state at the beginning of transmission network planning, which can effectively prevent the grid from entering the self-organized criticality, and reduce the system vulnerability[5].

Firstly, based on the traditional grid vulnerability assessment method, combining the power system uniformity theory and the Gini coefficient index, the global vulnerability factor model of power grid structure and state is put forward to assess the global vulnerability of power grid. Secondly, the multi-objective grid planning model is constructed on the basis of considering the economy of power grid, the global vulnerability factors of power grid structure and state , and the constraint set of grid security. Then, in the process of solving the problem, because of the dimensions of each objective are not uniform, the weight is not easy to obtain. In this paper, PCA is used to evaluate the individual population in each generation, and then the improved chaotic crossover and mutation genetic algorithm is used to optimize the planning scheme. Finally, the Garver-6 node system and the Garver-18 node system are used to analyze and prove the rationality and effectiveness of the proposed model.

2. Power Grid Planning Index and Model

2.1 Gini Coefficient

The Gini coefficient is an index used to judge the fairness of income distribution[6]. As shown in Figure 1.

Define Gini coefficient,

$$G = \frac{A}{A+B} \tag{1}$$

The larger the Gene coefficient is, the poorer the uniformity of the distribution of the system.

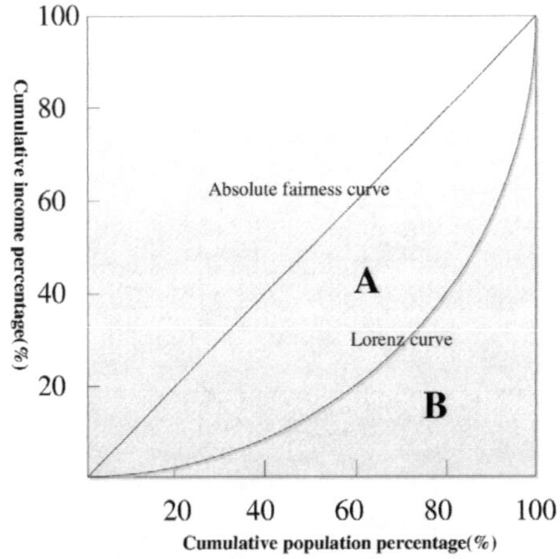

Figure 1. Gini coefficient curve

2.2 Global Vulnerability Factor of Structure

The electrical betweenness of nodes effectively quantifies the importance of nodes in the topology of the power grid, The model of electrical betweenness is available in the paper[7]。The uniform distribution of the node electrical betweenness can effectively reflect the uniformity of the topology of the power network. Therefore, based on the Gini coefficient theory, the index of the Gini coefficient of the nodes electrical betweenness G_n is established to measure the uniformity degree of the network topology.

2.3 Global Vulnerability Factor of State

The load rate of the power network branch can effectively measure the occupancy of the branches. The Gini coefficient of the branch flow load rate can effectively measure the uniformity of the system running state. Supposing the maximum active transport capacity of the branch i is $P_{i\max}$. When the system is running, the actual load flow of the branch i is P_i, and the load rate of the branch i can be as follows,

$$\eta_i = \left| P_i / P_{i.\max} \right| \ , i = 1, 2, \cdots, N \tag{2}$$

N is the number of branches

The Gini coefficient model for the load flow rate of the branches is established to measure the uniformity degree of the power grid state G_f.

2.4 Power Grid Planning Model

In this paper, the new lines is taken as the planning variable, Meeting the prerequisites of the number constraint of the new-built lines ω_i and grid security constraint set X, considering the investment of construction cost C, line operation loss cost P_{loss}, global structure vulnerability factor G_n, global state vulnerability factor G_f and safety constraint penalty O. The grid expansion planning model is as follows,

$$F = \min\{f_1(C), f_2(P_{loss}), f_3(\mathrm{G}_n), f_4(\mathrm{G}_f), f_5(O)\} \tag{3}$$

$$s.t.\ h(x) = 0 \tag{4}$$

$$g(x) \le 0 \tag{5}$$

Formula (3) is the objective function vector, formula (4) and (5) are the grid security constraint sets where x is running state variable.

3. Simulation Analysis

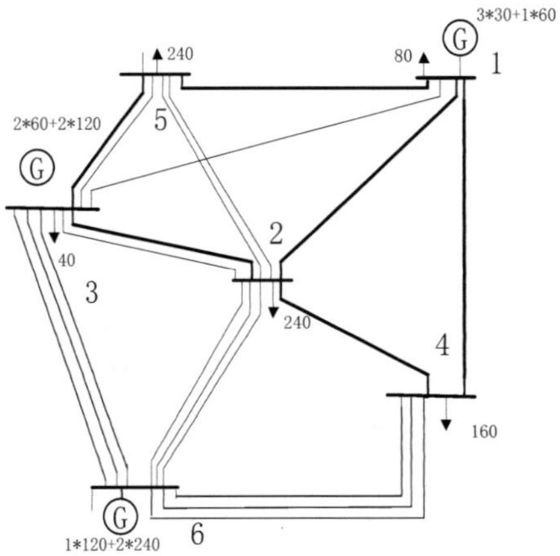

Figure 2. Planning route map

In this paper, the Garver-6 node system is taken as an example to simulate. It is considered that the 6 nodes can be connected with each other. In order to solve the problem that the weights of multi-objective programming are difficult to select, this paper uses PCA to evaluate the power grid planning comprehensively. Genetic algorithm is used to optimize the model. The wiring diagram of the optimization results is shown in Figure 2. The thin line is the added line.

It can be seen from table 1, first of all, the construction cost of model 1is 71097.95 thousand yuan, which is slightly larger than model 2. This is mainly because of the model 1 considers the uniformity of the structure and the state of the grid, more branches are needed to share the power flow of heavy load lines, and to make the topology of power network more reasonable. Then the vulnerability of planning model is reduced, and the ability of grid to resist various uncertain risks is enhanced. Simulation results also have proved this point; Secondly, the cost of grid loss of model 1 is obviously lower than that of the model 2, because of the rationality of the network topology and power flow distribution. In conclusion, the planning model established in this paper sacrifices the cost of a part of construction cost to achieve the optimal overall goal. Such a grid is more economical and reliable.

Table.1 The optimization simulation results comparison chart of Garver-6 node system

Parameter	Model 1	Model 2[8]
Number of expansion lines	2-3(1), 3-5(1)， 1-3(1)， 2-5(2)， 2-6(3)， 3-6(3)， 4-6(3)	2-3(1)， 2-6(4)， 3-5(2)， 4-6(3)， 3-6(2),
Construction expenses (Thousand yuan)	71097,95	63007,7
Global structure vulnerability factor	0.146	0.249
Global state vulnerability factor	0.315	0.386
Grid loss (Thousand yuan)	30472,76	31711,4

4. Conclusion

In this paper, a mathematical model of multi-objective power grid planning is proposed. Firstly, the global vulnerability factor model of power grid is constructed based on the uniformity of power grid topology and operation state. It can effectively measure the structure and state vulnerability of power grid. Secondly, the principal component analysis is used to comprehensively evaluate the economic, security constraints and global vulnerability of the planning model. Finally, the genetic algorithm is used to optimize the selected scheme, and the optimal scheme is obtained. Simulation results show the feasibility and effectiveness of the proposed model. At the same time, the planning model proposed in this paper can solve the problem that the constraint condition and the objective function of the power grid planning are difficult to deal with, meeting the needs of the actual project.

References

[1] Sun Weiqing，Wang Chengmin，Zeng Pingliang. Review on Evaluation method and index of power system homogeneity[J]. Power System Technology，2015，39(5)：1205-1212(in Chinese).W. Strunk Jr., E.B. White, The Elements of Style, third ed., Macmillan, New York, 1979.

[2] Xu Hang，Li Huaqiang，Zhao Xiangyun，et al. Assessment on nodal comprehensive vulnerability based on operational state and network structure[J]. Power System Technology，2014, 38(3)：731-735(in Chinese).G.R. Mettam, How to prepare an electronic version of your article, in: B.S. Jones, R.Z. Smith (Eds.), Introduction to the Electronic Age, E-Publishing Inc., New York, 1999, pp. 281-304.

[3] Cao Yijia, Wang Guangzeng, Cao Lihua, et al. An Identification Model for Self-organized Criticality of Power Grids Based on Power Flow Entropy[J]. Automation of Electric Power Systems, 2011, 35(7): 1-6(in chinese).

[4] Liu Wenying，Cai Wantong，Zhang Ning，et al. Evolution of self-organizing of grid critical state based on united weighted entropy[J]. Proceedings of the CSEE，2015, 6: 010(in Chinese).

[5] Yehesakul P D, Dabbaghchi I. A topology-based algorithm for tracking network connectivity[J]. IEEE Transactions on Power Systems, 1996,10:339-346.

[6] Zhang Jianhua. An convenient method to calculate Gini Coefficient[J]. Shanxi Agric. Univ. (Social Science Edition)，2007, 6(3)：275-278. (in Chinese)

[7] Xu Lin， Wang Xiuli， Wang Xifan. Electric betweenness and its application in vulnerable line identification in power grid[J]. Proceedings of the CSEE，2010，30(1)：33-39(in Chinese).
[8] Huang Ying， Li Yang， Weng Beibei， et al. A parameter estimation method for generator using trajectory sensitivity[J]. Automation of Electric Power Systems，2010 (23): 36-41(in Chinese).

Optimization Scheduling Model for Wind-thermal Power System Considering the Dynamic penalty factor

Siyu PENG[1],Jianchun LUO[2],Yunyu WANG[2],Jun YANG[2],Hong RAN[2],Xiaodong PENG[2],Ming HUANG[2],Wanyu LIU[3]

1.State Grid Chongqing Electric Power Co.Qi'nan Power Supply Branch Company, 401420Chongqing, China
2.State Grid Chongqing Electric Power Co.Wulong Power Supply Branch Company, 408500Chongqing, China
3.School of Electrical Engineering and Information, Sichuan University, 610065Chengdu, China

Email: 370379904@qq.com

Abstract. In this paper, a new dynamic economic dispatch model for power system is presented.Objective function of the proposed model presents a major novelty in the dynamic economic dispatch including wind farm: introduced the "Dynamic penalty factor", This factor could be computed by using fuzzy logic considering both the variable nature of active wind power and power demand, and it could change the wind curtailment cost according to the different state of the power system. Case studies were carried out on the IEEE30 system. Results show that the proposed optimization model could mitigate the wind curtailment and the total cost effectively, demonstrate the validity and effectiveness of the proposed model.

1. Introduction

With more and more large-scale wind farms integrated into power system, the uncertainty and volatility the wind power brings have caused many problems. Thus, how to reduce the wind curtailment is an important question[1-2]. In recent years, wind curtailment has become one of important factors restricting the development of wind power[3].

When system transmission capacity is enough, the wind curtailment is closely related to the active wind power and the power demand. To reduce the wind curtailment, extensive and deep researches have been carried on. And in most literature this problem was solved by considering the wind curtailment cost, but all the penalty factor introduced in the objective function is a fixed value[4]. All the methods do not consider the influence that the uncertainty and volatility the wind power brings, and they also don't consider the impact that the power demand brings to the power system. Cause the active wind power and the power demand is highly volatility, using the fixed wind curtailment penalty factor may influence the objectivity of the dynamic economic dispatch including the wind farm.

Thus, this paper introduces a new definition, namely"dynamic penalty factor". This factor considers the uncertainty of the active wind power and power demand comprehensively, and gets its value through the fuzzy logic. The penalty factor could change the wind curtailment cost according to the current state of the system so that to change the active wind power. And on this basis, this paper built an optimization scheduling model for wind –thermal power system considering the dynamic penalty factor. Finally, simulations were carried out with the modified IEEE 30-bus system using the cuckoo search optimization algorithm[5]. The results demonstrate the validity and effectiveness of the proposed model.

2. Dynamic penalty factor

2.1 Determination of the dynamic penalty factor

In high wind density region, wind power generation is of good support to the power system, especially at high demand situations. This support of wind power depends on the active wind power and the power demand of the system. Hence, the economic dispatch model needs some factors in addition to account for the support of wind energy while handling different situations of power demand scenarios. In order to manage such a situation, this paper introduces a new coefficient—"dynamic penalty factor", and its value depends on the active wind power and power demand, so it could influence the current state of the system.

2.2 Fuzzy logic

Both, power demand and wind velocity exhibit non-linear characteristics due to their uncertainty. They should therefore be presented as fuzzy quantities. Three variables are considered as fuzzy variables. They are active wind power, power demand and dynamic penalty factor.

For the design of dynamic cost coefficient, fuzzy logic[6] model implementation is shown in Fig. 1.

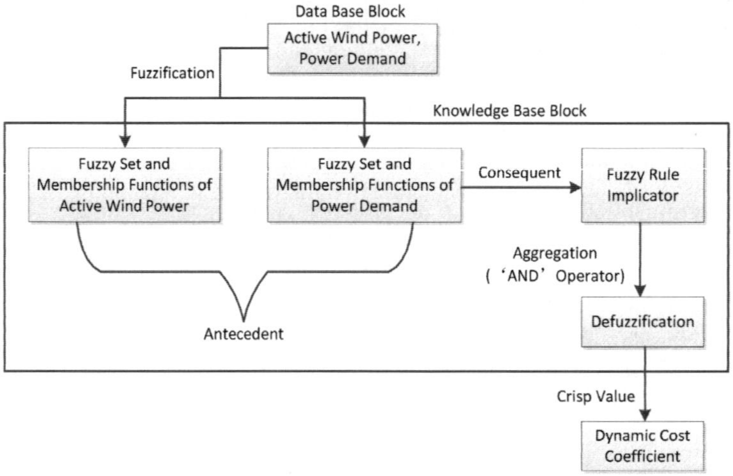

Figure 1. Fuzzy logic model of dynamic penalty factor

2.3 Demarcation for fuzzy parameters

In this model, three linguistic terms (e.g. high, low, medium) referred to as fuzzy set, are assigned to power demand, active wind power and dynamic penalty factor. The membership function of each variable are shown as follows :

(1) Power demand

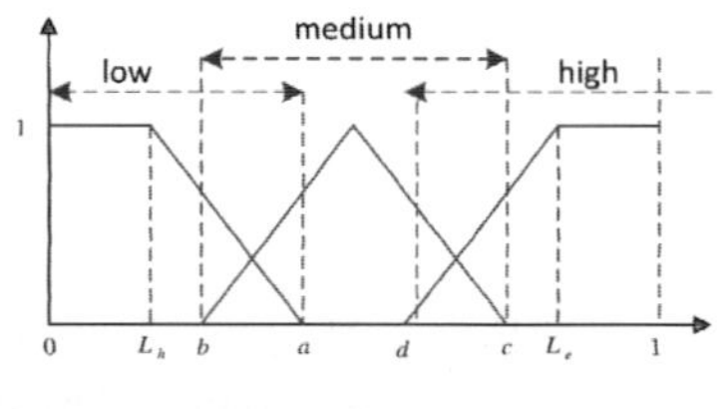

Figure 2. The membership function of the power demand

Where L_h is minimum power demand and 'a' is upper limit of low power demand. 'b' is lower limit of medium power demand and 'c' is upper limit of medium power demand. L_e is maximum power demand and 'd' is lower limit of high power demand.

(2) Active wind power

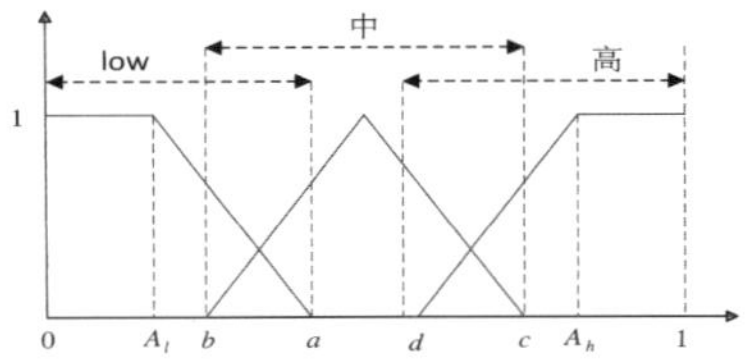

Figure 3. The membership function of the active wind power

Where A_l is minimum active wind power and 'a' is upper limit of low active wind power. 'b' is lower limit of medium active wind power and 'c' is upper limit of medium active wind power. A_h is maximum active wind power and 'd' is lower limit of high active wind power.

(3) Dynamic penalty factor

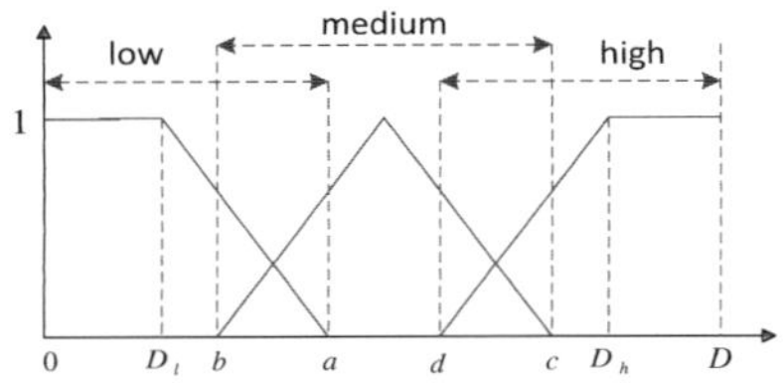

Figure 4. The membership function of the dynamic penalty factor

Where D_l is minimum dynamic penalty factor and 'a' is upper limit of low dynamic penalty factor. 'b' is lower limit of medium dynamic penalty factor and 'c' is upper limit of medium dynamic penalty factor. D_h is maximum dynamic penalty factor and 'd' is lower limit of high dynamic penalty factor.

2.4 Fuzzy rules

Case 1: When power demand is low, the wind curtailment is the only choice when the absorption capacity of the system is limited. Thus, whatever the active wind power is high or low, the value of the dynamic penalty factor is low.

Case 2: When the power demand is medium, system needs the support from wind farms, and in order to absorb wind power as much as possible, the value of the dynamic penalty factor is medium. But when the active wind power is very high and system could not absorb so much wind power, wind curtailment is the only choice, so the value will decrease while the active wind power increase.

Case 3: When the power demand is high, wind power will be a good support to the system, even if the active wind power is low. So in order to decrease the wind curtailment, the value of the dynamic penalty factor is high.

Table 1 shows a summary of the result from fuzzy logic system.

Table 1. Fuzzy rules

Power demand	Active wind power		
	L	M	H
L	L	L	L
M	M	—	L
H	H	—	M

3. Wind-thermal coordination optimization model

3.1 Objective function

With large scale wind farms integrated into power system, this paper builds the wind-thermal optimization model, considering power demand and active wind power comprehensively by introducing the dynamic penalty factor, and taking wind curtailment cost as one of the economic indicators, the objective function is as follows:

$$\min F_T = \sum_{t=1}^{T}\left\{\sum_{i=1}^{NG} f_c\left[P_i(t)\right] + \sum_{j=1}^{NW} f_c\left[P_j(t)\right]\right\} \tag{1}$$

Where F_T is the total cost of the power system generation including thermal and wind powered generators; NG is the number of thermal generators; NW is the number of wind powered generators; $P_i(t)$ is the power generated from bus i (thermal generator); $f_c\left[P_i(t)\right]$ is the cost of thermal power generation Pi; $P_j(t)$ is the power generated from bus j (wind powered generator); $f_c\left[P_j(t)\right]$ is the cost of wind power generation Pj.

Fuel cost of the thermal generators is as follows:

$$f_c\left[P_i(t)\right] = a_i P_i^2(t) + b_i P_i(t) + c_i \tag{2}$$

where ai, bi, and ci are constants of the fuel cost function.

Cost of the wind power generation is as follows:

$$f_c\left[P_j(t)\right] = d(t)\xi\left[P_{Nj}(t) - P_j(t)\right] \tag{3}$$

Where ξ is the wind curtailment coefficient; $P_{Nj}(t)$ is the predicted power of wind generator j for the time interval t; $d(t)$ is the dynamic penalty factor for the time interval t.

Subject to

(1) Real power balance constraint

$$\sum_{i=1}^{NG} P_i(t) + \sum_{j=1}^{NW} P_j(t) + P_{loss}(t) = P_l(t),\ t = 1,2,\cdots,T \tag{4}$$

Where $P_l(t)$ is the load forecasting for the time interval t; $P_{loss}(t)$ is the transmission losses that take place in the system during the time interval t.

(2) Real power operating limits

$$\underline{P_i} \le P_i(t) \le \overline{P_i},\quad i \in NG, t = 1,2,\cdots,T \tag{5}$$

Where $\underline{P_i}$ and $\overline{P_i}$ are the minimum and the maximum real power outputs of ith thermal generator, respectively.

(3) Total actual wind power generation limit

$$0 \le P_j(t) \le \tilde{P}_w(t),\ t = 1,2,\cdots T \tag{6}$$

Where $\tilde{P}_w$ is the predicted active wind power.

(4) Ramp up/down constraints

$$\begin{cases} P_i(t) - P_i(t-1) \le \overline{\eta_i}\Delta T \\ P_i(t-1) - P_i(t) \le \underline{\eta_i}\Delta T \end{cases},\ i \in I,\ t = 1,2,\cdots,T \tag{7}$$

Where $P_i(t-1)$ is the power generated from bus i (thermal generator) for the time interval t-1; $\overline{\eta_i}$ and $\underline{\eta_i}$ are the maximum ramping up rate and the maximum ramping down rate during normal operation of the ith thermal generator, respectively;

4. Methodology

4.1 The impact of the dynamic penalty factor on the wind curtailment

The cuckoo search optimization algorithm is adopted to solve equations established by objective function and constraints. IEEE30 system is used as the example system in this paper. The system includes 3 thermal units and 3 wind farms(A、B、C), and the installed capacity of 3 wind farms are all 50MW. Wind curtailment coefficient ξ is 975yuan/(MWh), vin、vN、vout are 3.5、13、25m/s respectively. In order to compare the 2 cases of with or without the dynamic penalty factor, simulations were carried out and the results show that after considering the dynamic penalty factor, there is a clearly decrease in the wind curtailment and the total operating cost.

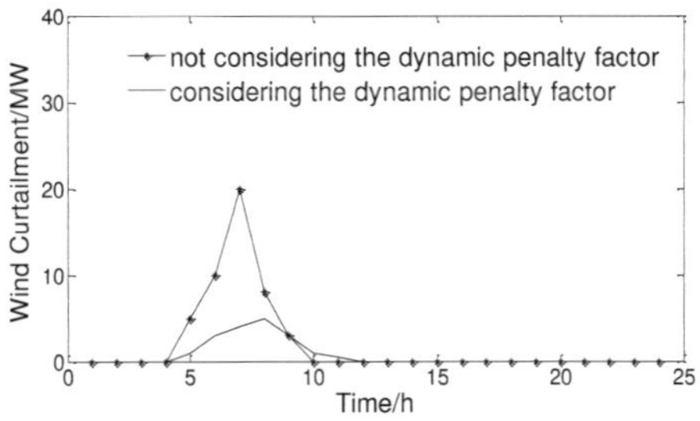

(a)wind farm A

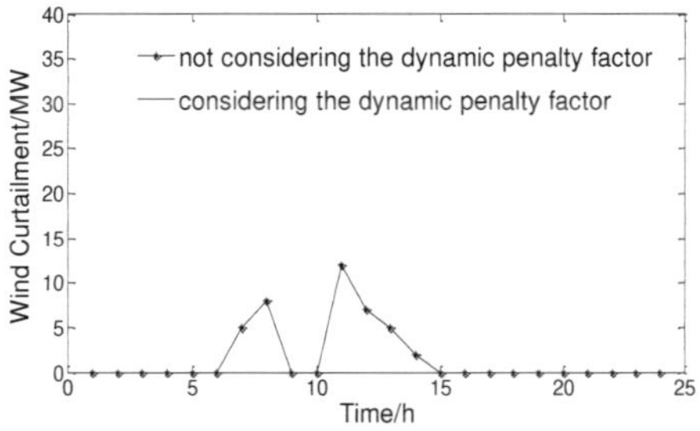

(b)wind farm B

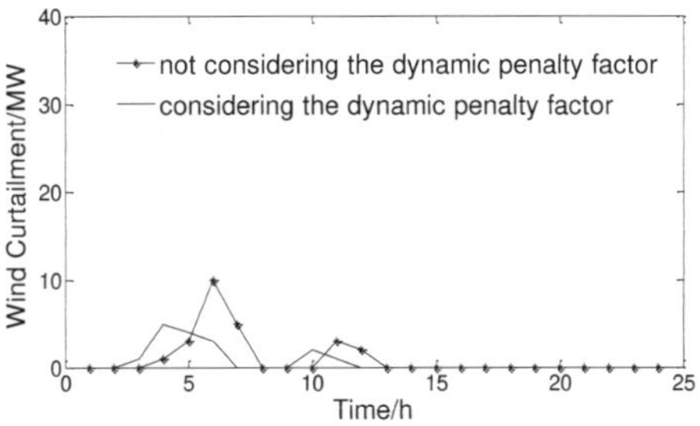

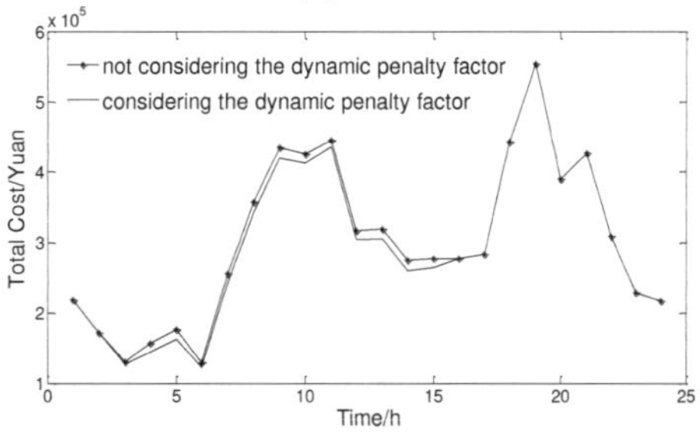

(c)wind farm C

Figure 5 Wind curtailment in three wind farms **Figure 6** Comparison of the power system total cost

The simulation results is shown in fig.5 and fig.6, and accroding to it, the wind curtailment has a obvious decrease after considering the dynamic penalty factor. And the wind curtailment of wind farm B is decreased to 0. The total cost of the system also decreased after considering the dynamic penalty factor when there is wind curtailment,and the decrease could up to 7.74% at most.

5. Conclusion

(1) The dynamic penalty factor will be changed with the power demand and the active wind power, so it could adjust its value in real time to change the wind curtailment cost.

(2) The introduction of the dynamic penalty factor could decrease the wind curtailment and the total operating cost effectively.

But the value of the dynamic penalty factor is subjective to some extent, and some limiting factor such as the environmental effect is important to the wind curtailment. So, how to make the value of the dynamic penalty factor more objective, and take the above factors into account in the model will be the follow-up research content.

References

[1] SUN Yuanzhang, WU Jun, LI Guojie. Influence Research of Wind Power Generation on Power Systems[J]. Power System Technology, 2007，31(20)：55-62.

[2] ZHANG Liying, YE Tinglu, XIN Yaozhong, et. al. Problems and Measures of Power Grid Accommodating Large Scale Wind Power[J]. Proceedings of the CSEE, 2010，34(1)：152-158.

[3] Fink S，Mudd C，Poter K，et al. Wind energy curtailment case studies[R]. Colorado：

National Renewable Energy Laboratory，2009.
[4] TROY N，ELEANOR D，O'MALLEY M. Base-load cycling on a system with significant wind penetration[J]. IEEE Trans on Power Systems，2010，25(2)：1088-1097.
[5] Yang Xinshe，Deb S. Engineering optimisation by cuckoosearch[J]. Mathematical Modelling and Numerical Optimisation，2010，1(4)：330-343.
[6] Homaifar A，Mccormick E. Simultaneous design of membership functions and rule sets for controllers using genetic algorithms IEEE Trans. Fuzzy Systems，1995，3(2)：129-139.

IOP Conf. Series: Materials Science and Engineering **339** (2018) 012017 doi:10.1088/1757-899X/339/1/012017

Load Forecasting of Central Urban Area Power Grid Based on Saturated Load Density Index

Yang Huping[1], Tang Chengyi[2] and Yu Meng[3]

[1,2,3]Information Engineering Institute, Nanchang University , Xinjian County,Nanchang,China

Email: yhping123@163.com

Abstract. In the current society, coordination between urban power grid development and city development has become more and more prominent. Electricity saturated load forecasting plays an important role in the planning and development of power grids. Electricity saturated load forecasting is a new concept put forward by China in recent years in the field of grid planning. Urban saturation load forecast is different from the traditional load forecasting method for specific years, the time span of it often relatively large, and involves a wide range of aspects. This study takes a county in eastern Jiangxi as an example, this paper chooses a variety of load forecasting methods to carry on the recent load forecasting calculation to central urban area. At the same time, this paper uses load density index method to predict the Long - term load forecasting of electric saturation load of central urban area lasted until 2030. And further study shows the general distribution of the urban saturation load in space.

1. Introduction

Urban electricity saturation load is a new concept put forward by power load forecasting and power network planning in recent years.Power load forecasting is one of the important work of the power sector. Accurate load forecasting can reasonably arrange the start and stop of the generator set inside the grid, keep the safety and stability of the grid operation, reduce the unnecessary reserve capacity and arrange the unit maintenance Plan to protect the normal production and life of society, effectively reduce power generation costs, improve economic efficiency and social benefits.The power load density is closely related to per capita GDP, population density, land use nature and regional economic planning layout. The trend of power load density increases with the growth of social and economic indicators.But load density can not grow indefinitely, regional electricity demand will become saturated after full development.

2. Electricity demand forecast

2.1. Common methods of load forecasting

Commonly used power load forecasting methods are broadly divided into two categories: The first method starts from the predicted power and is converted to the electricity load,such as integrated electricity level method, single consumption method, growth rate method and power elastic coefficient method; another type of method predicts the electricity load directly, such as regression analysis, gray system method and load density method.

2.2. The selection of Load forecasting methods

To forecast electrical load in the Central Urban Area, we mainly use the growth rate method, the regression method, the gray method, the per capita electricity consumption method, the average load method for calculating and Load density method. At the same time, we check with the number of hours of using the maximum load.

3. Load Forecasting

In this study, we selected a county in Jiangxi Province as an example to load forecast, the county is located in the eastern part of Jiangxi Province, it has a good industrial base. In the 12th five-year plan period, the economic and social development achievements of the central urban area are very significant. So it needs to further increase the demand for power network construction. This load forecast is a recent load forecast for the central urban area from the future to 2020, and the long-term load forecast from the future to 2030.

3.1. Recent load forecast

3.1.1. Electricity forecast.

According to central urban area historical load development level， a variety of methods are calculated by programming, we selected five of these methods for analysis.At the same time,we made a horizontal comparison of the development trend of meeting the maximum load utilization hours. The five methods are the growth rate method, the regression method, the gray method, the per capita electricity consumption method and the elastic coefficient method.

Table 3-1 Forecast of Total Electricity Consumption in Central Urban Area Unit: 100 million kWh

method of prediction	Year 2014	Year 2015	Year 2016	Year 2018	Year 2019	Year 2020
Gray method	6.9495	7.3392	8.7514	11.7464	13.6088	15.7665
Regression method	6.9495	7.3392	8.3205	10.0281	13.6088	11.7356
Growth rate method	6.9495	7.3392	8.3344	10.6064	11.8898	13.2749
Elastic coefficient method	6.9495	7.3392	8.253	10.4359	11.7351	13.1961
Per capita consumption method	6.9495	7.3392	8.3554	10.1178	11.0123	11.9374

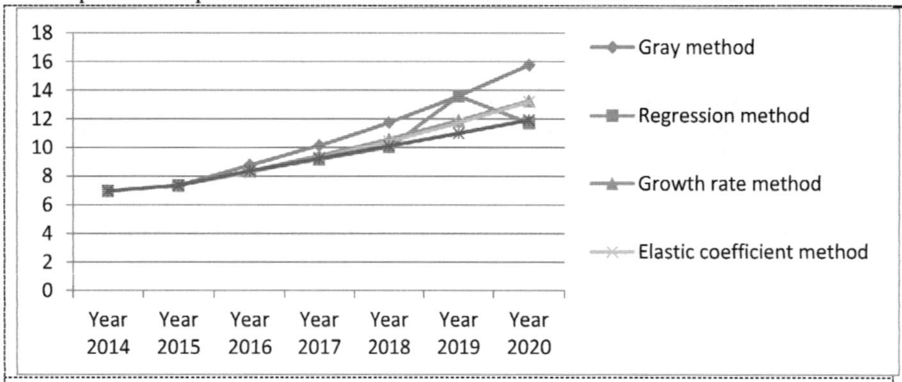

Figure 3-1 Forecast of Total Electricity Consumption in Central Urban Area

Unit: 100 million kWh

We choose high, medium and low three programs as the electricity forecast alternative,and recommend the program as the final plan for this power forecast,we can see from the table below, to 2020,Dongxiang County, the whole society electricity consumption is 13.2789 billion kwh.

Table 3-2 Central Urban Area, the whole society electricity consumption forecast results High and low units: 100 million kWh

IOP Conf. Series: Materials Science and Engineering **339** (2018) 012017 doi:10.1088/1757-899X/339/1/012017

Scheme comparison	Year 2014	Year 2015	Year 2016	Year 2018	Year 2019	Year 2020
High program	6.9495	7.3392	8.253	10.0281	10.8819	11.7356
Middle program	6.9495	7.3392	8.3344	10.6064	11.8898	13.2749
Low program	6.9495	7.3392	8.7514	11.7464	13.6088	15.7665

3.1.2. Load Forecasting.

According to load history data,combining with electricity consumption forecast results,we choose the appropriate forecasting method is used to forecast the power load. The five methods are the growth rate method, the gray method, the average load method and the maximum load utilization method. After a comprehensive analysis, we recommend a basic forecasting scheme.

Table 3-3 Highest load forecast for the whole society of Central Urban Area Unit:MW

method of prediction	Year 2014	Year 2015	Year 2016	Year 2017	Year 2018	Year 2019	Year 2020
Gray method	130.9	144.5	162.59	182.1	203.59	228.43	255.85
Growth rate method	130.9	144.5	157.51	168.53	178.64	187.57	195.08
Average load method	130.9	144.5	152.63	174.75	186.75	201.63	215.5
Maximum load utilization hours	130.9	144.5	157.32	186.69	191	207.83	227.09

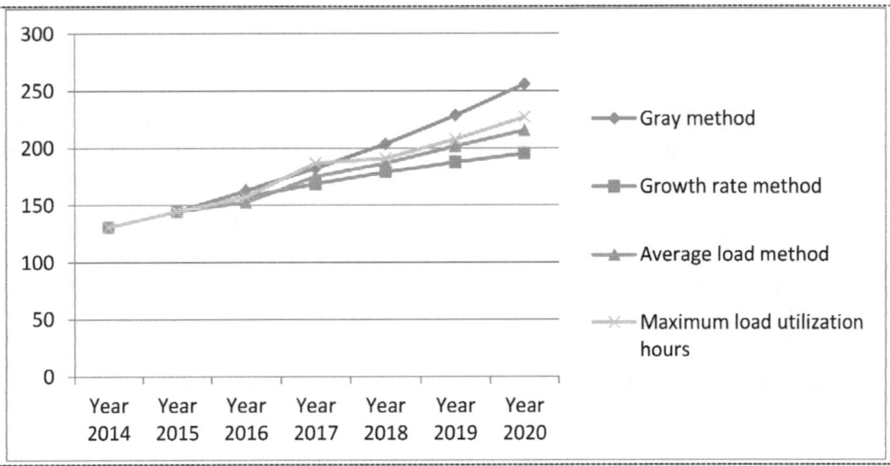

Figure 3-3 Highest load forecast for the whole society of Central Urban Area Unit:MW

We choose high, medium and low three programs as the load forecast alternative,and recommend the program as the final plan for this load forecast,we can see from the table below, to 2020, central urban area, the maximum social load forecast is 227.09MW.

Table 3-4 Central Urban Area, the total social load forecast results of high, medium and low programs Unit: MW

Scheme comparison	Year 2014	Year 2015	Year 2016	Year 2017	Year 2018	Year 2019
High program	130.9	144.5	162.59	186.69	203.59	228.43
Middle program	130.9	144.5	157.51	174.75	191	207.83
Low program	130.9	144.5	152.63	168.53	178.64	187.57

3.1.3. Maximum load utilization hours.

Maximum load typical day of this central urban area appears in the summer, summer temperatures are higher, residents living electricity consumption was affected by temperatures and increased. We combine the maximum load utilization hours in central urban area in recent years. Annual maximum load utilization hours of central urban area as shown in the table below.

Table 3-5 The recent electricity and load forecast results of Central Urban Area

Project	Year2014	Year 2015	Year 2016	Year 2017	Year 2018	Year 2019	Year 2020
The whole society electricity consumption (Billion kWh)	6.9495	7.3392	8.3344	9.4221	10.6064	11.8898	13.2749
The whole society the maximum load (MW)	130.9	144.5	157.51	174.75	191	207.83	227.09
Maximum load utilization hours (h)	5309.015	5079.0311	5291.347	5391.7597	5553.089	5720.926	5845.656

3.2. Forward load forecast

3.2.1. Load density index method.
Load Forecasting in this central urban area determine the load density indicator by investigating the load density of some of the same land properties in the case of saturation, considering plots development timing and strength. We use the cell load density index method to predict, and compare with other regions of the domestic load density and analysis, verify the rationality of the forecast results.According to the basic data collected in this plan, we determine load classification of load distribution prediction take the planning department classification method which according to land properties.Finally, we determine divide planning area load and land properties into eight categories. We divide Public management and public service facilities into administrative office land, cultural facilities land, educational research sites, sports land, medical and health land and so on. According to city planning results,the demand for electricity is different for different land properties, and therefore must be divided into blocks for analysis.

Table 3-6 Central Urban Area center city construction land balance table

Serial number	Land code		Land name	Planning (2030)		
				Area (ha)	proportion(%)	Per capita (m2 / person)
1	R		Residential land	961.3	30.6	32
2	A		Public management and public service facilities	194.1	6.2	6.5
	among them	A1	Administrative office land	40.4	1.3	1.3
		A2	Cultural facilities land	19.5	0.6	0.7
		A3	Education and Research Land	97.5	3.1	3.2
		A4	Sports land	16.5	0.5	0.6
		A5	Medical and health land	20.2	0.6	0.7
3	B		Commercial service facilities	155.8	5	5.2

IOP Conf. Series: Materials Science and Engineering **339** (2018) 012017 doi:10.1088/1757-899X/339/1/012017

4	M		Industrial land	840.1	26.7	28
5	W		Logistics warehousing land	89.8	2.9	3
6	S		Road and traffic facilities	444.5	14.1	14.8
	among them	S1	Urban road land	399.6	12.7	13.3
		S3	Transportation hub land	28.8	0.9	1
		S4	Traffic station land	33.6	1.1	1.1
7	U		Public facilities land	65.1	2.1	2.2
8	G		Green and square land	394.3	12.5	13.1
	among them	G1	Park green	288	9.2	9.6
		G2	Protection of green space	85.6	2.7	2.9
		G3	Square land	20.7	0.6	0.6
total			Urban construction land	3145	100	104.8

3.2.2. Selection of Load Density Index

Through the study of load density indicators, referring to a large number of similar types of urban load density indicators,and according to central urban area regional development, finally, we get the central urban area load density indicators of various types of land. As shown in the following table.

Table 3-7 Load density index of Central Urban Area

Land properties		Building Area Indicators（W/m2）	The required coefficient	Volume rate	Final load density index（W/m2）
Residential land（R）	Class II residential land	40	0.3-0.5	0.8-1.5	15
Public management and public service facilities（A)	Administrative office land（A1）	40	0.7~0.85	1.2	20
	Cultural facilities land（A2）	30	0.30~0.50	1.25	12
	Education and Research Land（A3）	30	0.4~0.6	1.2	12
	Sports land（A4）	20	0.3~0.5	1	10
	Medical and health land（A5）	35	0.5~0.7	1.2	12
	The land of cultural relics（A7）	10	0.5	0.75	5
	Religious land（A9）	8	0.5	0.7	4
Commercial service facilities (B)	Commercial and financial land	45	0.5~0.9	1.2-1.8	24
Industrial land（M）	Class II industrial land	50	0.4~0.7	0.7-1.2	18
Storage land（W）		10	0.3~0.5	1	5
Road and traffic facilities（S）		5	0.6~0.8	0.75	4.5
Utility land（U）		20	0.3	1.2	8
Green and square land（G)		1	0.2	1	0.2

3.2.3. Spatial load forecasting.

According to the selected results of the load density index for different types of land, combined with the actual situation of planning of planning area land, using the space load density index method, We get the load distribution prediction results of planning area land.

To target year(2030), prediction results of spatial load in central urban area are shown in the following table.

Table 3-8 Statistics of Central Urban Area load forecast results

Land properties	Area (km2)	Load（MW）	
Residential land	9.613	144.195	
Administrative office land（A1）	0.404	8.08	
Cultural facilities land（A2）	0.195	2.34	
Education and Research Land（A3）	0.975	11.7	
Sports land（A4）	0.165	1.65	
Medical and health land（A5）	0.202	2.424	
Commercial service facilities (B)	1.558	37.392	
Industrial land (M)	8.401	151.218	
Storage land（W）	0.898	4.49	
Road and traffic facilities（S）	4.445	20.0025	
Utility land（U）	0.651	5.208	
Green and square land（G）	3.943	0.7886	
Total	31.45	Not consider simultaneous rate	389.4881
		Considering simultaneous rate of 0.85	331.0649

As can be seen from the table above, to target year, central urban area total load is 331.0649MW, the city area is 31.45km², the load density is 10.53 MW / km².
Growth rate method is a more accurate way to predict long-term data, in this maximum load forecast of the whole society, we use the growth rate method.

Table 3-9 Total social load forecast of Central Urban Area in recent years

Years	2011	2012	2013	2014	2015
load	9.88	10.24	11.66	13.09	14.45
Forecast Year	2016	2017	2018	2019	2020
Predict load	15.7505	16.853035	17.864217	18.757428	19.507725

Based on previous data and nearly five years of forecast data, the forecast is 483.4458MW.
According to the previous description, the final load situation in this central urban area is shown in the table below.

Table 3-10 Central Urban Area Load Density Index +The load forecasting result of the household capacity method

Years	The target year
The maximum load of the whole society（MW）	483.4458
County urban area load （MW）	331.0649

From the table can be seen, to the vision of the whole society, the maximum load of 483.4458MW, central urban area load of 331.0649MW.Combined with urban positioning and development planning of this central urban area, the electricity consumption of central urban area is mainly composed with residential electricity and industrial electricity, to the maximum annual use of the number of hours to consider the number of hours 5700h, the total electricity consumption of 27.5564 billion kWh.

3.3. Load forecast results

From the short-term forecast results, we can see that by 2020, the total electricity consumption of Dongxiang County is 1.327 billion kWh, the maximum load of the whole society is 227.09MW, and the average annual growth rate of the "13th Five-Year Plan" period is 15.17% and 12%.From the long-term forecast calculation results, by 2030, Dongxiang County, the total electricity consumption of 2.756 billion kWh, the whole society, the maximum load of 483.45MW, the average annual growth rate of 18.54% and 16.83%.

4. Conclusion and analysis

Combined with the results and process of the load forecasting in this study, it can be found that the growth rate method, the gray system method and the regression analysis method in the load forecasting method have the advantages of accurately reflecting the overall level of the urban electricity load, but the disadvantage of these methods is that it can not reflect the spatial distribution of the load. In the construction of power grids, in order to more accurately promote the grid substation distribution and power grid construction, the total load and load growth position is the amount that we must know. The load density method is based on the acquired basic data and the predicted path, and it is more accurate than other prediction methods,the load density method satisfies our demand for the total load and the spatial load distribution. Therefore, the load density method is the better algorithm in our research process.

References:

[1] CHEN Dong. Discussion on Power Load Forecasting in Distribution Network Planning [J]. Technology and Innovation, 2014, (04): 27-28.

[2] Xiao Bai, Zhou Chao, Mu Gang.A Review and Prospect of Space Power Load Forecasting Methods [J]. Proceeding of the CSEE, 2013,33 (25): 78-92 (in Chinese)

[3] Yuan Yuan. Analysis of power load characteristics and long-term load forecasting method [D] .Shanghai Jiao Tong University, 2008.

[4] Wu JH, Yang JW. Several methods of power load forecasting and comparison [J] .Guangdong Electric Power, 2004, (01): 17-21.

[5] Xiao Yanhong, Mao Xiao, Zhou Jinglin, Jiang Huixia, Peng Yongjin.A REVIEW OF METHOD OF HARMONIC MEASUREMENT IN POWER SYSTEM [J] .Power System Technology, 2002, (06): 61-64.

[6] Zhang Mingli, Song Zhuoran, Liang Yi, Shi Zhe, Ye Peng.Analysis of urban long-term spatial load based on saturation load density [J]. Journal of Shenyang University of Technology:1-6(2017-10-25).http://kns.cnki.net/kcms/detail/21.1189.T.20171025.2113.044.html.

[7] TANG Xiao-ping. Research on Saturated Urban Aggregate Load Based on Urban Construction Land Area Ratio [A] .China Institute of Urban Planning, Nanjing Municipal Government.Transformation and Reconstruction - Proceedings of 2011 China Urban Planning Conference [C] .China City Planning Society, Nanjing Municipal Government:, 2011: 7.

[8] Huang Yuehua. Regional Power Grid Planning [D]. South China University of Technology, 2010.

Acknowledgments

This paper was supported by Major Program of National Natural Science Fundation of China(Grant No.51367014). We express our sincere gratitude to the professors and students who provided support and assistance for this pape.

IOP Conf. Series: Materials Science and Engineering **339** (2018) 012018 doi:10.1088/1757-899X/339/1/012018

Optimization of Power Generation Rights Under the Requirements of Energy Conservation and Emission Reduction

YANY Hu-ping[1],ZHONG Chong-wei[2] ,YAN Fei-fei[3], TANG Cheng-yi[4]

[1,4]School of Information Engineering, Nanchang University, Nanchang 330031, China
[2]Ganzhou Power Supply Company, State Grid Jiangxi Electric Power Company, Ganzhou 341000, China
[3]Nanchang Power Supply Company, State Grid Jiangxi Electric Power Company, Nanchang 330000, China

Email: yhping123@163.com

Abstract. In recent years, the energy crisis and greenhouse effect problem have caused wide public concern, if these issues cannot be resolved quickly, they will bring troubles to people's lives.In response, many countries around the world have implemented policies to reduce energy consumption and greenhouse gas emissions. In our country, the electric power industry has made great contribution to the daily life of people and the development of industry, but it is also an industry of high consumption and high emission.In order to realize the sustainable development of society, it is necessary to make energy conservation and emission reduction in the power industry as an important part of the realization of this goal.In this context, power generation trade has become a hot topic in energy conservation and emission reduction.Through the electricity consumption of the units with different power efficiency and coal consumption rate,it can achieve the target of reducing coal consumption, reducing network loss, reducing greenhouse gas emission, and increasing social benefit,and so on .This article put forward a optimal energy model on the basis of guaranteeing safety and environmental protection.In this paper, they used the IEEE30, IEEE39, IEEE57 and IEEE118 node system as an example, and set up the control groups to prove the practicality of the presented model.The solving method of this model was interior-point method.

1. Introduction

In recent years, in order to reduce the power industry's consumption of coal resources and make power generation trade can be carried out in the direction of energy-saving, many scholars put forward the relevant models. Early scholars have proposed the largest model of coal savings [1], while there are also scholars putting forward the smallest model of the network loss after transaction [2].For a medium-sized power system and assuming that the power transmission capacity is 1010kW · h / year, if the network loss reduces by 1%, then a year can save electricity 108kW · h. Also, assuming that the average unit coal consumption rate is 350g / kW · h, then the reduction of 1% loss means that a year will save 3.5 × 104t standard coal, which is a very impressive figure. Hence, reducing the network loss after transaction can also be used as a means of energy saving. However, there are some shortcomings in the two types of models: when the unit coal consumption after transaction is at minimum, the network loss may be larger. On the contrary, when the network loss after transaction network loss is at minimum, the unit coal consumption may be larger. So the results of the power generation trade under the guidance of the two models may fail to achieve the best energy savings.

Although a few scholars have proposed the optimal energy saving model to comprehensively consider the coal consumption and the network loss[3], but they did not fully consider the situation that when the system is at the maximum after transaction, the carbon emissions may exceed. They also neglected that the load margin may be too small to lead to the risk in operation. As a result, this paper presents a more comprehensive model, taking the coal consumption and the network loss after transaction into account. It also introduces the static voltage stability constraint and the carbon displacement constraint under the constraint condition. It aims to achieve the largest goal of saving energy after transaction in the premise of ensuring the safety and environmental protection of the system.

2. Research Methods

The model proposed in this paper comprehensively considers the change of coal consumption and the network loss after transaction. In order to verify if the model can achieve the largest energy savings, two sets of control groups need to be set up: one is the optimal model for the coal consumption after transaction and the other is the optimal model for the network loss after transaction. Then the energy savings of the two models after transaction will be obtained and be compared with the model of this paper. In order to verify the necessity of introducing double constraint conditions, two sets of control groups need to be set up to avoid the problem of the safety and environmental protection of the system. The static voltage stability constraints will be removed in the first group of models, and carbon emissions constraints will be removed in another. Then observe if the energy saving, load margins, carbon emissions and other parameters of the two groups of models after transaction have changed.

Apparently, the model of this paper deals with the nonlinear problem which covers a variety of constraints, and the interior point method can deal with such problems well. Here we adopt an improved interior point algorithm - the central correction interior point method based on the constraint relaxation strategy to solve the problem of this model. Compared with the traditional interior point method, the algorithm can solve the numerical problems caused by the complementary difference, and improve the convergence of the interior point method [4]. We use IEEE30, IEEE39, IEEE57, IEEE118 node system as an example of this paper, using MATLAB software to write the internal point of the program and calculate the results of various models.

3. Establishing the optimal power generation trade model with energy saving

3.1. Objective function

$$\min f_1 = K_s^T \cdot (P_{gt0} + P_{gp}) + \alpha \cdot [(e_1^T P_{g0} + e_2^T P_{gp}) - (e_3^T P_{d0} + e_4^T \lambda_P b)] \quad (1)$$

Table 1 Interpretation of variables

Variables	Meanings
KST	the coal consumption rate of the controllable generator, the row vector
Pgt0	the active output column vector of the controllable generator in the system before the start of the transaction
Pgp	the active output column vector of all controllable generators after transaction relative to that before the moment of transaction
α	the commutation factor of the network loss of coal consumption, whose value is equal to the power of the unit MW·h. It needs to consume the standard tons of coal
e1T	the row vector with M dimension is 1, here M is the number of all generators;
Pg0	the active output column vector of all the generators before the start of the transaction

e2T	the row vector with N dimension is 1, here N is the total number of controllable generators
e3T	the row vector that P dimension is 1, here P is for all load points
Pd0	the load column vector of the total load points at the beginning of the transaction
e4T	the row vector of Q dimension is 1, here Q is equal to the total number of real load points
b	the growth factor at the peak loadthe growing mode of the load , λp is the growth factor at the peak load
λp	the growth factor at the peak load
λpb	the load increment for all real load points from the starting time to the peak time.

3.2. Constraint condition

3.2. 1. Power flow equations.

$$L_0 + G(P_{gt0}) + l_p G(P_{gp}) - \lambda_p b - f(x_p) = 0 \quad (2)$$

Table 2 Interpretation of variables

Variables	Meanings
L0	the input power column vector of each node at the starting time
G(.)	the reflection of the controllable generator output vector to node input vector;
lp	the output factor of the controllable generator at the peak of the load
lpG(Pgp)	the active output increase of this node
f(.)	the calculation formula of the input power of the node
xp	the state variable of the node at the peak load, including the magnitude and phase angle of the voltage

3.2. 2. The active, reactive power constraints of generator.

$$P_{g\min} \le P_{g0} + P_{gp} \le P_{g\max} \quad (3)$$

$$Q_{g\min} \le Q_{gp} \le Q_{g\max} \quad (4)$$

Table 3 Interpretation of variables

Variables	Meanings
Pgmax	the upper limits of the active output of the generator at the peak load
Pgmin	the lower limits of the active output of the generator at the peak load
Qgmax	the upper limits of the reactive power output of the generator at the peak load
Qgmin	the lower limits of the reactive power output of the generator at the peak load
Qgp	the reactive power output of the generator at the peak load

3.2. 3 Node voltage amplitude constraint

$$V_{p\min} \le V_p \le V_{p\max} \quad (5)$$

Table 4 Interpretation of variables

Variables	Meanings

Vpmax	the upper limits of the reactive power output of the generator at the peak of the load
Vpmin	the lower limits of the reactive power output of the generator at the peak of the load
Vp	the amplitude of the node voltage at the peak load

3.2. 4 Branch current constraint

$$\left|I_{ij}(X_p)\right|^2 \le I_{ij\max}^2 \quad (6)$$

$$\left|I_{ji}(X_p)\right|^2 \le I_{ji\max}^2 \quad (7)$$

Table 5 Interpretation of variables

Variables	Meanings
Iij(·)	the branch current that flows from node i to node j
Iji(·)	the branch current that flows from node j to node i

3.2. 5. Static voltage stability constraint

The constraint of the power flow of voltage at the critical point:

$$L_0 + G(P_{g0}) + l_c G(P_{gp}) - \lambda_c b - f(x_c) = 0 \quad (8)$$

Table 6 Interpretation of variables

Variables	Meanings
lc	the output factor of the controllable generator at the voltage critical point
lcG(Pgp)	the active output increment of the system at that point
xc	the state variable of the node at the critical point

Load margin constraints:

$$\lambda_{ref} \le \lambda_c \quad (9)$$

$$\lambda_c = (p_{lc} - p_{l0}) / b_i \quad (10)$$

Table 7 Interpretation of variables

Variables	Meanings
Plc	the load of the voltage at the critical point
Pl0	the load at the current operating time
bi	the transition from the current operating point to the voltage at the critical point
λref	the given minimum load margin threshold
λc	the load margin of the system of the voltage at critical point

Load margin is generally used as a measure of static voltage stability, in which the greater the λc is, the better the stability of the voltage. Once the value is below the λref, the load will slightly increase, and the system is likely to face the risk of voltage collapse. We set λref of IEEE13, IEEE39, IEEE57, IEEE118, the four node systems relatively as 0.9, 1.35,0.95,1.05.

The reactive power constraints of generators of voltage at stable critical point:

$$Q_{g\min} \le Q_{gc} \le Q_{g\max} \quad (11)$$

Qgc is the reactive power output of generators of voltage at stable critical point.

3.2. 6. Carbon emissions constraints

The amount of carbon emissions in the system is set as Ce, which is calculated as follows:

$$C_e = C_t \cdot k_t \cdot P_d = 1.9774 \times 0.325 \cdot P_d \quad (12)$$

Table 8 Interpretation of variables

Variables	Meanings
Ct	the permissible carbon emissions of per unit of standard coal, which is 1.9774 t
Kt	the average coal consumption rate of power supply for the thermal power units, which indicates the amount of standard coal that is required to generate 1 MW·h of electricity, with the value of 0.325 t / MW·h
Pd	the total load of the system

Carbon emissions are defined as me, the formula is as follows:

$$m_e = k_e \cdot P_{gp} \tag{13}$$

In the formula: ke is the carbon flux intensity vector of the units, we set the carbon emission intensity of the large coal-fired unit, the small coal-fired unit and the gas turbine unit relatively as 0.75t / MW · h, 1.15t / MW · h, 0.4t / MW · h.

Therefore, the constrain of the system's carbon emission is:

$$K_e^T \cdot P_{gp} \le C_e \tag{14}$$

According to the above description, we set up the following power generation trade model to save energy to its limit :

$$\text{obj: } \min f_1 = K_s^T \cdot (P_{gt0} + P_{gp}) + \alpha \cdot [(e_1^T P_{g0} + e_2^T P_{gp}) - (e_3^T P_{d0} + e_4^T \lambda_p b)]$$

$$\text{s.t.:} \begin{cases} (2)\sim(9) \\ (11) \\ (14) \end{cases}$$

4. Case analysis

4.1 The necessity of considering the coal consumption and the change of the network loss in the meantime We decide to trade at the peak of the load.

Table 9 The raw data of the test system

Testing system	The coal consumption before transaction/t	The network loss before transaction/MW
IEEE30	116.21	9.75
IEEE39	2531.36	48.45

4.1.1 The optimal power generation trade model to save the coal consumption(model 1 for short in the following paper)

$$\text{obj: } \min\ f_2 = K_s^T \cdot (P_{gt0} + P_{gp})$$

$$\text{s.t.:} \begin{cases} (2) \sim (9) \\ (11) \\ (14) \end{cases}$$

Let Mt be the amount of coal after transaction, Mt = KST (Pgt0 + Pgp) t, t is for the settlement time, and t = 1h in this paper.

ΔM is the change of the coal consumption, which is the amount of coal before the transaction minus the amount of coal after the transaction. Ploss is for the network loss after transaction. Ploss=[(e1TPg0+e2TPgp)-(e3TPd0+e4TλPb)]·t ; ΔPloss is the change of network loss, which is the loss before the transaction minus the loss after the transaction; Q is the energy saving, Q=ΔM+αΔPloss，α=0.325t/MW·h. By trading the coal consumption after transaction will arrive at the minimum, so we can ensure that the best amount of coal will be achieved after transaction. According to this model, there are results as follows.

Table 10 The optimization results of model one

Testing system	Mt/t	ΔM/t	Ploss /MW	ΔPloss/MW	Q/t
IEEE30	115.87	0.34	8.1	1.65	0.88
IEEE39	2530.44	0.92	49.43	-0.98	0.6

4.1.2 The optimal power generation trade model to reduce the network loss(model 2 for short in the following paper)

obj: $\min f_3 = (e_1^T P_{g0} + e_2^T P_{gp}) - (e_3^T P_{d0} + e_4^T \lambda_p b)$

s.t.: $\begin{cases} (2) \sim (9) \\ (11) \\ (14) \end{cases}$

Table 11 The optimization results of model 2

Testing system	Mt/t	ΔM/t	Ploss /MW	ΔPloss/MW	Q/t
IEEE30	116.54	-0.33	5.85	3.9	0.94
IEEE39	2531.6	-0.23	48.28	0.17	-0.17

4.1.3 The optimal power generation trade model to save energy(model 3 for short in the following paper)

The model has been introduced in above two sections, so we do not discuss here. In the following table, C represents the carbon emissions after transaction, λc is the load margin of the system when the

Table 12 The optimization results of model 3

Testing system	Mt/t	ΔM/t	Ploss/MW	ΔPloss/MW	Q/t	C/(t/h)	Cref/(t/h)	λref	λc
IEEE30	116.31	-0.1	6.48	3.27	0.96	273.19	273.19	0.9	0.9
IEEE39	2530.51	0.85	48.28	0.17	0.91	5873.2	5919.72	1.35	1.35
IEEE57	497.86	-0.44	25.23	8.76	2.41	916.99	1152.73	0.95	0.95
IEEE118	1767.01	1.46	136.36	10.49	4.87	3752.63	3771.61	1.05	1.05

load is at the peak time after transaction.

From the analysis of the data in three tables above, it can be seen that the coal consumption of the power generation trade is ordered: Model 2> Model 3> Model 1, and the network loss after transaction is ordered: Model 1> Model 3> Model 2. Although Model 3 after transaction is not the best in the coal consumption and the network loss among the three models, the amount of its energy savings is the largest, which fully proves that Model 3 has the best energy-saving effect.

4.2 The necessity of introducing double constraint conditions

4.2.1 The optimal model of energy saving without the introducing of carbon emission constraints

The model removes the constraint (14) , and the result of the transaction is as follows:

Table 13 Trading results without the introduction of carbon emission constraints

Testing system	Mt/t	ΔM/t	Ploss /MW	ΔPloss/MW	Q/t	C /(t/h)	Cref/(t/h)	λref	λc
IEEE30	115.75	0.46	5.4	4.35	1.87	283.56	273.19	0.9	0.91
IEEE39	2530.44	0.92	45.43	3.02	1.9	6044.11	5919.72	1.35	1.35
IEEE57	497.52	-0.1	24.42	9.57	3.01	1236.98	1152.73	0.95	0.98
IEEE118	1759.43	9.04	125.99	20.87	15.82	3991.68	3771.61	1.05	1.05

Analysis: As can be seen from the data in Table 13, there is a slight increase in the load margin of the system of the IEEE30 and IEEE57 node at the peak load without introducing carbon emissions constraints. But the carbon emissions of the four nodes have increased. They are more than the provisions of the emission limit, hence, the environmental protection of this model greatly reduces.

4.2.2 The optimal model of energy saving without the introduction of static voltage stability constraints

The model removes (8), (9), (11), and the following is the result of the transaction in this case. λ0 is the load margin when the system does not introduce a single constraint condition.

Table 14 Trading results without the introduction of static voltage stability constraints

Testing system	Mt/t	ΔM/t	Ploss/MW	ΔPloss/MW	Q/t	C /(t/h)	Cref/(t/h)	λref	λc
IEEE30	116.93	-0.72	4.92	4.83	1.33	263.19	273.19	0.85	0.63
IEEE39	2527.67	3.69	38.13	10.32	7.04	5820.64	5919.72	0.67	0.29
IEEE57	497.15	0.27	23.93	10.06	3.54	908.16	1152.73	0.62	0.58
IEEE118	1758.49	9.98	101.75	45.1	24.64	36716.1	3771.61	1.35	0.32

From the data analysis of Table 14, it can be seen that under the condition of carbon emission constraint, the final carbon emission is less than the double constraint condition, and the result of the transaction is more environmentally friendly. However, the load margin of the whole system is significantly lower than that before the constraint, especially in IEEE39 and IEEE118 node systems, whose load margins at the peak of load are 19% and 22% respectively, which are most likely to lead to the serious consequences of voltage collapse, affecting the stable and safe operation of the power system.

In Table 12, Table 13 and Table 14, compared with that under the double constrains, the total amount of energy saving of the system increases under the single constraint. However, when the static voltage stability constraint condition is not introduced, the load margin of the system at the peak load point is small and there is a certain operational risk; if the carbon emission constraint is not introduced, the environmental protection of the model will reduce, both of which fail to achieve the requirements of "stability" and "environmental protection" of power generation trade. Therefore, it is necessary to introduce the two constraints of static voltage stability and carbon emissions in power generation trade.

5. Conclusion

In this paper, the optimal energy saving transaction model considering both static voltage stability and carbon emission constraints is proposed from the perspective of energy saving and emission reduction. It is proved that the largest amount of energy saving can be obtained by considering both the network loss and coal saving, the system is more in line with the requirements of safety and environmental after transaction. Using the model to guide the power generation trade can effectively alleviate China's current major issues of energy shortages and greenhouse gas emissions.

References

[1] Li Chungang. Research on power generation trade model for low-carbon development[D]. Beijing: North China Electric Power University, 2014.

[2] Yang Huping, Yan Feifei, Zhang Li, et al. Considering static voltage stability constraint low network loss power transaction optimization[J]. Power System Protection and Control,2017, 45(4): 46-48.

[3] Wang Yanling, Qiu Xiaoyan, Xu Chuanlong. The model of power generation trade blocking management based on energy saving and consumption reduction[J]. Power System Technology, 2012, 36(6):272-276.

[4] Xiong Ning, Zhang Wei, Huang Jinhai, et al.The central correction method based on the constraint relaxation variable strategy[J]. Power System Protection and Control,2012,40 (14):20-25.

[5] Xing Xiaoye. Research on power generation trading under the target of energy saving and consumption reduction[D]. Changchun: Northeast Electric Power University, 2014.

Acknowledgments

This paper was supported by Major Program of National Natural Science Fundation of China(Grant No.51367014). We express our sincere gratitude to the professors and students who provided support and assistance for this paper.

IOP Conf. Series: Materials Science and Engineering **339** (2018) 012019 doi:10.1088/1757-899X/339/1/012019

The Researches on Cycle-Changeable Generation Settlement Method

Jun XU, Suyan LONG, Jianhu LV
China Electric Power Research Institute(Nanjing), Nanjing, China 210003

Email: xujun3@epri.sgcc.com.cn

Abstract. Through the analysis of the business characteristics and problems of price adjustment, a cycle-changeable generation settlement method is proposed to support any time cycle settlement, and put forward a complete set of solutions, including the creation of settlement tasks, time power dismantle, generating fixed cycle of electricity, net energy split. At the same time, the overall design flow of cycle-changeable settlement is given. This method supports multiple price adjustments during the month, and also is an effective solution to the cost reduction of month-after price adjustment.

1. Introduction

Recently, the settlement cycle of power generation side in our country is fixed in month, that means power generation enterprises' net energy starts at zero clock of this month's first day and ends at zero clock of next month's first day, in month settlement, split the net energy and calculate every contract's energy and fee according to the priority of contracts. Price adjustment, which means the price adjustment of electricity contract, is a unique and common settlement business in China's long-term transaction contract model, usually the government departments issue a notice, including approved price, environmental protection price, contract price and so on in types, province and city Power Exchange Centers(PECs) execute in the monthly settlement.

This paper presents a cycle-changeable settlement method, which can solve the problem of month-forward price adjustment settlement and month-after price adjustment settlement by settlement task management, time energy process, etc., it also supports the random time price adjustment settlement. This method has clear thinking, and is operational, it can help to improve the level of price adjustment settlement.

2. Problems of Price Adjustment Settlement

In terms of price adjustment timeliness, there is a price adjustment in the month of the settlement, such as October 1 for September electricity settlement, the electricity measurement cycle is from September 1 to October 1, and September 12 to adjust the approved price, this type of price adjustment is usually called the month-forward price adjustment, The other type is adjust the history month contract which has completed settlement, such as the settlement on October 1, PEC receives a notice from the government that there is a power generation enterprises from May 12 began to adjust the price adjustment, the need for May 12 to September 1 has been the settlement of electricity to adjust the price, this type of price adjustment is usually called month-after price adjustment.

For month-forward price adjustment, PEC usually splits the net energy into sections from gathering to calculation, then deducts every contract's energy (including price adjust contract) from the total net energy according to contract's priority from high to low, at last splits the price changed contract's energy into sections according to the net energy segmentations' proportion, this method dealing with the price adjustment settlement is called "Settlement First, Split Second". For month-after price adjustment, because the history month settlement has been finished, and most history month energy could not get segmented energy, so the common method is calculating the effected energy from price adjustment time to this month, then calculating the compensation fee with effected energy and contract price, and put the compensation fee in this month settlement.

PEC usually deals with the price adjustment settlement manually, calculates the energy and fee offline, then uploads the results into system, this process has the following problems:

(1) Mostly calculate the effected energy of price adjustment manually or by Excel, this leads to heavy work and easily lead to errors.

(2) Possibly there are several price adjustments among a month, and power consumption side also has price adjustment, it leads to more complex operations.

(3) The compensation fee of month-after price adjustment is settled in current month, it cannot reflect the real purchase cost of the history month, also the current month purchase cost is distorted in analysis.

The existing researches mostly focus on the conventional settlement [1-3], the literature [4] and the literature [5] descript the conventional monthly settlement operation method from the perspective of power generation enterprises and regional grid, the literature [6] proposes a new method of settlement for month-after energy compensation, the literature [7] proposes a new settlement method in regional electricity markets, the literature [8] proposes a double settlement system to solve the problem of day ahead trade settlement and long term trade settlement. The above studies all study on the fixed monthly cycle of the settlement method, and some ones are not practical according to the actual business, lack the price adjustment method introduction of one or more times within a month, and lack the complete study and discussion on price adjustment settlement.

3. Cycle-Changeable Settlement

Cycle-changeable settlement means break the current settlement way which is fixed on monthly, by delimiting the start time and end time customized, to achieve any period of time settlement, In cycle-changeable settlement, the minimum time unit of all the metering energy is 15mins, and all the contracts energy also are broken down into 15mins, this means all the energy time cycle is 15mins. When there is a price adjustment, split the total month time into sections according to the price adjust time points, then net energy, contract energy is processed according to the sections, the net energy and contract in any section is calculated independently, and then get every section settlement result, at last collect the results and release the energy and fee which have been merged.

3.1. Settlement Task Creation

Current monthly settlement starts from the first day of month, the energy gathering, net energy calculation, contract energy confirms, net energy split and calculation all follow the settle day. In cycle-changeable settlement, it starts the settlement with creating a settlement task and setting a task management, and it runs following the task in all settlement parts. The settlement task defines all the attributes and configurations of the settlement, including task name, creator, involved participants, create time, start time of settlement data, end time of settlement data, through settlement task it achieves the random time definition of cycle-changeable settlement.

3.2. 15mins Energy Split

Divide an hour into 15mins for a time segment, then every hour has 4 segments, every day has 24 segments. Split net energy and contract energy into 15mins, manage all the energy in 96 segments per day.

Net energy calculation and split process: calculate every segment energy of the meters according to the start code and end code of every segment, then calculate the settle unit's every segment net energy through the net energy calculation formula.

Meter's segment energy calculation formulas:

$$\mathrm{EP}_n = (P_n - P_{n-1}) \times \mathrm{R} \qquad \mathrm{n} = 1,2 \ldots,96 \tag{1}$$

$$\mathrm{EQ}_n = (Q_n - Q_{n-1}) \times \mathrm{R} \qquad n = 1,2 \ldots,96 \qquad (2)$$

In formula (1) and (2): EP_n is the positive active power of meter at the n time segment; P_n and P_{n-1} are the positive codes of meter at n and n-1 time; EQ_n is the reverse active power of meter at the n time segment; Q_n and Q_{n-1} are the reverse codes of meter at n and n-1 time; R is the meter ratio.

Contract energy calculation and split process:

If the schedule plan has provided actually executed power curve data per 15 minutes of every contract, so it can calculate every time segment energy from the power curve data; if there is no power curve data per 15 minutes, the contract total energy can be extracted according to contract start and end time, the split the total energy into 15mins time segment on the basis of settle unit's 15mins net energy, as the following formula:

$$\mathrm{ConEng}_n = ConEng_{total} \times EngNet_n/\mathrm{EngNet}_{total} \qquad n = 1,2 \ldots,96 \qquad (3)$$

In formula (3) : $ConEng_{total}$ is the total energy from contract start time to end time; $EngNet_n$ is the settle unit net energy at time segment n; EngNet_{total} is the settle unit total net energy from contract start time to end time; ConEng_n is the contract energy at time segment n.

3.3. Fixed Cycle Energy Processing

According to the last part of the calculation of each time section of the net energy, the contract energy, in accordance with the day, week, month, quarterly four time periods of integration. In the daily zero clock, every Sunday zero clock, energy month first day zero clock, quarterly the first day of zero clock to calculate the generation of the previous day, the previous week, the previous month, the previous quarter of the total net energy and the total energy of each contract, when there are four time periods involved in the settlement, it can by read data quickly, so it can reduce data read and processing time.

3.4. Net Energy Split

In the process of net energy split, all the contract energy needs to split from net energy according to each settlement priority, the lowest priority contract is the last one to get the rest energy. At last calculate all contracts' actually settlement energy, price, fee. This step is implemented as follows:

Suppose that the settle unit net energy in specified time cycle is Eng_{net}, and it has m contracts, each contract has the settlement priority p_m (p_m to p_1 is in descending order), and the contract energy in the settlement cycle is EN_m. Now settle every contact according to p_m from high to low, the m-th contract has the highest priority, and the 1-st contract has lowest priority. Start from the m-th contract, so the i-th contract settlement energy as follows:

$$ES_i = \max(\min(\mathrm{EngNet} - \textstyle\sum_{t=i+1}^{m} ES_t \,, EN_i)\,, 0) \qquad (4)$$

In formula (4): EngNet is the settle unit net energy; ES_t is the t-th contract actually settlement energy; EN_i is the contract original energy in cycle; ES_i is the i-th contract final settlement energy.

4. Process Design and Application Characteristics

4.1. Overall Process Design

The cycle-changeable settlement overall process design is defined in Figure 1.

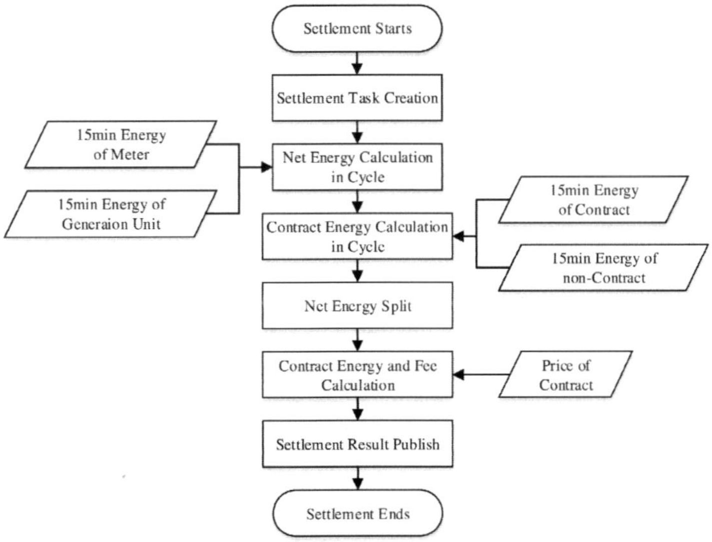

Figure 1 Overall Process Design of Cycle-Changeable Settlement

Description of significant steps:

a. Settlement task creation. Configurate parameters involved in cycle-changeable settlement, including participants, settlement data start time, settlement data end time.

b. Net energy calculation. Calculate the settle unit net energy according to 15min meter energy,15min generation unit energy.

c. Contract energy calculation. Calculate all contract energy in cycle according to 15min contract energy and 15min non-contract energy.

d. Contract price extract. Extract contract price from contract information.

e. Net energy split. Split and calculate contract actually energy from net energy in basis of every contract priority.

f. Fee calculation. Calculate every contract final settlement fee with the settlement energy and price.

4.2. Application Characteristics

a. Time irrelevantly

This method can define the settlement time from any start time to end time through the settlement task, the minimum time granularity is 15 minutes, and the maximum time granularity theory has no upper limit. It can meet the PECs of the current conventional monthly settlement, non-fixed time forecast settlement, quarterly settlement and other business requirements, but also can meet the day ahead and real time settlement needs in the future.

Suppose that the generation participant has price adjustment at day 10 and day 18 in this month, it only needs to set the settlement task at the corresponding point in time to complete the custom cycle of settlement.

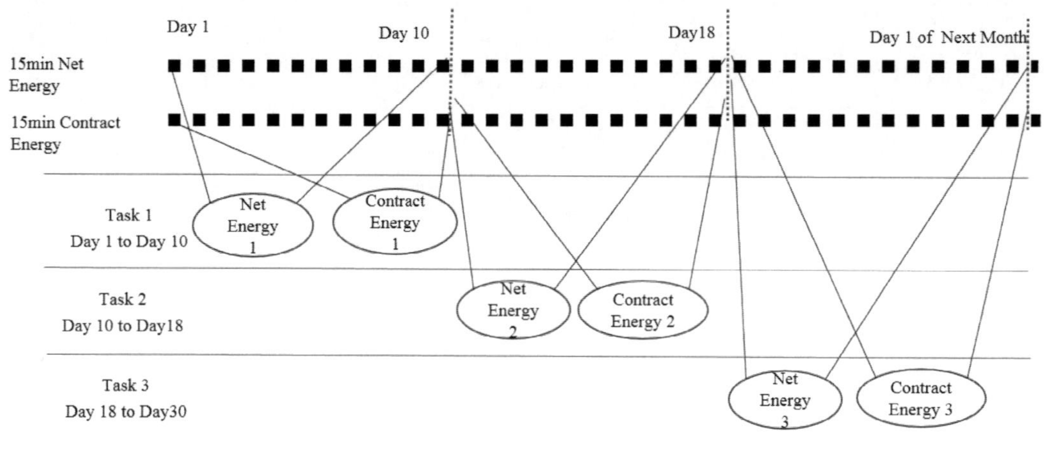

Figure 2 Radom Cycle Settlement

b. Multiple settlements in same cycle

Cycle-changeable settlement is driven by the settlement task, so the settlement can be repeated several times on same cycle. The results of the settlement can be saved independently and do not affect the other results. For multiple settlement results, you can choose any one of the settlement results to release.

c. Cost reduction of month-after price adjustment settlement

As the cycle-changeable settlement support for the historical data to re-settlement, and the data can be saved separately, so the energy influenced by month-after price adjustment can be re-settled, and the results can be compared with the historical data to generate deviation data. Statistical analysis of the deviation data will be accumulated in the historical data to restore, to solve the problem of the historical cost reduction and analysis.

The following is a case of the actual settlement of a power generation company, the case described as: the current time is October 1, PEC starts September settlement, and receives a notice from the government that the company in May approved price increase of 40 RMB/MWH. The traditional settlement results from January to September in Table 1:

Table 1 Settlement Results of Traditional Method

Month	Net Energy (MWH)	Price (RMB/MWH)	Purchase Fee(RMB)	Compensation Fee(RMB)	Sum Fee(10 Thousands RMB)	Average Price of Purchase(RMB/MWH)
5	254300	400	10172		10172	400
6	255800	400	10232		10232	400
7	263500	400	10540		10540	400
8	249800	400	9992		9992	400
9	274200	440	12064.8	4093.6	16158.4	589.29

From columns Sum Fee and Average Price of Purchase, it can be found that if put the compensation fee from March to August in September, it will lead to a significant increase in the cost of purchase in September, the history of the settlement date and the current September purchase costs are seriously distorted.

Resettle the history month energy by cycle-changeable settlement method, compare the resettlement results with history month settlement results, the deviation data is saved separately, and restore the deviation when analyze. The results from resettled by cycle-changeable settlement method as Table 2:

IOP Conf. Series: Materials Science and Engineering **339** (2018) 012019 doi:10.1088/1757-899X/339/1/012019

Table 2 Settlement Results of Cycle-Changeable Method

Month	Net Energy (MWH)	Price (RMB/MWH)	Purchase Fee(RMB)	Compensation Fee(RMB)	Sum Fee(10 Thousands RMB)	Average Price of Purchase(RMB/MWH)
5	254300	400	10172	1017.2	11189.2	440
6	255800	400	10232	1023.2	11255.2	440
7	263500	400	10540	1054	11594	440
8	249800	400	9992	99.2	10991.2	440
9	274200	440	12064.8	0	12064.8	440

5. Conclusion

This paper presents a cycle-changeable settlement method to support any period of time in order to solve the problems existing in the month-forward price adjustment and month-after adjustment in the current business. The method starts with settlement task, through split any energy into 15mins time segment to meet the finest time granularity of 15 minutes' settlement, in order to adapt to a variety of price adjustment of the system to achieve and after the price adjustment of the real cost of tracking to provide strong support.

References

[1] Cheng Haihua, Geng Jian, Hu Jun, et al 2011 A Settlement System Design for Provincial Power Generation Enterprises Based on Algorithm Library[J]. Automation of Electric Power System, **35(23)**:99-101.

[2] Long Suyan, Zhang Xian, Yang Zhenglin, et al 2013 A Component Library Based Flexible Electricity Settlement Method[J]. Automation of Electric Power System, **37(14)**:89-94.

[3] Zhu Changjiang 2015 Research on Electricity Settlement Method[J]. Science & Technology Information, **30(194)**: 194-195.

[4] Zhang Senlin 2008 Practical Research on Settlement Mode of Transacted Electricity Quantity in Regional Electricity Market[J]. Power System Technology, **32(3)**: 56-61.

[5] Zhu Xinglin 2007 Discussion on Settlement Method in East China Power Market [J]. Automation of Electric Power System, **31(2)**: 98-101.

[6] Cheng Haihua, Zhang Xian, Zheng Yaxian, et al 2013 A New Method of Complementary Settlement for Provincial Power Plants [J]. Automation of Electric Power System, **37(22)**: 118-121.

[7] Jiang Jianjian, Chen Wei, Huang Tao, et al 2008 New Settlement Method in Regional Electricity Markets [J]. Automation of Electric Power System, **32(13)**: 40-44.

[8] Zeng Ming, Cao Fang, Cui Tuowei, et al 2010 Study on Coordination Electricity Market System Based on Double Settlement System [J]. Water Resources and Power, **28(4)**: 154-156.

IOP Conf. Series: Materials Science and Engineering **339** (2018) 012020 doi:10.1088/1757-899X/339/1/012020

Research on the Automatic Fusion Strategy of Fixed Value Boundary Based on the Weak Coupling Condition of Grid Partition

X Y Wang[1,2], J M Dou[1,2],H Shen[1,2] ,J Li[3],G S Yang[1,2], R Q Fan[3],Q Shen[4]

1. China Electric Power Research Institute, Beijing 100192, China;
2. The State Key Laboratory for security and energy saving, Beijing 100192, China;
3. State Grid Shandong Electric Power Company, Jinan 250001,China;
4. State Grid Jinan Power Supply Company, Jinan 250001, China

Email: 93537158@qq.com

Abstract. With the continuous strengthening of power grids, the network structure is becoming more and more complicated. An open and regional data modeling is used to complete the calculation of the protection fixed value based on the local region. At the same time, a high precision, quasi real-time boundary fusion technique is needed to seamlessly integrate the various regions so as to constitute an integrated fault computing platform which can conduct transient stability analysis of covering the whole network with high accuracy and multiple modes, deal with the impact results of non-single fault, interlocking fault and build "the first line of defense" of the power grid. The boundary fusion algorithm in this paper is an automatic fusion algorithm based on the boundary accurate coupling of the networking power grid partition, which takes the actual operation mode for qualification, complete the boundary coupling algorithm of various weak coupling partition based on open-loop mode, improving the fusion efficiency, truly reflecting its transient stability level, and effectively solving the problems of too much data, too many difficulties of partition fusion, and no effective fusion due to mutually exclusive conditions. In this paper, the basic principle of fusion process is introduced firstly, and then the method of boundary fusion customization is introduced by scene description. Finally, an example is given to illustrate the specific algorithm on how it effectively implements the boundary fusion after grid partition and to verify the accuracy and efficiency of the algorithm.

1. Introduction

In recent years, the rapid development of UHV AC and DC grid and renewable energy power plants has made great significance to optimize the allocation of resources and promote the efficient use of energy in China, and also makes the connection of regional grid more tight and frequent, and the power grid structure more complicated. For the regional grid, on the one hand, a set of open, regionalized data is needed to complete the calculation of protect set point based on the conservation value of the region; on the other hand, a high-precision, quasi-real-time boundary fusion technique is required to constitute an integrated fault computing platform by stitching each region seamlessly and to accomplish the transient stability analysis with high-precision which can cover whole grid and multiple mode, so as to cope with the hazard for non-single failure and interlocking failure, and to build a strong "first defending line" for the power grid.

During the normal operation and maintenance of the grid, the set point calculation personnel usually maintenance the grid model by incrementing or modifying the local new-built power plant on it, and receive the latest equivalent boundary figure from the supervisor dispatching center half yearly or yearly. This mode is convenient and fast for the calculation of model of the partition grid, but the tight and frequent connection of the regional power grid highlights the drawback of the modeling method above. And the rapid development of renewable distributed energies and the energy interconnection of China has increased the difficulties of power grid analysis and set point calculation. So how to keep real-time synchronization of the change speed of the grid, while maintaining the original efficiency based on regional calculation is the goal of this paper to be solved.

The research on the boundary set point fusion strategy based on distributed autonomous agent is developed and applied accordingly [1-3]. It implements an open, shared architecture for relay protection setting calculation of power system. The system allows the updating based on the local distributed parameters and protection schemes among the boundaries and the fusion of strategies and the coordination between them. Its main contents are:

- Replace simplified modeling with global parameter modeling to reduce data maintenance difficulty;
- To determine the labeled technical specifications based on relay protection settings to facilitate the standard, transparent management of protection algorithm of the various manufacturers at the bottom layer, in order to achieve the fusion at the bottom layer;
- For each voltage level, a unified rule base is specified for the whole calculation, providing technical support for the setting calculation.

From the above it can be seen that in dealing with the boundary set point fusion strategy, the core two aspects are as follows: one is standardized design, which mainly reflects in the application of the label technology and the unification of the rule base; the other is the realization of complete modeling, whose deep meaning and realization rely on open data platform, the partition accurate modeling, and finally, the boundary fusion technology. Therefore, it can be seen that the labeling, standardization as the foundation, and the boundary fusion as the means, the ultimate goal is to achieve accurate transient calculation of the whole grid with an open platform and quasi-real-time splicing of data of high capacity [4-5]. As there exists real-time grid parameters quasi-synchronous with the development of the grid, the corresponding operation mode can also fully be reflected. When the short-circuit current (i.e., transient stability) reaches the critical point, the engineers can adjust the operation mode. For a districting power supply mode as an example, engineers can adjust the operation mode to one with sufficient margin of stability. And the modeling platform concerning complete-parameter gives most timely warning information by short-circuit current calculation method, which is the most simple and effective.

This complete-parameter real-time calculation platform, due to the usage of minimum parameters (super-transient impedance) of the short-circuit calculation for protection value setting, as well as the maximum possible operation mode, when the global short-circuit current reaches the critical point of transient stability, the system will guide the operator to re-adjust the operation mode, in which the transient stability can be strengthened substantially. Such mechanism is absolutely positive and effective, where early predictive transient stability level information is directly used to adjust operation mode, keeping the grid operating at a safe level.

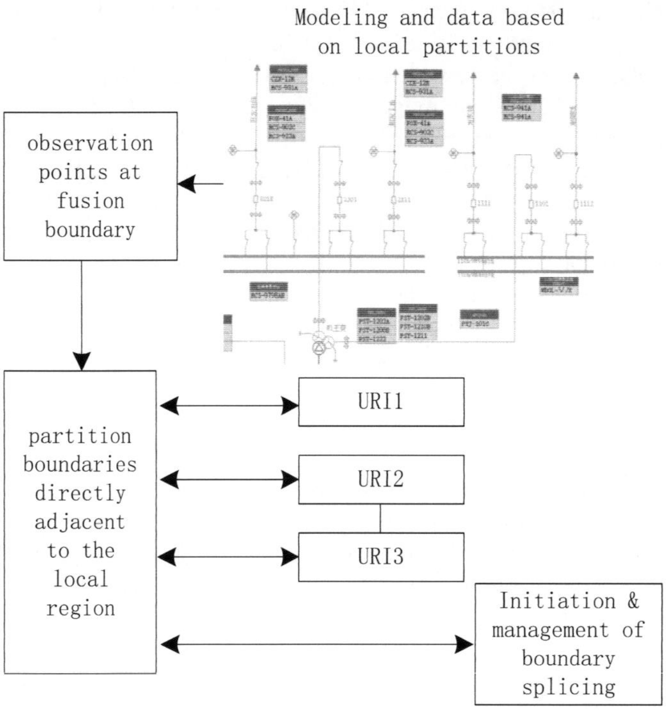

Figure. 1 A Scheme of Unique Identification Based on URI Partition

It can be seen from figure 1 that this algorithm is essentially a distributed method, and adopts splicing and fusion method afterwards [7-8]. In order to realize this method, we adopt the technique of data identification based on the URI partition. According to the scheduling scope of the distributed system, we configure the URI prefix for the data of scheduling unit, guaranteeing the uniqueness of the URI of each data object in the whole grid. In the progress of distribute data modeling, when a data object is to be generated, the URI of the data object is made up of the URI prefix of the scheduling unit stitching on ID of the data object. Such is how the object-based URI-stitched data works, as during the modeling of distributed data, the association between data objects is recorded by their URI. Once the URI of the data object is created, it will never be modified, realizing the seamless stitching of data objects.

2. Decision system of boundary fusion of BDI main objects based on multi - agent technology

There are many concrete implementations of border fusion based decision-making. In this paper, we firstly submitted data of scheduling unit, commit the confidence calculation afterwards, and submit for the manual review finally when the confidence value exceeds the threshold, see in figure 2. The confidence mentioned above is divided into one, two, three levels according to the source of the parameters and the verification situation, so as to measure the reliability of the parameters. The evaluation progress of the confidence of the parameters include the pre-assessment and post-evaluation steps in reality. The pre-assessment corresponds to the evaluation before the parameter by the means of software and artificial comprehensive assessment so as to test the integrity and correctness of the parameters. After the parameters being used, the post-evaluation verify the correctness of the parameters using the actual operation data, and with the state estimation and fault recovery as the main means, being the most important criterion for parameter error prevention. The parameters passing the pre-assessed process are at the premier and secondary level, which are lower, while parameters passing the post-evaluation verification by fault-recovery data are at the third level.

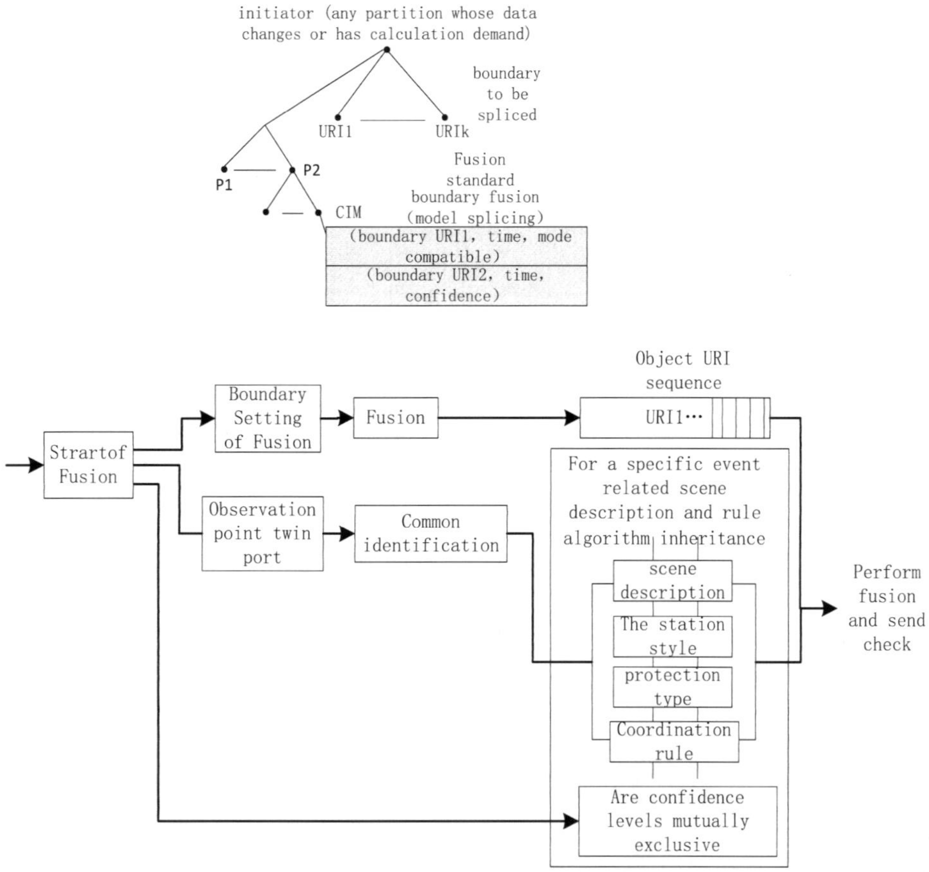

Figure. 2 Structure of log tree based on BDI boundary fusion and queue formation method diagram

It can be seen from figure 2, that sometimes it's necessary to judge whether the scenes are exclusive mutually. When the mutual exclusion of the scenes is distinguished, the border fusion with limited condition will be needed. The specific flow is shown in figure 3.

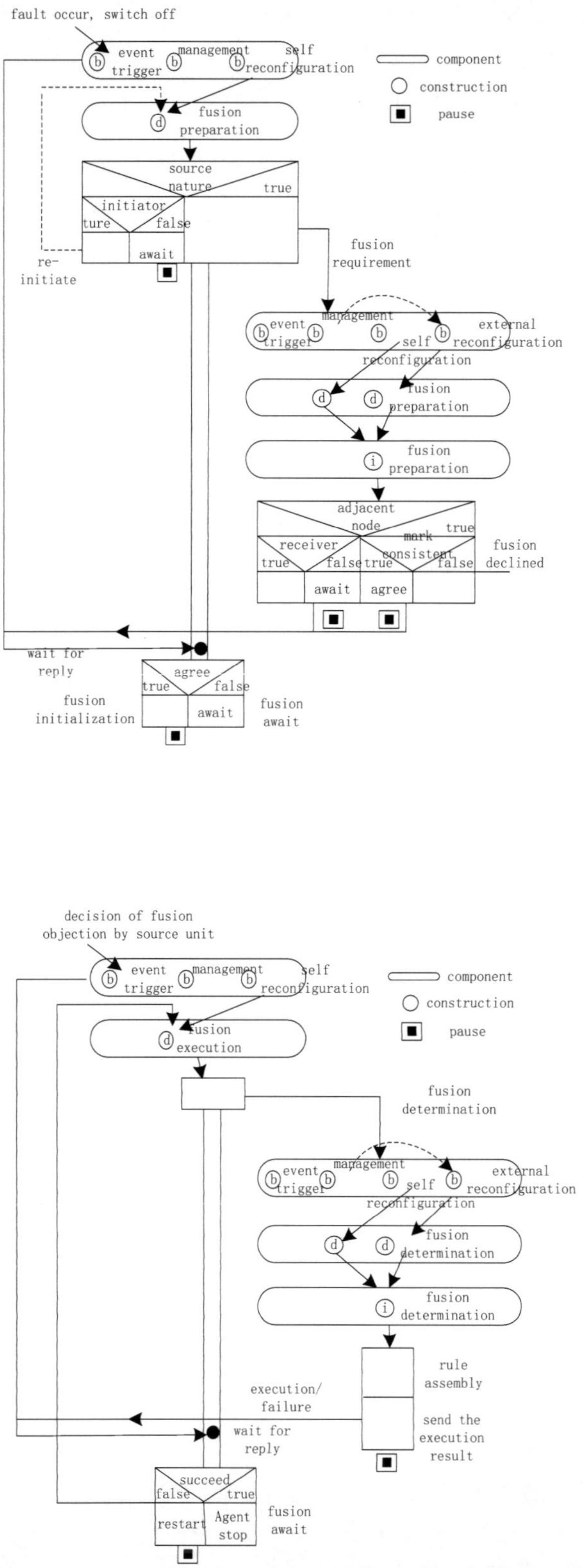

Figure. 3 Boundary integration process based on mutual exclusion audit

The realization of the method above is based on the Multi-agent technology, including the manual audit process in the method. According to the assumptions made by the main objects of component

BDI, the behavior of the Agent is determined entirely by the spirit (belief, desire, goal, and intent) component it possesses. Because of this, the exclusion conditions of the boundary fusion can be reflected proactively. The decision-making and implementation of the Agent's fusion also reflects on the changes in the spiritual components. What is important is that Agent uses its perceptual component to understand the environment. While information provided by the perceptual component of the agent is not static, each agent analyzes the environment by establishing its individual environment model. And whether the relay protection setting calculation is reasonable depends on whether the scene settings are reasonable and the model of Agent meets the actual application on site. As for the protection rules and the relationship between the settings, it is derived from the practical application set by the scene. If the scene changes, the corresponding rules of fault calculation will change accordingly.

3. Boundary Fusion Algorithm Based on Mesh-operation of Power Grid

3.1. The principle of algorithm based on scene setting

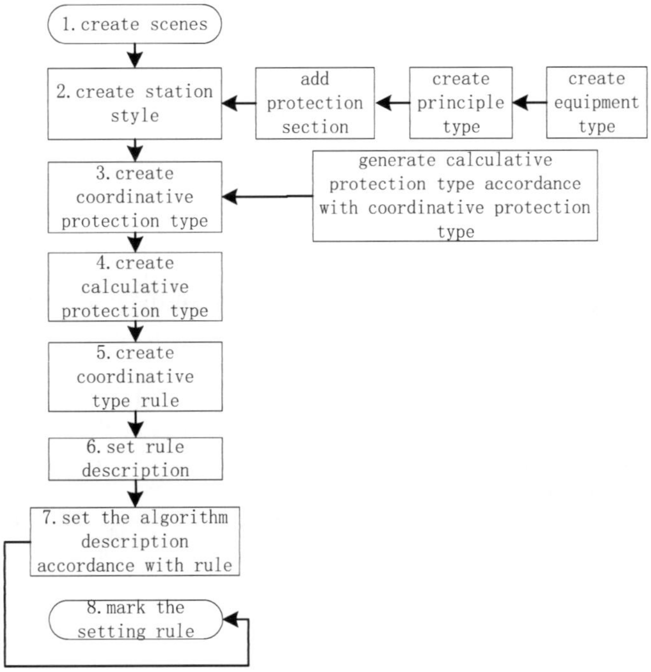

Figure. 4 Protection coordination settings based on scenario settings

It can be seen from figure 4 that through the scene setting, the observation point, cyclical mode of the station, operation mode of the line during the fault are set globally, and the boundary fusion must be one that meets the needs of this scene, otherwise it will lose its meaning. Because the final optimization direction of the protection scheme is actually the one that is at a level of the grid health assessment and able to maintain the rationality of the protection action when the multi-level fault occurs, and this scheme is in fact more from the judgment on the safety degree of the grid subjectively as well as the comparison of advantages and disadvantages among various options. So, it can be said that all protection strategies are relatively better ones after trade-off. Similarly, sometimes boundary fusion computing is a specific application of the way in which dispatching partitions are interconnected. As the excellent operation mode are actually such ideal situations as partition management, local balance, sharing of resources. Meanwhile, in the event of cascading accidents, it is best to divide power supply scope of partitions in view of balancing the supply and demand naturally, so as to limit the scope of the failure. Therefore, this paper divides the power supply partition naturally to balance the supply and demand in each partition, while limit the fusion rules on the boundary by the

IOP Conf. Series: Materials Science and Engineering **339** (2018) 012020 doi:10.1088/1757-899X/339/1/012020

operation mode, so as to limit the short-circuit current level of the power grid and improve the transient stability. Meanwhile, the boundary fusion set a limited operating conditions, corresponding to limited calculation scenes, so that the calculation is greatly simplified, while at the same time, evaluate the transient stability conforming to the operating conditions, as shown in figure 5.

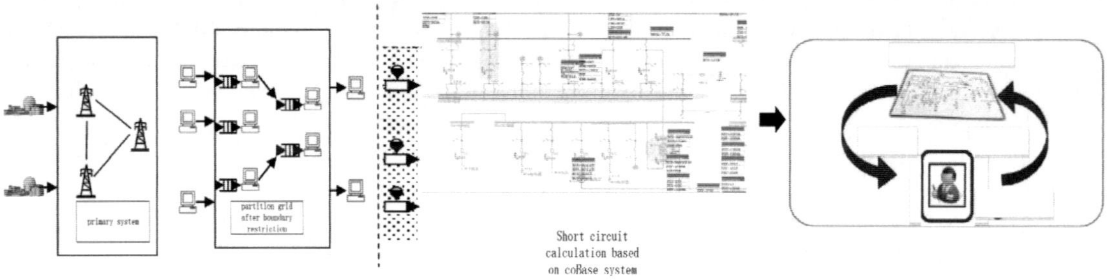

Figure. 5 The transient analysis diagram of grid power grid

3.2. Boundary conditions setting of fusion based on grid meshing

The overall trend of the power flow of the 220kV, 500kV system is from the west to east. With the reinforcement of the grid structure, the incensement of the power generation, and the completion of special high voltage grid, the power transmission capacity and the level of static stability rose dramatically, but also brought about problems such as power grid short circuit current over-limited at 220kV, conformation of electromagnetic loop, and larger power supply area than usual. In order to take into account of the level of power grid transient stability, try the following adjustment of operation mode: take the balance line between the generation and the local load at normal power flow as flow gates, divide the grid into mesh partitions, the tie-line of the grid keeps open-loop operation. Each meshing grid supply the power dependent on its regional generation, while in weak connection situations the open-loop line turn to direct-supply mode in order to realize power support. However, due to the variation of the operation mode of the open-loop lines and the plant at the ends of the lines, and the rejection between the original operation modes, the result of the approach above is that the cyclical combination of simple station mode will result in errors in the coordinative calculation of boundary fusion. To avoid this, in the previous boundary fusion process, integrate the relevant circumstances of the scene, and give the conditions to avoid the mutual exclusion, so as to give a reasonable fusion result to accord with the operation mode.

The setting principle of the plant on both sides of the boundary line of sub-regions is to ensure that the combination of cyclical modes of the plant at the two sides of the line avoid the one that makes the open-loop line operate in ring network mode. Therefore, the boundary fusion is actually one that is defined on the weak interconnection between partitions.

3.3. Condition setting of boundary fusion based on weak coupling connection mode

IOP Conf. Series: Materials Science and Engineering **339** (2018) 012020 doi:10.1088/1757-899X/339/1/012020

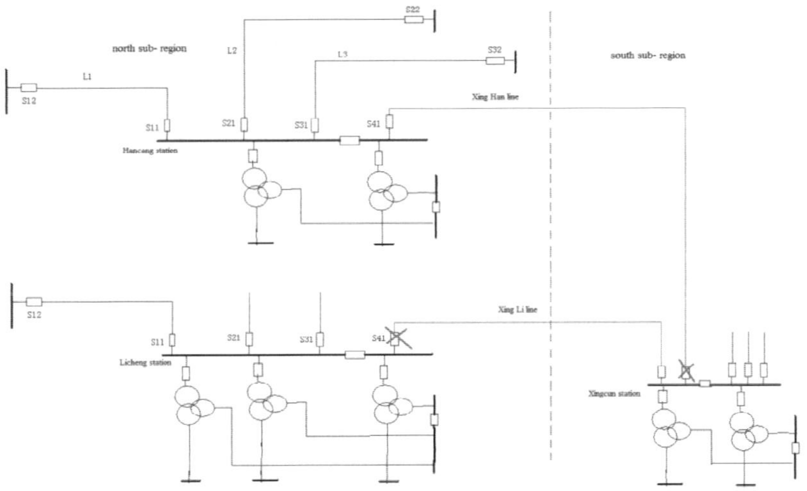

Figure.6 Open-loop operation schematic diagram of Han Xing Line, Xing-Li Line

Take the setting principal of the operation mode of a station connecting two sets of open-circuit lines as the example. In Fig. 6, the red cross mark denotes the intermittent point of the normal operation. The boundary fusion process sets the Xing Han line and Xing Li line as open-loop, and as direct distribution line when the local power supply encounter shortage or at the accident mode. In the latter situation, Xing Han line supply Hancang station (north sub- region) or a main transformer of Xingcun station (south sub- region) directly, while Xing Li line supply Licheng station (north sub-region) or a main transformer of Xingcun station (south sub- region) directly.

Therefore, basing on the considerations above, during the basic mode setting of Xingcun station on boundary fusion, the following should be considered. Normally, the normal open-loop operation should be considered, i.e. Xing Han line switch open, Xing Li line switch open, # 1,2 stations operate, # 1 station is grounded. When the local power supply encounter shortage or at the accident mode, direct distribution line operation is needed, i.e. Xing Han line switch off, Xing Li line switch open, Xing Han line only take load of # 1 main transformer of Xingcun station. At the same time, the high, medium and low voltage side of the bus-coupler of Xingcun station is open, to split the Xingcun station from the original grid, whose load is taken by the lines of adjacent partition in open-loop mode. Similarly, Xing Han line directly takes the load of # 2 main transformer of Xingcun station, the high, medium and low voltage side of the bus-coupler of Xingcun station are open. Or Xing Li line only take the load # 1 main transformer (or # 2 main transformer) of Xingcun station, and the high, medium and low voltage side of the bus-coupler of Xingcun station are open. The specific diagram is shown in figure 7.

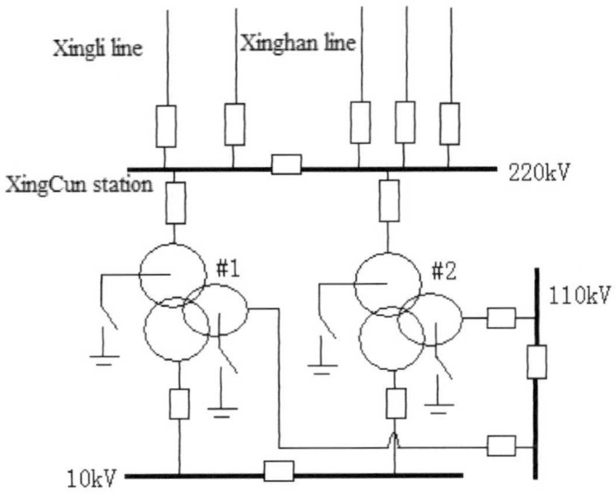

Figure.7 Main wiring diagram of Xing Cun Station

IOP Conf. Series: Materials Science and Engineering **339** (2018) 012020 doi:10.1088/1757-899X/339/1/012020

4. Conclusions

This paper introduces a boundary fusion algorithm suitable for mesh operation and weak coupling of modern grid. This algorithm takes the actual operation mode for qualification, complete the boundary coupling algorithm of various weak coupling partition based on open-loop mode, improving the fusion efficiency, truly reflecting its transient stability level, and effectively solving the problems of too much data, too many difficulties of partition fusion, and no effective fusion due to mutually exclusive conditions. This article provides an effective method for realizing the integrated modeling and calculation of grid splicing.

References

[1] Yang Y, Zhou T J and Yu R J 2015 General Matching Method of Large-Scale Power System Complex Data *Power & Energy* **36** 612-617.

[2] Dai W, Lian X H and Geng L S 2010 Distributed Relay Protection Setting Calculation System *Shandong Electric Power* **6** 34-37.

[3] LUO Jian, ZHU Bo-tong, CAI Ming, et al.Interoperability of CIM and SCL model based on CIMXML[J]. Power System Protection and Control, 2011,25(7): 18-22.

[4] LUO Jie, CHENG Hong-bo, CHEN Chi-pei. Research and design of data exchange platform based on CIM in power system[J]. Power System Protection and Control,2008, 36(13): 74-77.

[5] IEC61970 energy management system application program interface part 301： common information modelbase[S]. 2007.

[6] CHEN Guoyan, ZHANGZhe, YIN Xianggen, et al. A new algorithm of wide area backup protection based on evidence theory fusion[J]. Transactions of China Electrotechnical Society, 2014, 29(10): 246-256.

[7] SHAO Yaning, TANG Fei, LIU Dichen, et al. An approach of transient angle stability assessment in power system for WAMS measured data[J]. Power System Protection and Control, 2015, 43(6): 33-39.

[8] CUI Wei, SHI Yong, SUN Bing. The construction and in-tegration of grid model based on IEC61970/61968[J].Power System Protection and Control, 2011, 39(17):60-63.

IOP Conf. Series: Materials Science and Engineering **339** (2018) 012021 doi:10.1088/1757-899X/339/1/012021

The Researching on Evaluation of Automatic Voltage Control Based on Improved Zoning Methodology

ZHU Xiao-jun, FU Ang, DONG Guang-de, WANG Rui-miao, ZHU De-fen

State Power Research Institute of Chongqing Electric Power Company, Chongqing 421120, China

Email: 1654663144@qq.com

Abstract: According to the present serious phenomenon of increasing size and structure of power system, hierarchically structured automatic voltage control(AVC) has been the researching spot. In the paper, the reduced control model is built and the adaptive reduced control model is researched to improve the voltage control effect .The theories of HCSD, HCVS, SKC and FCM are introduced and the effect on coordinated voltage regulation caused by different zoning methodologies is also researched. The generic framework for evaluating performance of coordinated voltage regulation is built. Finally, the IEEE-96 stsyem is used to divide the network. The 2383-bus Polish system is built to verify that the selection of a zoning methodology affects not only the coordinated voltage regulation operation, but also its robustness to erroneous data and proposes a comprehensive generic framework for evaluating its performance. The New England 39-bus network is used to verify the adaptive reduced control models' performance.

1. Introduction

With the continuous development of society science and technology and economy, the scale and structure of power system is becoming more and more complex, the difficulty by controlling the voltage balance of each grid node voltage to control the entire power grid is also increasing. At present, the research direction of automatic voltage control (AVC) has been extended to the hierarchical and sub regional control of network [1-2].

France is one of countries that implement AVC layering earlier, and voltage control mode implemented by EDF company is one of the most advanced control systems in the world. The whole control system is divided into three levels, voltage control [3]. Germany RWE company is the leader in the voltage control system, but its application is two level voltage control, optimal power flow is the core of the optimal control, but because many control variables and high computational complexity, computational cycle is longer, so there are higher requirements on the reliability of [4]. At present, Fujian is the first implementation of AVC, system also uses two level control mode, because the regional controller in order to adapt to the operation state of the power system change is difficult, so the three level control and France compared to the lack of two level voltage control area, control bus hub selection and generator the control layer [5]; Taizhou city has also developed a voltage optimization control system to node voltage as constraint conditions, in order to minimize the power loss, the function of "four remote" to realize the optimization of voltage [6].

According to the AVC partition problem, now has a large number of scholars to conduct research, literature [7] presented a multi population genetic algorithm to improve the voltage control partitioning algorithm of fuzzy c-means clustering of the C, AGC was established in [8] and AVC integrated optimization model based on layered in the time dimension for the frequency and voltage of

IOP Conf. Series: Materials Science and Engineering **339** (2018) 012021 doi:10.1088/1757-899X/339/1/012021

decomposition and coordination. The [9] introduces evaluation based on optimal power flow, the power system is divided into several regions, the formation of three level voltage control, but can not reflect the real-time power network topology for partitioning the effects of Shandong power grid AVC system introduced by the literature [10], isolated and EMS means no establishment of communication links, to ensure the stability and reliability of AVC.

In view of the current research situation and development direction, this paper has done a few work: This paper has established a general framework to evaluate the effect of different partition methods on voltage order control, and evaluate the adaptive model in voltage order control In the applicability. This paper introduces the principle and application of single-distance stratification clustering (HCSD), VAr control space stratification clustering (HCVS), spectral K clustering (SKC) and fuzzy C clustering method (FCM) The application of the method is evaluated on the problem of voltage order control. The spatial region is divided and the selection of the dominant node is based on different examples. The adaptability of the adaptive model in the topology change is studied, and the various emergencies are considered The feasibility of each clustering method is verified.

2. Overview of 1 and AVC hierarchical control architecture

At present, hierarchical voltage control can be divided into three stages including voltage control, power system, voltage level control system refers to the voltage change when first need to response, when the system load fluctuation, power grid accident, a first voltage response, the response time of a few seconds in the range of two level voltage control; set in the hub system, is based on regional control, closed-loop control, response time in a minute level, at the level of voltage control, the partition problem is the key to realize the problem of two level control, is also the focus of this paper; three level voltage control system is arranged in the dispatch center, is a kind of prevention and control the response time is about tens of minutes, the target three level voltage control is to ensure the safe operation of the whole network, but also to ensure the economic operation .

The AVC system partitioning general framework diagram is shown in figure 1:

In order to realize the real-time voltage automatic control, it is necessary to optimize all the variables of the whole network in order to obtain the distribution of the optimized power flow. However, the reality is not feasible due to the large and complex network, network partition partition voltage control is the orderly development direction and trends of the day, the first is to obtain the control partition coupled within the region, to reduce the mutual influence between regions, each region has second is the requirement of reactive power balance the basic ability. The current partitioning studies are based on the concept of electrical distance, which can be represented by the sensitivity of the load node, i.e., the Jacobi matrix can be described . After controlling the partition, a dominant node is set so that it can represent the change of the load node in the region. Two regulation as a level of voltage regulation and two regulation of the bridge, the received signal level three voltage control, the voltage amplitude of dominant nodes at a given value, if the shift, by changing a voltage controller parameters of offset voltage offset.

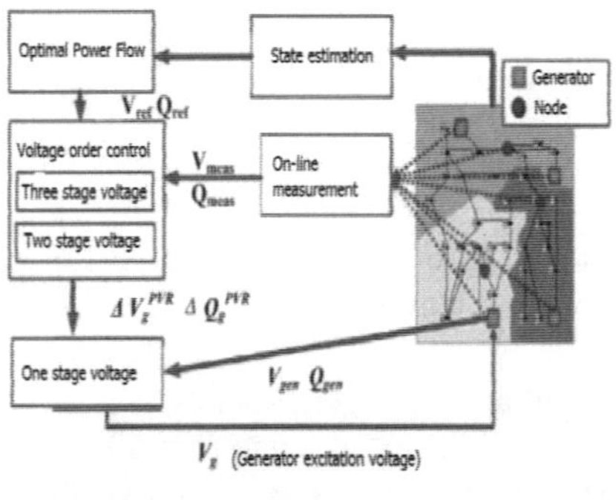

Figure 1 general framework diagram for AVC system partition

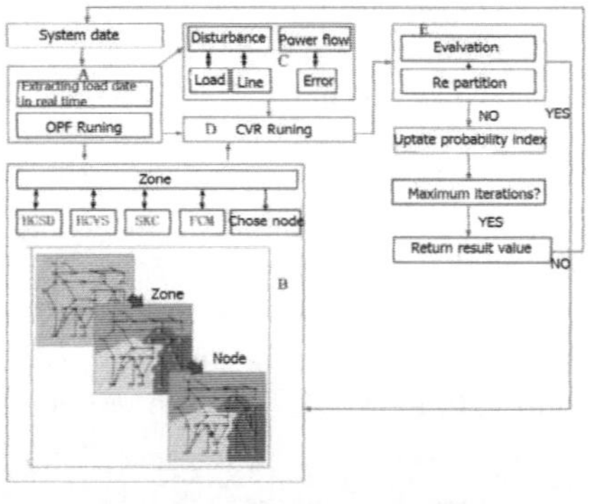

Fig. 2 general framework flow diagram

In order to realize the real-time voltage automatic control, it is necessary to optimize all the variables of the whole network in order to obtain the distribution of the optimized power flow [14]. However, the reality is not feasible due to the large and complex network, network partition partition voltage control is the orderly development direction and trends of the day, the first is to obtain the control partition coupled within the region, to reduce the mutual influence between regions, each region has second is requirement of reactive power balance the basic ability.The current partitioning studies are based on the concept of electrical distance, which can be represented by the sensitivity of the load node, i.e., the Jacobi matrix can be described.After controlling the partition, a dominant node is set so that it can represent change of the load node in the region. Two regulation as a level of voltage regulation and two regulation of bridge, the received signal level three voltage control, the voltage amplitude of dominant nodes at a given value, if the shift, by changing a voltage controller parameters of offset voltage offset.

3. Voltage order control frame diagram

In this section, a general framework will be proposed to evaluate the performance of voltage order control. Module A and C are responsible for data acquisition,B, D and E are responsible for evaluating system status, and general framework flow diagram is shown in figure 2.

Fig. 2 flow chart of general framework, load curve provides basic load data for the system, and then considers the safety constraints to optimize the power flow. The module is module B and module C provides the active, reactive and voltage data (P, Q, V) of the power system, providing the voltage reference value and reactive power reference value for the module D,Control range (Vref, Qref, Vmax, Vmin, Qmax, Qmin).

B. module B

The module will be integrated into the different cluster method frame model, the whole network is divided into several weak coupling region, and then select the dominant node for each region, (with methods will be discussed in the fourth part of this article). The system status information collected by module A is transmitted through the module to the D module provides the basis for planning voltage control in the area.

C. module C

Through calculation of the module, module C can obtain the system load and line loss, voltage offset, and provide the basis for the voltage control of the module D.No probability random perturbations in the near normal values by the load fluctuation, the disturbance is temporary, but also will cause random line tripping of reactive load fluctuations, emergency situations should be taken into consideration, so as to control the voltage of the order.

It is assumed that the transient response of the generator reaches a steady state when the voltage is controlled orderly, and the voltage in the middle section can be calculated by the steady-state alternating current flow equation. The number of nodes in the N network, a lot of V voltage offset is a N*1 matrix calculation method, as shown in formula (1):

$$\Delta V = V_{ref} - V_{ad} \quad (1)$$

Among them, the Vre f is the voltage standard value obtained by the module A, and the Vad is the voltage value with perturbation.

From Figure 1, it is assumed that voltage order control is generally based on normal controller can obtain voltage displacement vector Vp of leading node and the reactive power generated by the controllable device such as the generator in the voltage ordered control process Qg.

D. module D

The module includes a voltage control strategy of orderly control, synchronous equipment by adjusting the voltage in each partition to regulate the orderly adjustment of bus voltage of PNset, the ability to control the area of synchronous generator as shown in formula (2):

$$QR_g \cdot C_{pg} \geq a_{C_k} \quad (2)$$

Among them, QRg is reactive capacity of the generator, and Cpg is dominant node in the region.To control the sensitivity of the generator, AC is the smallest controllable volume of K in the first k region.The primary goal of voltage order control is to allow the voltage of the leading node.

PNset can follow the optimal calculated by optimal power flow forecasting module A reference value, a second goal is the rational allocation of regional reactive power distribution, so as to improve the safety of power area, regional reactive power as the reference value of formula (3) shown:

$$Qref_J = \frac{\sum_{g \in SVR_J} Q_g}{\sum_{g \in SVR_J} Q_{g,\max}} \tag{3}$$

Among them, the SVRJ is a group of generators that participate in the two stage voltage control in the regional J.
The objective function of the two programming problem is shown in equation (4):

$$\min\{ \sum_{i \in PN_{set}} (\Delta V_P)^2 + \mu \sum_{J \in k} (\sum_{g \in SVR_J} (Q_{refJ} - \frac{Q_g + \Delta Q_g}{Q_{g,\max}})^2)\} \tag{4}$$

The constraints are as follows:

$$V_{P,\min} \le V_p + C_{pg} \cdot \Delta Q_g \le V_{P,\max} \tag{5}$$

$$V_{G,\min} \le V_G + C_{Vg} \cdot \Delta Q_g \le V_{G,\max} \tag{6}$$

$$Q_{g,\min} \le Q_G + \Delta Q_g \le V_{g,\max} \tag{7}$$

Among them, is the weight factor of the two objective functions, and Qg is the generator without Work power output, Qg refers to the amount of adjustment obtained by iterative calculation, VP and VG, respectively, bus voltage and generator voltage, Cpg and Cvg respectively refers to voltage sensitivity matrix. Equations (5) and (6) are upper and lower bounds on voltage operation. (7) the upper and lower limits of reactive power operation.
E. module E
The module evaluates the control effect and calculates the correction Qg in each voltage ordered control loop.
In a N bus system, the PI value is calculated based on the mean relative error absolute value in each voltage ordered control iteration, as shown in (8):

$$PI_i = \frac{1}{N} \cdot \sum_{i=1}^{N} \left| 1 - \frac{V_{ac}(i) - V_{ad}(i)}{\Delta V(i)} \right| \tag{8}$$

Among them, subscript AC is the numerical value after the order control. Ad is the value after the voltage disturbance, and V is calculated by formula (1). That between different voltage levels of the orderly control is a dynamic decoupling, the reactive power control loop time constant is greater than a constant voltage control circuit of power plant, lower than the two level voltage control loop constant, therefore, the dynamic characteristics of voltage can be ignored. If the adaptive simplified control model can be applied to the partitioning method, when PIi is less than the threshold PIthree, you need to reconfigure the partition and select the dominant node. Then, go back to the B module and update the calculation PIi.

$$OPI = 1 - E\{PI\} \tag{9}$$

After each iteration, a non sequential Monte Carlo method is used to update the data. The lower the OPI value, the worse the order control performance of the voltage. Through the comparison of the OPI, we can measure the degree of voltage control scheme for different combinations of modules A, B, C, D and E. In this paper, the influence of the change of the precision of data on the method of partitioning is considered, and then the influence of different partition methods on the performance of voltage order control is studied.

3. Zoning method

In this section, four partitioning methods based on clustering optimization are introduced. In principle, the partition method includes two steps: (1) using the partition algorithm, the area is divided into several voltage for orderly control of weak coupling region; (2) for each small area leading node and bus, the load voltage variation can best represent the region.

IOP Conf. Series: Materials Science and Engineering **339** (2018) 012021 doi:10.1088/1757-899X/339/1/012021

3.1 Network partitioning algorithm

Engineering and partition problem is a clustering optimization problem, existing the following definition: A. similarity measure: electrical distance represents the similarity degree of two nodes, most of the electrical distance definition is similar to the actual distance, namely the coupling is stronger, the smaller the distance; B. clustering standard in clustering and classification, need consider the economic objective function; C. clustering evaluation: through evaluation algorithm applicability related clustering method.

1) single distance hierarchical clustering algorithm (HCSD)

The concept of the algorithm and the electrical distance was first proposed .

A. similarity measurement: the degree of coupling between the M and the N nodes is affected by the degree of voltage change, as shown in equation (10):

$$a_{mn} = \frac{\left[\dfrac{\partial V_m}{\partial Q_m}\right]}{\left[\dfrac{\partial V_n}{\partial Q_n}\right]} \tag{10}$$

However, the use of AMN matrix is not appropriate, the concept of size and electrical distance aspects of AMN is the reverse, on the other hand, amn is not equal to anm, which is not logical, therefore, need to construct new matrix to accomplish the transformation of electrical distance, as shown in Figure 11:

$$D_{mn} = -\log(a_{mn} \cdot a_{nm}) \tag{11}$$

B. clustering criteria: a clustering algorithm is merged into the node group, in each iteration, the two cluster CI, CJ similar characteristics such as formula (12) shows, the standard of clustering between clusters as shown in formula (13):

$$ICD_{I,J} = \max\{D_{mn} : m \in C_I \ and \ n \in C_J\} \tag{12}$$

$$C_I \cup C_J : \min\{ICD_{I,J}\} \tag{13}$$

The result of the iterative algorithm is a clustering tree, called a tree graph, which shows the correlation between clusters and clusters.

C. clustering verifies that the diameter of a cluster is the maximum distance between any two nodes in a cluster, as shown in equation (14). The relative diameter of the first k cluster can be obtained by formula (15):

$$diam_c = \max\{D_{mn} : m, n \in C\} \tag{14}$$

$$RD = \frac{\sum_{i=1}^{k} diam_{C_i}}{k} \tag{15}$$

The quality of the grouping corresponds to the slope of the RD curve. Finally, the optimal number of partitions K can be obtained by formula (16):

$$k : \max\left\{|RD_2 - RD_1|, \cdots, |RD_{k_{\max}} - RD_{k_{\max}-1}|\right\} \tag{16}$$

2) VAr control space hierarchical clustering (HCVS)

For g a reactive power source and a network of L nodes classification, Sij is the node of the I node of J voltage reactive power output sensitivity, VAr control space is defined as the G dimension Euclidean space , each node load are defined by the ordered vector xij:

$$x_{ij} = -\log\left(|S_{ij}|\right) \tag{17}$$

Based on the above definition, the ordered vector of the node represents the degree of coupling between the node and the specified reactive power source.

A. similarity measurement: for two load nodes, m{xm1, xm2,... Xmg} and n{xn1, xn2,... Xng}, the power distance is shown in equation (18):

$$D_{mn} = \sqrt{|x_{m1} - x_{n1}|^2 + \cdots + |x_{mg} - x_{ng}|^2} \tag{18}$$

B. clustering criteria: single distance similarity and hierarchical clustering algorithm, also using the agglomerative clustering algorithm, single clustering by iteration, node and a set of passive resources strong coupling of the same will be placed in the same cluster.

$$ICD_{IJ} = \frac{1}{\sum^{memb}(C_I)\cdot\sum^{memb}(C_J)}\sum_{m\in C_I}\sum_{n\in C_J}(\mathrm{D}_{mn}) \tag{19}$$

C. cluster verification: using the average intra cluster distance (AD) can determine the most suitable class number in the range of [1 and K kmax]:

$$AD = \sqrt{\frac{\sum \mathrm{ICD}_{IJ}^2}{k}} \quad I = 1,\cdots,\mathrm{k};\; J = I+1,\cdots,k \tag{20}$$

The larger the AD, the weaker the coupling between the clusters, and the K definition is shown in formula (21):

$$k:/\max\{\mathrm{AD}\} \tag{21}$$

3) spectral K clustering (SKC)
The spectral K clustering method is based on spectral analysis, and extracts the global information about the structure of the graph from the eigenvalues of the graph matrix .
A. similarity measurement: the relation between weighted proximal matrix and network G is A (G) =[wij]. Wij stands for the weights of nodes I and j placed in different clusters, such as the formula (11): (assuming that nodes I and j are adjacent to the G diagram).
The metric matrix of figure D is shown in equation 22:

$$d_{ij} = \begin{cases} 0 & if\ \mathrm{i} \neq \mathrm{j} \\ \sum_{k=1}^{n} W_{ik} & if\ \mathrm{i} = \mathrm{j} \end{cases} \tag{22}$$

The Laplasse operator of G is a N*N symmetric matrix, L (G), =D (G), -A (G), and the normalized Blass equation is presented in this paper as follows:

$$L_n = D^{-(1/2)}\cdot L\cdot D^{-(1/2)} \tag{23}$$

Laplasse G is the operator for a symmetric matrix N*N, L (G) =D (G) -A (G), the standard proposed La Blass equation are shown as follows: Blass operator standard is singular, its characteristic value contains 0, the rest of the eigenvalues are also positive. For spectral classification, K characteristic values for nodes in Rk coordinate arrangements. Vector x standardization publicity, as shown in (24), equivalent to the epitome of k-1 space.

$$U_i = \frac{x_i}{\|x_i\|} \quad 1 \le i \le N \tag{24}$$

B. clustering criteria: the K mean is used to assign N nodes to K clusters. The K mean method iteratively iteratively reduces the objective function value:

$$J = \sum_{j=1}^{k}\sum_{i=1}^{N}\|U_i^j - C_j^2\| \tag{25}$$

Among them, the U I j at C J refers to the node to the geometric center of the distance, the optimization iteration until the center point of the movement down to a certain minimum threshold or below the maximum number of iterations is reached to stop.
C. clustering verification: in this paper, eigenvalue analysis is used to determine the most appropriate clustering decision, while feature difference refers to the difference between two consecutive eigenvalues. The values of the eigenvalues Ln are arranged in descending order, relative to eigenvalues, as shown in equation 26:

$$\gamma = \frac{v_{k+1} - v_k}{v_k} \tag{26}$$

When the gamma reaches the maximum value, it represents the subarea and the result is optimal. It is worth noting that the relative eigenvalue criterion allows the most appropriate number of clusters to be confirmed prior to the K mean optimization. This will significantly reduce the computation time of the SKC method.

4) fuzzy K means clustering (FCM)

This paper adds voltage sensitivity research based on similarity measurement, and uses fuzzy statistical method to cluster verification.

A. similarity measurement: Xij refers to the voltage coupling of load nodes I and J. The formula is shown in equation 11.

B. clustering criterion: the fuzzy K mean method is used to assign N load nodes to K clusters, and the optimization function is as follows

$$J_m = \sum_{i=1}^{N}\sum_{j=1}^{k} u_{ij}^{m} \cdot \left\| x_i - C_j^2 \right\| \quad 1 \le m \le \infty \tag{27}$$

Among them, Uij is the membership degree of Xi in J clustering. M determines the ambiguity of clustering. In the simple K mean, Uij converges to 0 or 1 in m=1. The fuzzy clustering is iteratively optimized by formula (27), and the formulas updated by Uij and Cj are shown in formula (28):

$$u_{ij} = \frac{1}{\sum_{t=1}^{k}\left(\frac{\left\| x_i - C_j \right\|}{\left\| x_i - C_t \right\|}\right)^{2/(m-1)}} \qquad C_j = \frac{\sum_{i=1}^{N} u_{ij}^{m} \cdot x_i}{\sum_{i=1}^{N} u_{ij}^{m}} \tag{28}$$

When the conditions shown in the formula (29) are satisfied, the result converges to the minimum or saddle point Jm: 0, epsilon 1.

$$\left\{ \left| u_{ij}^{(iter+1)} - u_{ij}^{(iter)} \right| \right\} < \varepsilon \ \text{ or } \ \max\{iterations\} \tag{29}$$

C. clustering verification: Based on the value of K, the quality of the clustering decision can be verified by formula (30). The smaller the VXB, the better the clustering results.

$$V_{XB}(k) = \frac{\sum_{j=1}^{k}\sum_{i=1}^{N} u_{ij}^{m} \cdot \left\| x_i - C_j^2 \right\|}{N \cdot \min_{ij} \left\| C_i - C_j \right\|} \tag{30}$$

3.2 Leading node selection method

After partitioning the area, it should start looking for the dominant node in each region, and the selected leading node needs to be able to replace the voltage level [22] of an area. In the same region, the similarity between the two nodes in the C is shown in listing 31:

$$L_m = \sum_{m,n \in C} D_{mn} \tag{31}$$

Among them, Dmn refers to the electrical distance between nodes m and n.

The system needs to be able to monitor load disturbances accurately and to keep the voltage from within the proper range. The number of dominant nodes best and clustering the same number of dominant nodes too much though will make cluster more uniform, but also increase the coupling between regions, so they need to control more complex and dynamic interactions to handle loop instability, a good partition method must be able to do a good job in the balance two hand.

4. Simulation results and discussions are presented

4.1 Partition method results are discussed

All partitioning methods use similarity measurement, and only the input Jacobi matrix is needed, and the Jacobi matrix can be periodically updated with conditional changes. Because of the existence of the Jacobi matrix, all partitioning methods are related to state and structure. A good partitioning method can identify different boundary, based on the IEEE-96 system as an example, through the 72 mile 230KV line second and third regions connected by 67 mile 230KV line first and third regions are connected, the system grouping result is shown in Figure 3, the system is divided into three clusters, consistent the partitioning method clustering results, this method can also verify the partition basic ability to confirm the clear boundary.

4.2 Adaptive simplified model zoning method adaptive discussion

This section mainly verifies whether the partitioning method can be applied to the adaptive simplified model, and studies the main factors that affect the applicability of the simplified model partitioning method.

A simplified model of adaptive partition method and the advantage of Jacobi is based on the matrix update can quickly update the voltage control ordered partition model, in order to allow the orderly control voltage can quickly work in the first cycle, the simplified model running time need less than 1min control. In this paper, Poland 2383 node is used as an example, HCDS and HCVS are all based on agglomerate hierarchical clustering method, and their computational complexity depends on the number of cluster nodes. In order to reduce the computation time, the network needs to reduce the tree nodes and reduce the nodes to 1733 by reducing the tree nodes before clustering..

The computational complexity of SKC and FCM methods is represented by O (NKI) and O (nk2i), respectively. N is the number of wires to be classified, K is the number of clusters, and I is the maximum number of iterations. For fast clustering, K needs to be limited to range {10151}, and the reactive margin should be above 20MVAr. For the SKC method, the relative eigenvalue calculation has been introduced in the fourth part of this paper. The SKC method can select the most suitable classification K value, and also increase the computing time.

The results of the relative eigenvalues of the Poland system example are shown in Figure 4, and 17 regions are divided based on the SKC method.

Figure 5 FCM running time curve decided to simplify the control model, HCSD and HCVS are based on agglomerative clustering analysis algorithm, the most potential, SKC can quickly solve the classification problem in about 8 seconds, but in the eigenvalue calculation but increase in computation time.

Figure 6 is the different classification methods of operation of computing time, in the FCM method, the clustering of different K value in this range of {10151} can be processed in parallel, unlike FCM, HCSD, HCVS and SKC method can simplify the control online decision model, HCSD, HCVS analysis algorithm based on agglomerative clustering, with the most development the potential SKC can quickly solve the classification problem in about 8 seconds, but in the eigenvalue calculation but increase in computation time.

4.3 example comparison

This paper studied 39 UK bus network based on classic model of network voltage control is ordered, it has 9 synchronous generator, HCSD, HCVS, SKC and FCM methods are to solve a single network state simplified control model problem the most common method, clustering validation parameters are shown in Table 1, each partition the algorithm partitions are shown in Figure 7.

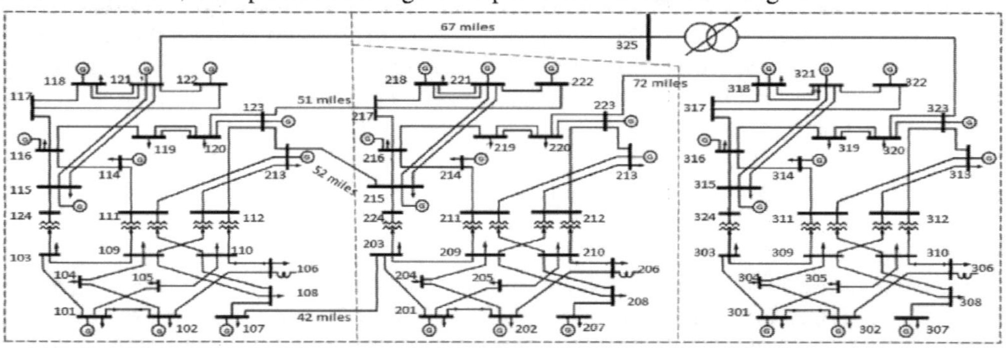

Figure. 3. Regional system topology diagram

IOP Conf. Series: Materials Science and Engineering **339** (2018) 012021 doi:10.1088/1757-899X/339/1/012021

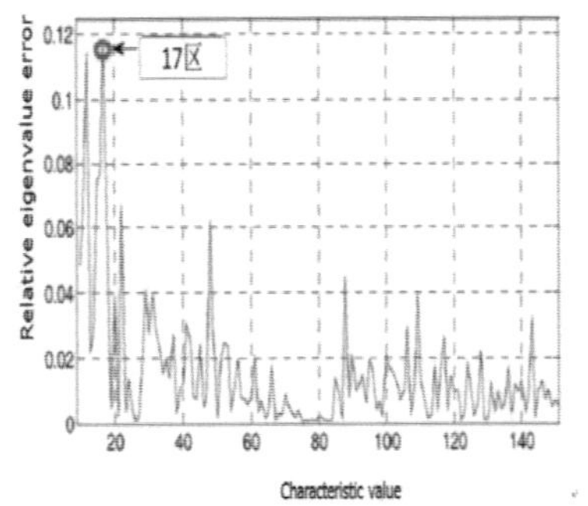

Figure. 4. Calculation results of eigenvalues

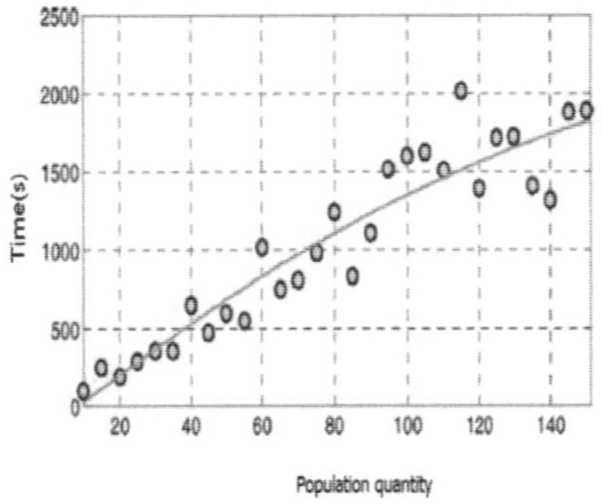

Figure 5. FCM run time diagram

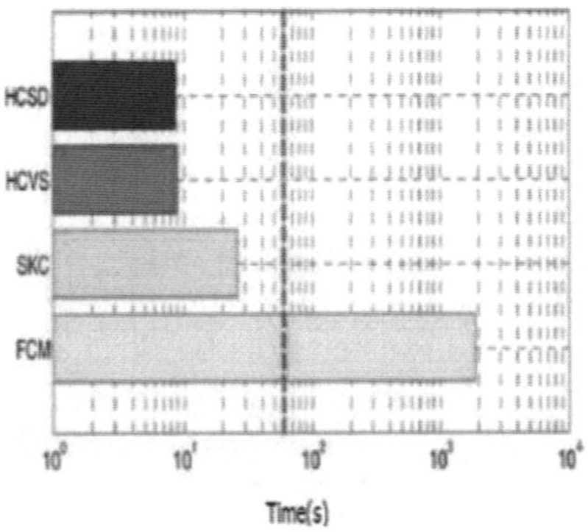

Figure.6. 2383 partitioning time consumption diagram of bus Poland power system

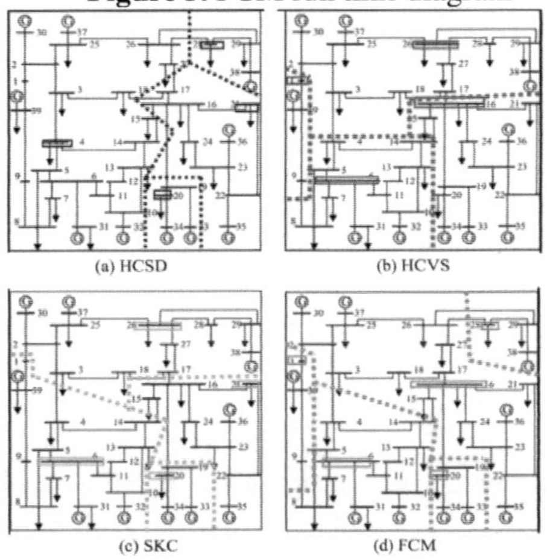

Figure 7. partitioning and leading node validation diagrams

Tab.1 Features of Clustering Validation

Zoning method	Verification standardf(x)	optimal value{f}	X:/optimal value
HCSD	max{ΔRD(k)}	0.0265	4
HCVS	max{ΔAD(k)}	7.0612	4
SKC	max{γ(k)}	0.4498	4
FCM	Min{V_{XB}(k)}	0.0961	5

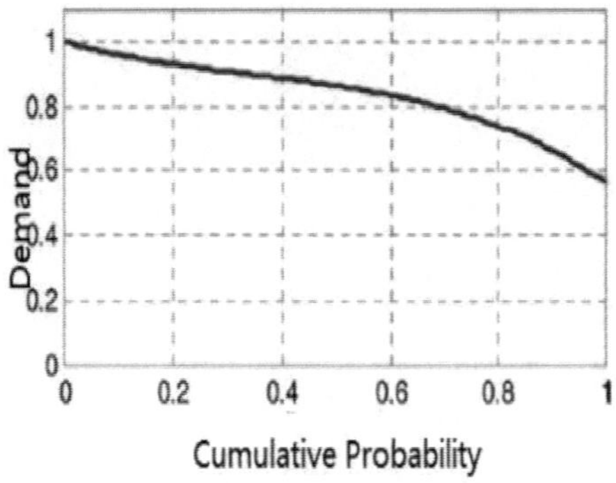

Figure. 8 .load cumulative probability distribution

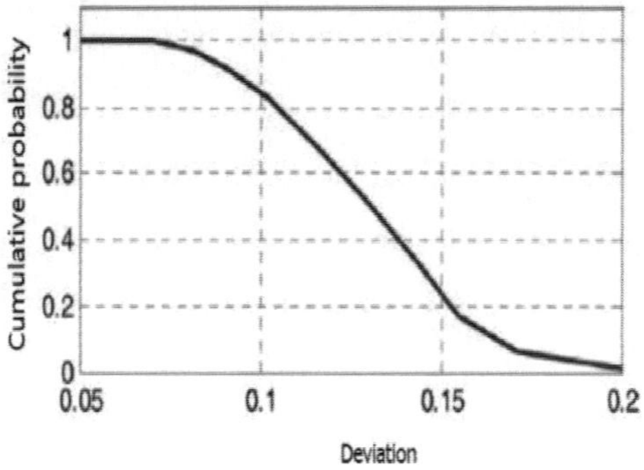

Figure. 9. cumulative probability distribution of load deviation

The results are highlighted to partition state markers: {HCSD:[4, 20, 21, 28]}, {HCVS:[1, 6, 16, 26]}, {SKC:[6, 20, 21, 26]}, {FCM:[1, 6, 16, 20, 28]}, is not difficult to see, although the network and control objective function is the same, but the results obtained using different classification methods the different, the problem is how to select the most suitable method for voltage control and orderly.

4.4 Static simplified control model and precise measurement evaluation

This section compares the performance of four partitioning methods in voltage order control, and studies the static simplified control based on the new England 39 bus test network

Assume that load obeys the cumulative probability distribution shown in Fig. 8, assuming that the base load deviation obeys the cumulative probability distribution as shown in figure 9:

The accurate measurement of the simulation results are shown in Table 2, the performance results show that the classification method affect the choice of order voltage control method, from table 2 it is not difficult to see, when the spatial classification all can meet the requirements of the time, HCVS and SKC can reduce the loss and improve the voltage distribution, show better performance.

Tab.2 Zoning Methodology Evaluation

Spatial classification	HCSD	HCVS	SKC	FCM
OPI(%)	86.91	92.09	91.23	89.88

4.5 Static simplified control models and dominant bus noise measurements are evaluated

This section studies the robustness of the measurement error of the spatial classification method, as shown in equation (32):

$$\Delta\hat{V}_P = |1-\varepsilon| \cdot \Delta V_P \qquad (32)$$

The error of initial value is set to 2%, gradually increasing to find 10% different optimal power flow under the condition of error results as shown in Figure 10, with the increase of the error, the OPI approximation decreases linearly, the slope is small, the robustness of the method of regional classification is stronger, the slope of HCSD decreased significantly, FCM slope is the most slowly, when the error increases, the SKC method is better than HCVS. From the above conclusions, it is easy to see that the selection of the partitioning method needs to be selected according to the expected accuracy of the voltage order control.

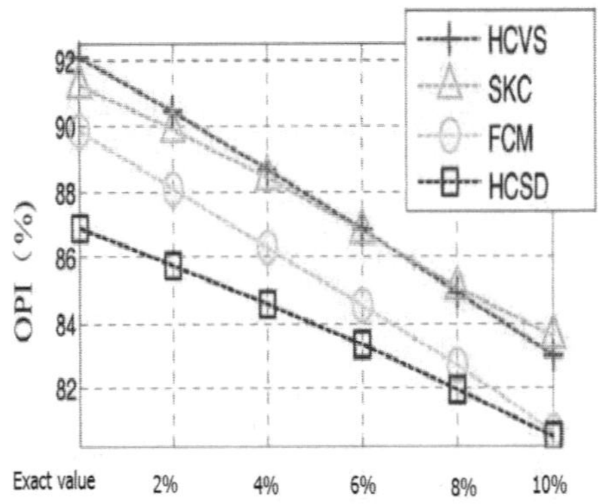

Figure. 10. optimization results of different partitioning methods based on different measurement errors

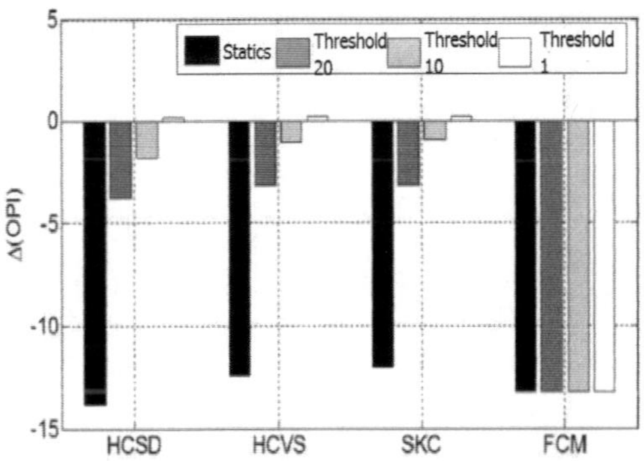

Figure. 11. performance analysis of adaptive RCM adapted to topology changes

Figure 10 based on different measurement error under different partition method based on adaptive topology optimization results of the 4.6 simplified control model will change the adaptive simplified control model is applied to the new England 39 bus network, when the network topology changes, the coupling of space inside and outside will change, the simplified model of adaptive control can maintain good performance. The combination of the partitioning method and the ordered voltage control method, coupled with accurate measurements, can be used to evaluate the most unstable states (maximum load, disturbance reaching 20%). In the actual situation, the accidental occurrence of voltage control in order to control the result will not be ignored. Once the case is lower than the minimum threshold for the E module introduced in the third section of this article, a simplified control model needs to be re run to achieve a steady state for the HCDS, HCVS, and SKC methods with an adaptive simplified model.

Figure 11compares the performance bias shown by the four partitioning methods in the presence of topological changes OPI.

Partitioning method with adaptive simplified control model has the ability to reduce the attenuation, the effect depends on (PI) thres, the lower the threshold, the better the performance, but the higher the reliability performance requirements. When (PI) thres=1%, the adaptive simplified control model is adapted to HCSD, HCVS and SKC, the network topology has changed, and the corresponding spatial correlation and control characteristics will be improved immediately.

5. Conclusion

In order to improve the voltage coordinated control, a new framework is proposed to evaluate the performance of voltage order control. The framework can select appropriate simplified model configurations for different candidate solutions and for the first time take into account robust characteristics. Although the test results mainly focus on HCSD, HCVS, SKC and FCM four kinds of partition method, but the framework has universal applicability, and can simplify the control model, and any data acquisition technology and matched control scheme. Among the four spatial classification methods, HCSD, HCVS and SKC can quickly and easily decide the simplified control model online. Meanwhile, when the network topology changes, it can improve its status quickly and improve its performance. For the SKC method, the relative eigenvalues have important value, and he needs to calculate the number of clusters by optimizing algorithm before the eigenvalue calculation, so as to improve the operation efficiency. In principle, adaptive simplified control models rely mainly on increased sensing and communication capabilities, and therefore need to be improved in the context of smart grids to enhance online telemetry and infrastructure construction.

References

[1] Wen J Y, Wu Q H, Turner D R, et al. Optimal coordinated voltage control for power system voltage stability[J]. IEEE Transactions on Power Systems,

2004, 19(2): 1115-1122.
[2] Tao Zhidong, Ding Xiaoqun, Deng Jixiang, et al. Reactive power control subarea based on SAGA clustering analysis [J]. electrical measurement and instrumentation, 2015, 52 (8):85-90.
[3] Brenna M, Berardinis E D, Carpini L D, et al. Automatic Distributed Voltage Control Algorithm in Smart Grids Applications[J]. IEEE Transactions on Smart Grid, 2013, 4(2): 877-885.
[4] Corsi S, Pozzi M, Sabelli C, et al. The coordinated automatic voltage
[5] control of the Italian transmission Grid-part I: reasons of the choice and overvi ew of the consolidated hierarchi cal system[J]. IEEE Transactions on Power Systems, 2004, 19(4): 1723-1732.
[6] Jiang Pengfei. Research on partitioning algorithm of reactive power and voltage control based on clustering and AVC in local power grid Applied [D]. South China University of Technology, 2014.Zhou Xiaohua, Wang Lifang, Qin Jin woo, et al. Voltage regulator SVC improved single neuron based on PID [J]. electrical measurement and instrumentation, 2011, 48 (9):73-77.
[7] Wang Ying. Two stage voltage control subarea based on fuzzy clustering method [D]. Hunan University, 2008.
[8] Zhang hang, Ding Xiaoqun, Deng Jixiang, et al. Reactive power / voltage control division based on multiple population genetic improvement [J]., FCM electrical measurement and instrumentation, 2015, 52 (7):71-75.
[9] Hu Wei, Wang Shuying, Xu Fei, et al. Automatic control strategy for coordinated optimization of AGC and AVC based on hierarchical control [J]. automation of electric power systems, 2011, 35 (15):40-45.
[9] Yonezawa S, Takayama S, Ishigame A, et al. Automatic Voltage Control by SVC using Moving Average Control in Distribution System[J] . Ieej Transactions on Power & Energy, 2016, 136(3): 302-310.
[10] Wei Wei, Zhu Tao, Zhao Chuan, et al. Three stage voltage division method based on clustering analysis. [J]. automation of electric power systems, 2016 (5): 127-132.

IOP Conf. Series: Materials Science and Engineering **339** (2018) 012022 doi:10.1088/1757-899X/339/1/012022

Comprehensive vulnerability assessment method for nodes considering anti-interference ability and influence

Jianchun LUO [1], Yunyu WANG [1], Jun YANG [1], hong RAN [1], Xiaodong PENG [1], Ming HUANG [1], Hao FENG [2], Meijun LIU [3]

[1]State Grid Chongqing Electric Power Co.Wulong Power Supply Branch Company, Chongqing, China
[2]State Grid Sichuan Electric Power Co.Tianfu New Area Electric Power Supply Company, chengdu,China
[3]School of Electrical Engineering and Information, Sichuan University, Chengdu, China

Email: 564851778@qq.com

Abstract. The vulnerability assessment of power grid is of great significance in the current research. Power system faces many kinds of uncertainty factors, and the disturbance caused by them has become one of the main factors which restrict the safe operation of power grid. To solve this problem, considering the anti-interference ability of the system when the system is disturbed and the effect of the system when the node is out of operation, a set of index to reflect the anti-interference ability and the influence of nodes are set up. On this basis, a new comprehensive vulnerability assessment method of nodes is put forward by using super efficiency data envelopment analysis to scientific integration. Finally, the simulative results of IEEE30-bus system indicated that the proposed model is rational and valid.

1. Introduction

There have been a series of large-scale power outages at home and abroad in recent years, which have brought serious challenges to the safe and stable operation of power system[1,2]. Compared with the open line, the result of large power outage caused by bus trip is more serious, And not easy to take effective measures to control short-term emergency. Therefore, fast and accurate identification of the weak nodes of the system, can provide valuable theoretical reference for the power system safe and stable operation, and early warning, is of great significance.

Power grid vulnerability research, as the safety and reliability of the depth and extension of electricity has been a focus of attention of academic and engineering. Traditional vulnerability assessment methods of power grid are mainly Energy function method based on system operation state, risk theory, complex network theory, etc. However, any system and component is not isolated and there is a relationship, power grid vulnerability assessments should take into account the interaction between system and components.

This paper proposes the indexes to reflect the anti-interference ability and the influence of nodes by four steps, which can recognize the weak nodes of the system effectively.

(1).Application energy margin, sensitivity and components Outage probability model build the anti-interference ability indexes.

(2) According to the electrical between and load loss rate build node influence indexes.

(3).And on this basis, the indexes can be generated scientifically by the super efficiency data envelopment analysis, thus, a new comprehensive vulnerability assessment method of nodes was introduced. Relatively, this method can select the weight of the index objectively, so that it can recognize the weak nodes of the system comprehensively and effectively.

(4).The results of the IEEE30-bus system simulation can verify the rationality and feasibility of the method.

2. The anti-interference ability of nodes

2.1 The anti-interference ability of nodes under load fluctuation

Static energy function is an effective tool to assessment nodes vulnerability under load fluctuation[3]. The energy function of nodes is expressed as follow:

$$E_i = \int_{(\delta_i^s, V_i^s)}^{(\delta_i, V_i)} [f_i(\delta_i, V_i), g_i(\delta_i, V_i)] \cdot \begin{bmatrix} d\delta_i \\ dV_i \end{bmatrix} \tag{1}$$

where V_i is the voltage amplitude of node i ; δ_i is the phase angle of node i ; f_i is the active power of node i ; g_i is the reactive power of node i .

This paper take into account energy margin and sensitivity to assessment the anti-interference ability of nodes under load fluctuation. The energy margin of nodes is defined as follow:

$$\lambda(i) = \frac{|E_i - E_{icr}|}{|E_{i0} - E_{icr}|} \tag{2}$$

Where E_{i0} is the initial static energy of nodes; E_i is the static energy of nodes in the current running state; E_{icr} is critical energy corresponding to the critical point of voltage stability. The sensitivity of nodes is expressed as follow:

$$\beta(i) = \frac{\partial E_i}{\partial V_i} \tag{3}$$

with the aggravation of the load, voltage amplitude will drop continuously, the energy changes of each node show different sensitivity.

Considering the energy margin and sensitivity under load fluctuation, the anti-interference ability index of nodes is defined as follow:

$$C_n(i) = \begin{cases} \beta(i) \times \lambda(i), \beta(i) \geq 0 \\ \beta(i) \times \lambda(i)^{-1}, \beta(i) < 0 \end{cases} \tag{4}$$

2.2 The anti-interference ability of nodes under fault condition

In this paper, the voltage offset of the node after the line is out of operation to measure the anti-interference ability of the node in the failure mode. Because of the difference of the power flow, the probability of failure of each line is different[4], this will have a certain effect on the node voltage offset, therefore, the fault probability of the line is introduced to the anti-interference ability index of nodes under fault condition. The curve of the line fault probability changes with the line power flow is shown in Figure1.

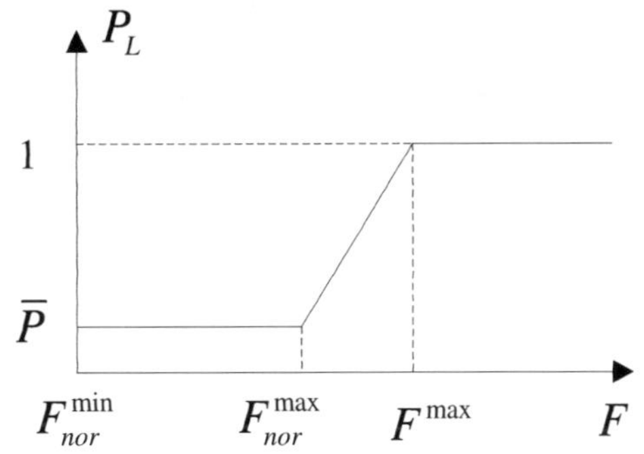

Figure1. Failure probability model for transmission line

In Figure 1, $F_{nor}^{\max}$ and $F_{nor}^{\min}$ which respectively indicates the maximum and minimum values of the allowable flow during the normal operation of the line; $F_{\max}$ is the limit value of line transmission capacity. when the power flow of line is in the rated transmission range, the protection movement is almost no effect by the tidal current. Take the statistical value $\overline{P}$ The failure probability caused by internal factors is defined as follow:

$$P_L = \begin{cases} \overline{P}, & F_{nor}^{\min} < F \le F_{nor}^{\max} \\ \dfrac{F - F_{nor}^{\max}}{F^{\max} - F_{nor}^{\max}} \times \left(1 - \overline{P}\right) + \overline{P}, & F_{nor}^{\max} < F \le F^{\max} \\ 1, & F > F^{\max} \end{cases} \tag{5}$$

The anti-interference ability of nodes under fault condition is defined as follow:

$$\Delta V_a = \sum_{b=1}^{N_l} \frac{\omega_b (V^a - V_N^a)}{V_N^a} \tag{6}$$

Where ω_b is probabilistic weight of line b , $\omega_b = p_b \Big/ \sum_{i=1}^{N_l} p_i$; p_b is failure probability of line b ; V^a is actual voltage of node a ; V_N^a is rated voltage of node a .

3. Influence of nodes

In order to study the influence of nodes in power system, it should be considered from two aspects: the importance of nodes in the topology and the impact on the system after the nodes out of running.

3.1 The electric betweenness model of nodes

The electric betweenness of nodes[5] is expressed as follow:

$$B(i) = \sum_{m \in G, n \in L} \sqrt{W_m W_n} \left| B_{mn}(i) \right| \tag{7}$$

$$B_{mn}(i) = \begin{cases} \dfrac{1}{2} \sum_j \left| I_{mn}(i,j) \right|, i \ne m,n \\ 1 \qquad ,i = m,n \end{cases} \tag{8}$$

where G and L which respectively indicates the set of generators and load nodes; W_m is the weight factor of generator node m , take actual output or rated capacity of the generator ; W_n is the weight factor of load node n , take actual or peak load; $I_{mn}(i, j)$ is the current generated on branch $i-j$ after the "generator-load" nodes is injected current element.

3.2 Load loss rate

Loss of load ratio is the ratio of the load loss after failure and the initial total load of power grid. This paper the influence of component is expressed as the effect of component failure on the terminal loads, we take the load loss rate to measure the degree of influence of the system after node failures. The load loss rate is expressed as follow:

$$S_i = \frac{P_i}{\sum_{j \in L} P_j} \times 100\% \tag{9}$$

where P_i is the load loss after node i failure; P_j, the load capacity of load node j in normal operation; and L is the load nodes set.

4. The model of SEDEA

DMUi（i=1,2,…,n）is expressed the decision making unit of i [6], each decision making unit has m input indexes which belong to cost index, the smaller the better; each decision making unit also has s output indexes, it is belong to profit index, the bigger the better. Efficiency index is expressed as follow:

$$E_i = \frac{u^T y_i}{v^T x_i} = \frac{\sum_{r=1}^{s} u_r y_{ri}}{\sum_{k=1}^{m} v_k x_{ki}} \tag{10}$$

where x_{ki} is the number k input index of DMUi , $x_{ki} > 0$, $x_i = \left(x_{1i}, x_{2i}, \ldots, x_{mi}\right)^{\mathrm{T}}$ is input vector of DMUi ; v_k is the weight of number k input index, $v = \left(v_1, v_2, \ldots, v_m\right)^{\mathrm{T}}$ is the weight vector of input index; γ_{ri} is the number r output index of DMUi , $\gamma_{ri} > 0$, $y_i = \left(y_{1i}, y_{2i}, \ldots, y_{si}\right)^{\mathrm{T}}$ is output vector of DMUi ; μ_r is the weight of number r output index, $u = \left(u_1, u_2, \ldots, u_s\right)^{\mathrm{T}}$ is the weight vector of output index.

When evaluating the efficiency of decision making units DMUi , we take the weight vector of the input and output indexes as variables, with the efficiency index E_i as the goal, and all DMU efficiency index E_a $(a = 1,2,...,n, a \neq i)$ as constraint except DMUi ,then we get fractional programming model as follows:

$$\max E_i = \frac{u^T y_i}{v^T x_i}$$

$$s.t. \quad E_a = \frac{u^T y_a}{v^T x_a} \leq 1 \tag{11}$$

$$v \geq 0, u \geq 0$$

Based on the fractional programming model, solving each decision making unit respectively, we can get the corresponding maximum value of E, and then evaluate the efficiency of each decision making unit. The model of SEDEA is more objective to select the weight, thus making the result more practical.

5. The comprehensive vulnerability assessment method of nodes based on SEDEA

This paper proposes a method to evaluate the vulnerability of nodes, which is based on the SEDEA model, and carries out a comprehensive consideration from anti-interference ability and influence these two aspects. In this method, we can regard each node as a decision making unit (DMU), and make the anti- interference ability index into cost type as the input, where the index is smaller, the anti- interference ability of the node is weaker. Meanwhile, making the influence index into efficiency as the output, where the index is bigger, the influence of nodes is greater. The efficiency index, E, is defined as the comprehensive vulnerability index of nodes, where the higher the index, the weaker the anti- interference ability of nodes and the greater the influence in the system. The proposed method in this paper has a clear physical meaning and a certain practical significance.

6. Test scenario

This paper takes the IEEE 30-bus system as an example. Load fluctuation mode is selected as the load increased by 30% to calculate the anti-interference ability of nodes under load fluctuation. Branch outage probability statistics is set to 1.81×10-4, limit value of branch power flow normal value is set to 0.95-1.05 times of the rated capacity to calculate the anti-interference ability of nodes under fault conditions.

Using the traditional DEA and super efficiency data envelopment analysis (SEDEA) model to calculate the comprehensive vulnerability index of nodes. The result is shown in Figure 2.

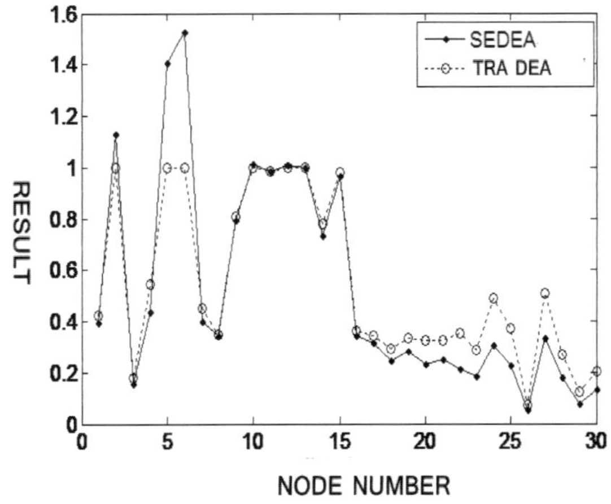

Figure2. node comprehensive vulnerability index with different models

As shown in Figure2, the node comprehensive vulnerability index curve trend is very close with traditional DEA and SEDEA model , so the rationality of the method in this paper is demonstrated. But the traditional DEA model can not identify the nodes which comprehensive vulnerability indicators are 1, and this paper can effectively distinguish the vulnerability of each node. Finally, the comprehensive vulnerability of all nodes is sorted.

Node 6 is the most vulnerability in this paper, further analysis showed that node 6 is the most important links in the system, it undertake the task of power injection from generators to end of load, this node needs to add transport capacity immediately to maintain the power flow balance when the load changes, So it is more susceptible to be disturbed than other nodes, and the anti-interference ability is more weak; furthermore, node 6 is the node of main transformer branch, it will lead 5 and 8 generator node power cannot transport to the central section of system if node 6 fault or outage, it is very easy to cause the whole power flow to move significantly, eventually lead to large-scale blackout, it has great influence to the power system. Node 5 is similar to node 6, node 5 is one of few generator nodes which affected by load fluctuations greatly, and the ability to maintain voltage stability is weak, in addition, it will cause large-scale blackout if node 5 quit running, so node 5 shows a strong vulnerability.

Through the simulation results and analysis above, the vulnerability nodes are mostly generator nodes or important links in the system. Results show that the presented method is feasible and reasonable.

7. Conclusion

In this paper, a comprehensive vulnerability index set of nodes is proposed, which is based on the ability of anti-interference and influence. Then, the SEDEA model is used to integrate multiple indexes, and a new method for evaluating the vulnerability of nodes is obtained, which is more objective to select the index weight, and it can effectively identify the weak and influential nodes in the system. The comprehensive vulnerability index resulted in this paper is expected to be applied in engineering practice, and it can provide valuable theoretical reference for early-warning and stable operation of power system.

References

[1] Bai Jia lin, Liu Tianqi, Cao Guoyun, et al. A survey on vulnerability assessment method for power system[J]. Power System Technology, 2008, 32(S2): 26-30(in Chinese).

[2] Zhou Yanheng, Wu Junyong, Zhang Guangtao, et al. Assessment on power system vulnerability considering cascading failure[J]. Power System Technology, 2013, 37(2): 444-449(in Chinese).

[3] Huang Zhaomeng, Li Huaqiang, Du Tao, et al. Vulnerable branch assessment based on branch energy function[J]. Power System Protection and Control, 2012, 40(15): 7-11(in Chinese).

[4] Sun Yuanzhang, Cheng Lin, Liu Haitao. Power system operating reliability evaluation based on real-time operating state[J]. Power System Technology, 2005, 29(15): 6-12(in Chinese).

[5] Xu Lin, Wang Xiuli, Wang Xifan. Electric betweenness and its application in vulnerable line identification in power system[J]. Proceedings of the CSEE, 2010, 30(1): 33-39(in Chinese).

[6] Wu Yuhua, Zeng Xiangyun, Song Jiwang. A DEA model with AHP restraint cone[J]. Journal of Systems Engineering, 1999, 14(4): 330-333(in Chinese).

Analysis and Research on the effect of the Operation of Small Hydropower in the Regional Power Grid

Fu Ang, Dong Guangde, Zhu Xiaojun,Wang Ruimiao and Zhu Shengyi1

State Grid Chongqing Electric Power Research Institute,Yubei District, 401120, Chongqing, China

Email: 1654663144@qq.com

Abstract: The analysis of reactive power balance and voltage of power network not only affects the system voltage quality, but also affects the economic operation of power grid. In the calculation of reactive power balance and voltage analysis in the past, the problem of low power and low system voltage has been the concern of people. When small hydropower stations in the wet period of low load, the analysis of reactive power surplus and high voltage for the system, if small hydropower unit the capability of running in phase is considered, it can effectively solve the system low operation voltage of the key point on the high side.

1. Introduction

For the reactive power in the grid is too high, the grid voltage is too low, the usual solution to the operating state of the generator to adjust the phase into the running state, with the working principle of the phase to solve the power grid in the absence of Power power is too much, the grid voltage to reduce the problem. The specific approach is: First of all, to fully understand the power system voltage levels on the reactive power of what are the main aspects. Secondly, the characteristics of the phase of the operation and the working principle of mastery, and finally, the two fully integrated, play their respective roles, so that the hydropower station to get a better run.

With the construction and development of modern power grid with high voltage, large capacity and long distance transportation, the problem of high voltage at the peak point of the power system is becoming more and more serious, which seriously affects the safe operation of the transmission and distribution equipment. Commonly used in the grid regulation mode, the generator into the phase with a smooth operation, no additional investment, easy to real.Commonly used voltage regulating method in power grid， the generator into the phase with a smooth operation, without additional investment, easy to implement and other unique advantages and gradually applied.

The phase-running operation and the late-phase operation together constitute the operation of the entire generator. The phase-running operation and the late-phase operation of the generator have their own characteristics, both between each other, but also promote each other and mutual conversion, mutual cooperation, the generous function of the generator to play out. Among them, the late operation means that the generator does not issue active power, only to the grid output inductive reactive power running state, which play a regulatory system power, to maintain the role of system voltage level. And the phase of the operation and the late phase of the opposite operation, the phase is running in the power system running reactive power over the process, the system voltage is high, the normal operation of the system blocked, etc., when the generator running to adjust the state Phase operation of the turntable, phase operation can be issued active power, and active power to the excess reactive power can be absorbed, so you can be too high in the power system to reduce the voltage, so that the power system to resume normal operation. But the phase of the operation also will be subject to a number of factors, limiting the function of play into play.

2. Theoretical Analysis on the Operation of Small Hydropower Development

1.1 Definition of generator phase advance operation

With the increasing number of hydropower stations in China, the scale of hydropower stations has been expanding, and some problems in hydropower systems have also changed. It is very difficult for traditional single generators to run a relatively satisfactory answer to some new situations and problems. And some of the advantages of the phase of the operation is being constantly highlighted, in the solution of these new situations and new problems have certain advantages, so the generator phase is very necessary to run. Because the current number of reactive power in China's power grid is increasing, but the voltage anomaly in the power grid is more serious, more serious situation is the voltage on the wire exceeds the upper limit of the specified value, which is more serious reactive power increase. The increase of reactive power is a common phenomenon in the field of power grid in China .Because with the continuous improvement of China's motor manufacturing technology and the rapid development of the power industry, the current power industry, the main direction of development of the main direction of the transition to the direction of large units, combined with high-voltage transmission lines, large power grid system, these two constitute a common The general characteristics of the current power system. These features make the reactive power in the grid continue to accumulate, increase, which is an urgent need to run into the orderly implementation.

1.2 generator into the basic principles of the operation

The generator operating conditions are running at a later time, where the stator current lags the terminal voltage and the generator is in the over excitation state. The phase-to-phase operation is relative to the generator's late-running operation, where the stator current is ahead of the terminal voltage and the generator is in underexcitation operation state . When the generator is running at the same time, the active power and inductive reactive power are supplied to the system. The active power table and the reactive power meter are all positive. The active power and capacitive reactive power are supplied to the system. The power meter indicates a positive value and the reactive power meter indicates a negative value, where the generator absorbs inductive reactive power from the system. The electrical parameters of the generator are still symmetrical when the generator enters the

IOP Conf. Series: Materials Science and Engineering **339** (2018) 012023 doi:10.1088/1757-899X/339/1/012023

phase, and the generator still maintains the synchronous speed, which is an operating condition when the power factor changes in the normal operation mode of the generator, but only widens the normal operation of the generator range. Therefore, in the allowable phase of the operating range, as long as the grid needs, the generator can run for a long time into the phase. As shown in Figure 1, respectively, for the phasor diagram of lagging and leading phase.

If the generator G is directly connected to the infinite system S, the terminal voltage U_G is constant, the generator potential is E_q, the stator current is I, the power factor angle is φ, the excitation current is If, the synchronous reactance is X_d, Angle is δ.

When I_f increases, E_q then increases, I produce demagnetization armature reaction, then I lag behind U_G, φ at the same time change, is lagging, G issued active power P and reactive power Q, then G For the late phase of operation, as shown in Figure 1 (1); on the contrary, when reducing I_f, so that E_q decreases, φ ahead of the U_G, I generate the armature reaction, At this point, G is converted to Q and G is running in leading phase, as shown in fig.1 (2).

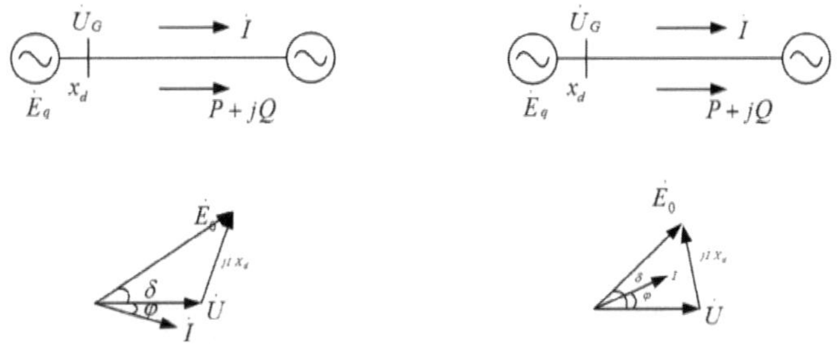

Figure 1. Phasor diagram of lagging and leading phase

3. Small hydropower access grid operation characteristics

Small hydropower refers to a single capacity of 5 MW below the transmission station to use power generation. It changed the load of grid and was susceptible to seasonal changes during seasonal changes. After a long period of research, the heavy load of the wet season and low load of the dry season can run normally. The voltage of the grid system does not change greatly. The whole grid system is very stable. The dam head of dam type hydropower station depends on dam height. At present, the maximum head of dam type hydropower station is not more than 300 meters. The reference flow of dam type hydropower station is large, the scale of power station is large, and the utilization of water power is sufficient . (Due to dams, the formation of reservoir upstream, can be used to adjust the flow) At present the world's installed capacity of more than 2000 megawatts of large hydropower stations are mostly dam-type hydropower station. In addition, the comprehensive utilization of reservoirs of dam-type hydropower stations is high, which can meet the requirements of flood control, power generation and water supply. The dam type hydropower station has large investment and long construction period. Due to the large scale of the project, the flooded area caused by the reservoir is large and the migration population is too large. Applicable to the river slope is slow, the flow is bigger, and has the condition of dam construction.

When the grid is in the low trough load, it will cause some nodes to have the problem of excessive voltage. Through personnel research and analysis, through a variety of means to conduct a comprehensive study of the system and make in-depth study on the voltage and reactive power. Through the theory of reactive power flow and voltage and power grid, combined with the network to select the appropriate reactive power control mode, the characteristics and performance of the reactive power compensation equipment are used to determine the influence caused by the operation of the small hydropower access grid. Dam-type power station is the water for the storage medium, from the regulatory role. Mainly to solve the power system peak peaking problem. When system load is low, the excess power of the system will be used to drive the pumping station to the upper reservoir (motor + water pump) and store it in the form of water potential energy. When the system load is high, the water on the reservoir will be pushed down to propel hydroelectric generator (turbine + generator) to generate electricity to supplement the lack of electricity in the system. With the reform of the power industry, the implementation of load peak high price, low load low price, the economic benefit of pumped storage power station will be remarkable.

(1) scattered spots, local development, the nearest power supply;

Due to the control of river flow and fall, small hydropower stations have small installed capacity, from thousands of kilowatts to tens of thousands of kilowatts, small storage capacity, poor regulation of power plants, and most of them are runoff type power generation. The power station is scattered and distributed, developed locally and supplied by the nearest power supply. As a result, small hydropower stations are small in size and wide in distribution.

(2) small hydropower near the network, into a piece of power supply

As the small hydropower station installed capacity is small, stand-alone capacity is also small, the corresponding lower power supply voltage level, generally the following voltage near power supply, power supply radius is short. In order to improve the reliability of power supply and the stability of the generator set, the small hydropower stations are connected to each other, and the electric power is complement each other and the mutual adjustment is carried out to form a small distribution network.

(3) small hydropower operation and maintenance is simple

Small power station, such as the distribution of power, as the installed capacity is small, the output line voltage level is low, there are 110kV, 35kV, 10kV, etc., the output line of the wire cross-sectional area is small, pole and tower relative to 220kV and 500kV tower The. So the line operation and maintenance is relatively easy to ultra-high pressure, do not need a lot of manpower and large machinery and equipment, encountered natural disasters easy to repair. In 2008, China's snow disaster in the south, the large power supply, ultra-high voltage line 500kV line down tower, broken line, power station to send electricity, restore 500V line to mobilize a lot of manpower and resources, equipment support was able to recover, Time is relatively long, under the same circumstances, the distribution of small hydropower than the reliability of large power grid is also high.

(4) small hydropower station as a black power supply for large power grids

In January 2008, the southern provinces of China suffered 50 years of extreme frozen disasters, large power grid 500kV and 220kV lines have broken down the tower, the external power to send into a county or even a city to become an island, power outages, The whole network crash all black, large power supply due to disconnection can not send, can not start the grid, only to small hydropower station as a black start power, start the grid, and gradually restore the grid to run.

4. Reactive power balance and voltage stability of small hydropower networks

4.1 Reactive power

When the voltage and current are sinusoidal waveforms and the load is linear, the instantaneous value expression of voltage and current can be written as

$$u = \sqrt{2}U \sin \omega_1 t \quad (1)$$

$$\begin{aligned} i &= \sqrt{2}I \sin(\omega_1 t - \varphi) \\ &= \sqrt{2}I \cos\varphi \sin\omega_1 t - \sqrt{2}I \sin\varphi \cos\omega_1 t \end{aligned} \quad (2)$$

If the current i is decomposed into an active component ip in phase with the voltage and a reactive component iq which is 90 ° out of phase with the voltage. , Can be written

$$\begin{aligned} i_p &= \sqrt{2}I \cos\varphi \sin\omega_1 t \\ i_q &= \sqrt{2}I \sin\varphi \cos\omega_1 t \end{aligned} \quad (3)$$

The instantaneous power P is:

$$\begin{aligned} p &= ui = 2UI \sin\omega_1 t \sin(\omega_1 t - \varphi) \\ &= UI\cos\varphi(1-\cos 2\omega_1 t) - UI \sin\varphi \sin 2\omega_1 t \end{aligned} \quad (4)$$

The instantaneous power can be divided into two parts that are non-sinusoidal, $UI\cos\varphi(1-\cos 2\omega_1 t)$, which is the irreversible component of the input or output instantaneous power; $UI\sin\varphi\sin 2\omega_1 t$,the sine quantity is the reversible component of the instantaneous power, which alternates alternately in one cycle, To exchange energy.

Reactive power Q

$$Q = UI \sin\varphi \quad (5)$$

It is the magnitude of the reversible component in the equation, when $\sin\varphi > 0$ is considered that the port absorbs reactive power when $\sin\varphi < 0$ is assumed that the port is generating reactive power. , Respectively, $I\sin\varphi$、$U\sin\varphi$ for the current, voltage reactive components, reactive power reflects the internal and external exchange of energy between the situation. For a single component is also applicable, that is $Q_R = 0, Q_L = U_L I_L > 0, Q_C = U_C < 0$.

The reactive power loss in the transformer is divided into two parts, namely, the loss of the excitation branch and the leakage in the winding leakage. Wherein the percentage of the loss of the excitation branch is substantially equal to the percentage of no-load current, about 1% to 2%, and the percentage loss in the leakage of the winding is substantially equal to the short-circuit voltage when the transformer is fully loaded Percentage, about 10%.

$$\begin{aligned} Q_{LT} &= \Delta Q_0 + \Delta Q_T \\ &= U^2 B_T + \frac{P^2+Q^2}{U^2} X_T \\ &\approx \frac{I_0\%}{100} S_N + \frac{U_k\% S^2}{100 S_N} (\frac{U_N}{U})^2 \end{aligned} \quad (6)$$

$$\Delta Q_T = \frac{I_0\%}{100} S_N + \frac{P^2+Q^2}{U^2} X_T \quad (7)$$

The excitation power is roughly proportional to the square of the voltage. When the apparent power

through the transformer does not change, the reactive power loss in the leakage resistance is proportional to the square of the voltage. For a transformer or a transformer network, the transformer reactive power loss is not large, full load is about a few percent of its rated capacity. But the multi-voltage level network, the transformer reactive power loss is considerable.
Transmission line with the equivalent circuit Π that the line series reactance in the reactive power loss ΔQ_L is proportional to the square of the current through, that is,

$$\Delta Q_L = \frac{P_1^2 + Q_1^2}{U_1^2} X = \frac{P_2^2 + Q_2^2}{U_2^2} X \tag{8}$$

B / 2 is the equivalent charge in the shape circuit. The total reactive loss of the line is:

$$\Delta Q_L + \Delta Q_B = \frac{P_1^2 + Q_1^2}{U_1^2} X - \frac{U_1^2 + U_2^2}{2} B \tag{9}$$

4.2 reactive power balance
Reactive power balance means that at every moment the grid is running, the reactive power of all reactive power sources is equal to the reactive power consumed by all loads and the sum of reactive power losses in each link in the system. According to the distribution of the trend, the statistics of the data, you can determine whether the system reactive power balance. The voltage level of the power system is the specific performance of the system reactive power supply and demand balance, but the system frequency and the system active power supply and demand balance situation is very different. The transmission of reactive power not only produces a large amount of active loss but also has a large voltage drop along the transmission path, so the voltage characteristics of each hub in the system are more regional in nature. As the reactive power supply and demand distribution varies, the same time at different points of the voltage level varies. The voltage regulation at each point in the system is mainly dependent on the local balance of reactive power to achieve the mutual support and adjustment of the reactive power supply and demand between different points, often limited, especially in the distance between the electrical distance .

4.3 Voltage control of power systems
(1) to change the grid reactive power distribution. If you can change the lines, transformers and other power components on the voltage loss, it will change the voltage of the power grid nodes. In practice, the reactive power of the generator alone is far from meeting the needs of the reactive power of the grid, and it is necessary to configure various reactive power compensation devices. Reactive power compensation principle is to do reactive power hierarchical partition balance, and to leave enough accident reactive power reserve. Application of reactive power compensation device to adjust the voltage. In the appropriate location of the power grid access to the parallel reactive power compensation device, can reduce the line and transformer reactive power transmission, which can reduce the line and transformer voltage loss and improve the power grid voltage level, while reducing the power grid Power loss, improve economic efficiency. When the system load changes, by adjusting the reactive power compensation device output reactive power, you can control the power grid voltage. Commonly used reactive power compensation equipment is a parallel capacitor and shunt reactor, in the peak load into the parallel capacitor can improve the voltage level of the whole network, the load is low, you can remove some of the capacitors, and even all the cut and then into the parallel reactor,

IOP Conf. Series: Materials Science and Engineering **339** (2018) 012023 doi:10.1088/1757-899X/339/1/012023

To prevent the voltage level is too high.
(2) change the transformer variable ratio regulator. By switching the transformer tap to change the ratio, you can adjust the transformer low voltage side or high voltage side of the voltage, only when the system reactive power supply capacity is sufficient, with the transformer to change the ratio of pressure to work. Otherwise, not only by the voltage regulator point change is not, but also cause a further decline in its voltage, which will lead to the entire system voltage collapse. So when the system reactive power is insufficient, the first should be installed reactive power compensation equipment, the system reactive power capacity has a certain margin.
(3) change the generator terminal voltage. Application of generator voltage regulator is no need to increase the investment regulator. The generator terminal voltage is controlled by the excitation regulator, changing the voltage setting of the regulator to change the terminal voltage. The voltage of the generator is closely related to the reactive power output of the generator. When increasing the terminal voltage of the generator, it also increases the reactive power output of the generator. Instead, the terminal voltage of the generator is reduced, and the reactive power output of the generator is reduced. Therefore, the generator terminal voltage adjustment by the generator reactive power limit limit, when the generator output reactive power reaches its upper or lower limit, the generator can not continue to pressure regulator. The reactive power output limit of the generator is related to the active output of the generator. When the active output is small, the range of the reactive power adjustment will be larger and the capacity of the regulator will be stronger.

4.4 Voltage stability of power systems
Voltage stabilization is initially considered a static problem, so the mechanism of voltage instability is also explained from a static point of view. As the former Soviet Union, Markovic proposed the first voltage stability criterion, the dQ / dU criterion, on a single load-infinite system as a classic and intuitive physical interpretation of voltage stability problems. When the reactive power is excessive, the voltage rise also has a great impact on the stability of the grid. Through the trend calculation can be observed and analyzed.

Figure 2 shows a schematic diagram of the power supply wiring for a single load infinite system. The system is affected by the terminal voltage power characteristics of the system by the power supply electromotive force E by the transmission impedance jX to a load by the end of the system operating voltage characteristics.

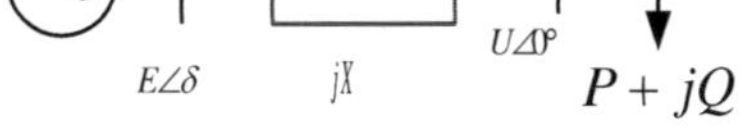

Figure 2. single load and infinite system diagram

$$\frac{PX}{E^2} = \frac{U}{E}\sin\delta \qquad (10)$$

$$\frac{QX}{E^2} = \frac{U}{E}\cos\delta - (\frac{U}{E})^2 \qquad (11)$$

$$(\frac{PX}{E^2})^2 + [\frac{QX}{E^2} + (\frac{U}{E})^2] = (\frac{U}{E})^2 \qquad (12)$$

$$(\frac{U}{E})^2=\frac{1}{2}[1-2\frac{QX}{E^2}\pm\sqrt{1-4\frac{QX}{E^2}-4(\frac{PX}{E^2})^2}\,] \tag{13}$$

In the above expression, each quantity is the value of a voltage (kV) and a certain capacity (MVA). In particular, when the load is absorbed by the system reactive power, Q is positive; and when the load to the system to send reactive power, Q is negative. All P and Q values correspond to the value of the corresponding voltage U value, rather than the value of the rated voltage.

The voltage stability problem is related to the power generation system, the transmission system and the load system. Transmission network strength and power transmission level, the generator reactive power voltage control restrictions, the load voltage characteristics, reactive power compensation equipment characteristics and the role of voltage control equipment together determine the level of system voltage stability.

The power supply line produces both reactive and reactive power, and its net reactive value must be equal to the amount of power that the two ports of the line absorb or emit from the system. The shunt capacitor or the charging capacitor sends a reactive power proportional to the square of the voltage. Since the node voltage must remain within the rated voltage (1 ± 5%), the reactive power generated by the line is relatively constant. Transmission line series inductance consumption and the current square is proportional to the reactive power. Because the current from heavy to light load changes greatly, power line reactive power consumption also will have a big change. In this way, the net reactive power of the transmission line changes in a load cycle.

5 Analysis of Feed - in Operation and Stability of Hydro - generator

5.1 Water turbine generator working principle

The hydro-generator is a salient-pole synchronous generator. The stator is the stationary part of the hydro-generator, mainly composed of the base, the core and the armature winding stator winding. The rotor is composed of the magnetic pole, the rotor winding and the rotating shaft. They constitute two main parts of the hydro-generator. As the two parts are equipped with windings, the current generated by the cohesive magnetic field. The structure of the hydro-generator must be such that the energy dissipation of the combined magnetic field is a function of the angular displacement of the rotor in order to produce a continuous motor energy conversion during rotation. Specifically, the stator, the rotor winding is a field winding that is the rotor winding, through the shaft on the collector ring and brush device to introduce DC current, the introduction of the current in the rotor magnetic pole to produce the main magnetic flux and the formation of the main pole magnetic field, And the other is the armature winding stator winding, which is mounted on the other side of the air gap to have relative movement between the main magnetic field and the main magnetic field. When the turbine drags the generator shaft rotor rotation, the rotor main pole magnetic field rotation, cutting armature winding, so that the induction in the winding potential, when the armature winding and the outside three symmetrical load connected, the armature winding will have AC current, the whole process of work is the basic working principle of hydro-generators.

Excitation system by the excitation of the main circuit and excitation circuit composed of two parts. The excitation main circuit includes the excitation winding of the synchronous generator, the excitation power supply, the rectifying device and the demagnetizing circuit. The excitation circuit includes automatic and manual excitation adjustment device.

IOP Conf. Series: Materials Science and Engineering **339** (2018) 012023 doi:10.1088/1757-899X/339/1/012023

5.2 Static stability analysis of hydroelectric generator in phase

The principle of the hydrodynamic phase of the hydro-generator is described in detail in Section 1.2. The hydro-generator is a salient-pole synchronous generator. The static stability limit is the same as the power angle between the static stability limit and the variable angle. At the moment, the relationship between the active power and the reactive power of the salient-pole generator in the system when the static stability limit is reached:

$$P^2 = \frac{(Q+\frac{U^2}{X_q})^3}{Q+\frac{U^2}{X_d}} = \frac{(Q+\frac{U^2}{X_q})^3}{U^2(\frac{1}{X_q}-\frac{1}{X_d})-(Q+\frac{U^2}{X_q})} = \frac{(Q+\frac{U^2}{X_q})^3}{U_s^2(\frac{1}{X_q}-\frac{1}{X_d})-(Q+\frac{U_s^2}{X_q})} \tag{14}$$

$$P^2 = \frac{(Q+\frac{U^2}{X_q})^3}{U_s^2(\frac{1}{X_{qs}}-\frac{1}{X_{ds}})-(Q+\frac{U_s^2}{X_{qs}})} \tag{15}$$

The above equation shows the relationship between the active power and the reactive power of the system when the static stability limit is obtained after the system impedance. After taking into account the effect, the same P value is allowed to enter the phase of the reactive power reduction. If the generator to maintain a certain stability reserve, the phase of the operation of the safety limit should be further reduced. At this point, static stability will be the main limit for the output of the generator to allow the output of the generator.

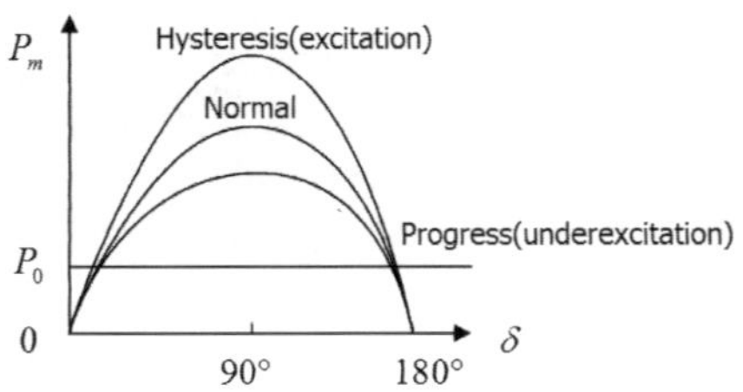

Figure 3. Power-angle diagram in different excitation

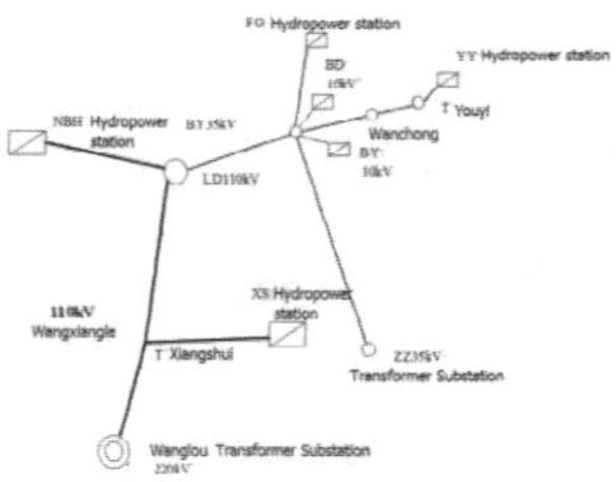

Figure 4. LD power grid topology

6. Case study

For a region of Hainan LD power grid, the regional power grid is rich in a large number of small hydropower stations, in the event of natural disasters, can effectively provide electricity to support it, to ensure that the county power supply, but when the low load period, often make LD power grid node voltage Too high, this paper through the trend of the calculation and analysis, in the technical conditions of small hydropower units can enter phase operation. In the actual investigation of the situation of hydropower stations also proved that small hydropower units can enter the phase to run.

The small hydropower output and load values in LD grid are shown in Tables B1 and B2.

The voltage of each node before and after the operation of small hydropower is shown in Fig:

IOP Conf. Series: Materials Science and Engineering **339** (2018) 012023 doi:10.1088/1757-899X/339/1/012023

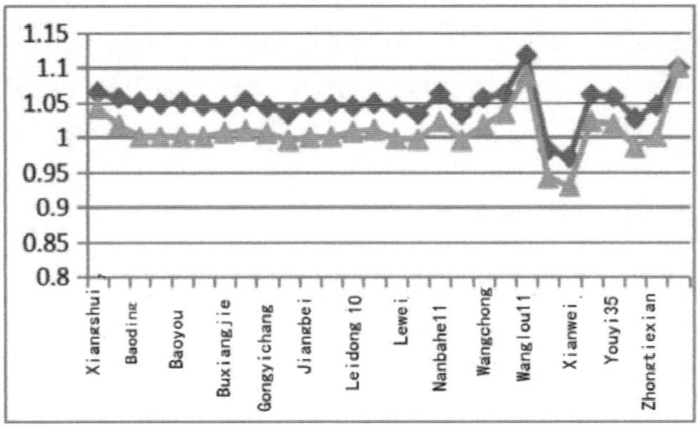

Figure 5. Comparison of voltage before and after the operation of small hydropower

As can be seen from the figure, in the hydropower overdue period to avoid the network node voltage is too high, you can consider the small hydropower phase into the run, and can be a good drop in a node voltage.

7. Conclusion

In summary, through the above experimental study, the grid system voltage adjustment method, for the small hydropower network in the wet period of low load, the grid voltage changes and the reasons for the increase in the grid has a comprehensive understanding. The non - power balance of the power grid has a certain control effect on the voltage stability of the power grid system, and can effectively stabilize the voltage of the grid in the low water load during the wet period. As long as the control voltage within the specified range, it is necessary to control the power of the grid, and then the grid voltage control in a reasonable range.

In the representative small hydropower station carried out on-site tuning magnetic operation, showing the phase of the operation of the power grid has a certain effect, and that the small hydropower units have the ability to enter the phase. In the water during the full period to avoid the network node voltage is too high, you can test

Consider small hydropower into phase operation, and can be a good way to reduce the node voltage.

References

[1] YAN Wei,CHEN Jun, SHENQuan-rong.Discussion onlarge non-salient pole generatorphase-advancement operation[J].Automation of Electric Power Systems,2007, 31(2): 94-97.

[2] SHI Jia-yan, SHI Yuan-su,ZHAO Xiao-min, et al. The new method of determining the capability of leading phase of generator and realizing the on-line monitoring[J]. Proceedings of the CSEE, 2006, 26(11): 139-143.

[3] ZHANG Jian-zhong, WAN Li,LIU Hong-zhi, et al.Study on leading phase operation and voltage regulation testof power network for large steam turbine- generator[J]. Electric Power, 2006, 39(12): 11-15.

[4] WANG Zhenhao, LIU Jinlong, LI Guoqing, et al. Power and voltage regulation of wind farm based on EDLC energy storage[J].Electric Power Automation Equipment, 2011, 31(3):

113-116.
[5] TELEKE S, BARAN M E, HUANG A Q,et al. Control strategies for battery energy storage for wind farm dispatching ［J］. IEEE Trans on Energy Conversion,2009,24（3）：725-732.
[6] TELEKE S,BARA N M E,BHATTACHARYA S,et al. Optimal control of battery energy storage for wind farm dispatching [J].IEEE Trans on Power Systems,2010,25（3）：787-794
[7] TELEKE S,BARAN M E,BHATTACHARYA S,et al. Rule-based control of battery energy storage for dispatching intermittent renewable sources ［J］. IEEE Trans on Sustainable Energy,2010,25（3）：117-124.
[8] CHEN Xun. Analysis on several problems in carrying outgenerator under-excitation operation. Guangdong ElectricPower, 2000, 13(1): 1-4.
[9] LI Jun. Practice on 300 MW non salient-pole generator phase-advancement operation. Electric Equipment, 2003, 4(1): 1-4.
[10] SHI Shiwen.The theory of large generator and transformerprotection. Beijing: Hydraulic and Electrical Power Press,1987.
[11] WANG Weijian.The theory and application of electric mainequipment protection. Beijing: China Electrical Power Press,1996.
[12] SUN Ziping, GAO Chunru.Summary of microcomputer-basedprotection operation experience of large-sized generator-transformer unit. Automation of Electric Power Systems, 1999,23(11): 55-58.

IOP Conf. Series: Materials Science and Engineering **339** (2018) 012024 doi:10.1088/1757-899X/339/1/012024

The Fatigue Life Prediction of Train Wheel Rims Containing Spherical Inclusions

Yajie Li*[1,2], Huanguo Chen[1,2], Li Cai[1,2], Pei Chen[1,2], Jiacheng Qian[1,2] and Jianwei Wu[1,2]

[1]Zhejiang Province's Key Laboratory of Reliability Technology for Mechanical and Electrical Product
[2]Zhejiang Sci-Tech University, Hangzhou, 310018, China

Email: ppluoxi@163.com

Abstract. It is a common phenomenon that fatigue crack initiation occurs frequently in the inclusions of wheel rims. Research on the fatigue life of wheel rims with spherical inclusions is of great significance on the reliability of wheels. To find the danger point and working condition of a wheel, the stress state of the wheel rim with spherical inclusions was analyzed using the finite element method. Results revealed that curve conditions are dangerous. The critical plane method, based on the cumulative fatigue damage theory, was used to predict the fatigue life of the wheel rim and whether it contained spherical inclusions or not under curve conditions. It was found that the fatigue life of the wheel rim is significantly shorter when the wheel rim contains spherical inclusions. Analysis of the results can provide a theoretical basis and technical support for train operations and maintenance.

1. Introduction

Wheels are a key component to guarantee vehicle dynamics performance and the safety of trains. Compared with trucks and medium or low speed trains, Electric Multiple Units (EMU) can easily have safety problems when running because of their higher speed and more intense vertical vibration. To ensure the safe operation of trains, it's essential to put forward higher requirements on wheel fatigue life and reliability. At present, the estimation methods for fatigue life mainly include the nominal stress method [1], the local stress strain method [2,3] and the critical plane method [4,5].

Numerous researchers have studied fatigue life prediction methods and related theories, and achieved a lot of results. In order to calculate the fatigue life of the train wheel, Li Shulin [6] proposed a new method based on the static strength analytical method and the nominal stress method to analyze the strength of train wheels. The main purpose is to calculate the fatigue life of the EMU wheels more realistically. Ekberg [7] studied rolling wheel fatigue and pointed out that rim crack can occur whether it contains inclusions or not, but the probability of rim crack occurrence was bigger when it contained inclusions. He proposed that the fatigue life of the railway wheel included crack initiation life and crack propagation life. American Brant Stratman [8] proposed a method to estimate the high cycle fatigue life of railway wheels. He applied the cumulative fatigue damage theory to the problem of wheel rail contact and established a 3D elastic-plastic finite element model which considers rail wheel contact. Fatigue wheel damage was calculated by the time load history of 360 degree wheel rotation, which was loaded into the finite element analysis.

Although there are many efficiency and high precision analytical methods for estimating contact fatigue life of wheels, there is a of lack of related guides for research on the life of wheel rims, especially in estimation of the fatigue life of wheel rims containing inclusions. Firstly, a finite element model was created to get the stress of the wheel rim around spherical inclusions under different working conditions. Then the dangerous working condition was found. Lastly, fatigue life of the wheel rim containing spherical inclusions under dangerous working conditions based on the fatigue prediction model was predicted.

2. Fatigue life prediction model

For the three methods to provide for the fatigue life of parts, the premise of the nominal stress method and the local stress strain method requires that the material is uniform with no defects and no initial cracks. The object of this paper is a wheel rim containing spherical inclusions 10 mm below the tread. Inclusions in the rim are defective. So the life of the wheel rim with inclusions can be predicted by using the critical plane method. Among the many critical plane criteria, McDiarmid's criterion [9,10], which considers two forms of fatigue crack initiation proposed by Brown and Miller is the most widely used.

McDiarmid proposed that the maximum fatigue damage is in the maximum shear stress amplitude plane. The McDiarmid model suggests that the shear stress amplitude and the maximum normal stress at the critical plane are two parameters that affect the fatigue crack growth. The McDiarmid model criterion is expressed as follows.

$$\tau_{ns} + \frac{\tau_{-1}}{2\sigma_b}\sigma_{n,\max} = \tau_f^{'}(2N_f)^b \tag{1}$$

On the wheel rim material, τ_{ns} is the shear stress amplitude on the critical plane and the unit is MPa; τ_{-1} is the torsional fatigue limit strength of the material under a symmetrical load and the unit is MPa. When the torsion fatigue limit of the material is unknown, it can be estimated by using the empirical formula $\tau_{-1} = 0.58\sigma_{-1}$, where σ_{-1} is the symmetric bending fatigue limit which can be obtained according to the material list. $\sigma_{n,\max}$ is the normal maximum stress of the critical plane and the unit is MPa; $\tau_f^{'}$ indicates the torsional fatigue strength coefficient and the unit is MPa. Normally $\tau_f^{'} = \sigma_f^{'} / \sqrt{3}$ and the tensile strength coefficient is $\sigma_f^{'} = \sigma_b + 350$. The high cycle fatigue life of the wheel rim with spherical inclusions can be predicted by McDiarmid's critical plane criterion. The procedure is expressed as follows in Fig. 1.

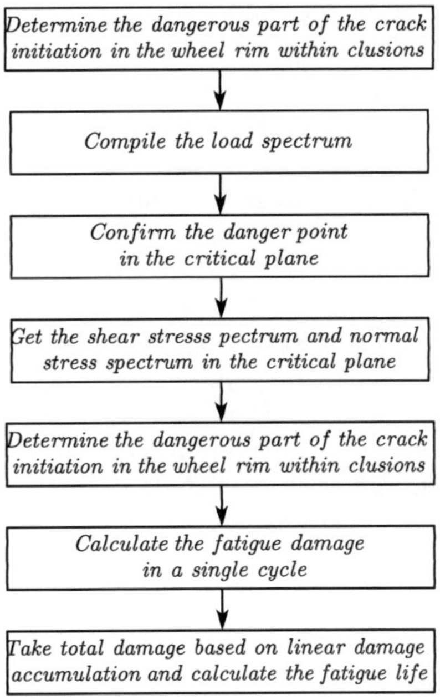

Figure 1. Steps of fatigue life prediction based on the McDiarmid model

3. Compiling the wheel load spectrum under dangerous conditions

3.1. Determine the dangerous conditions

According to the symmetry characteristics of an axle-wheel system, and the load and constraint condition of the axle-wheel, the finite element model of the 1/4 axle-wheel is established and shown in Fig. 2. Using 3D solid unit SOLID187 when meshing. In order to calculate the accuracy of the results, refine the inclusion and surrounding basal material which are shown in Fig. 3[11].

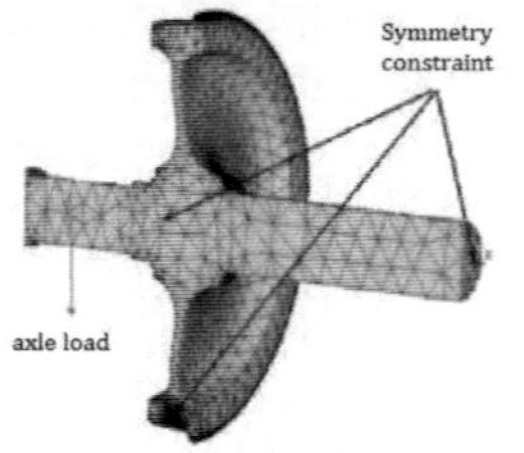

Figure 2. Finite element model of a wheel

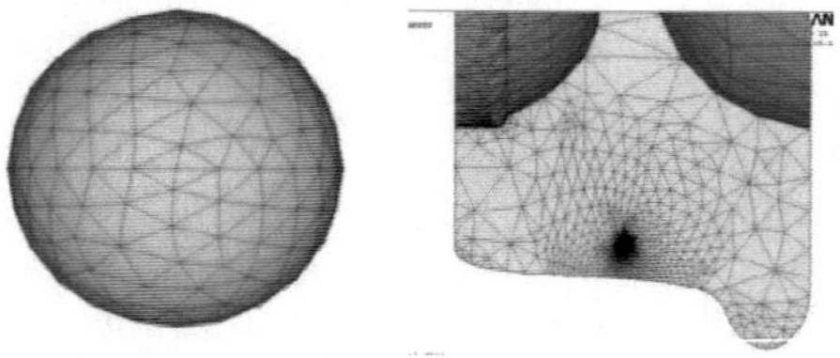

Figure3. Mesh of inclusion and its surrounding basal material

CRH5 EMU usually run on straight, curved or turnout rails, so the contact forces on the wheel are quite different according to different conditions. The finite element method is used to analyze the distribution of stress around the spherical inclusions in the wheel rim under three different operating conditions. The analysis results are presented in Table 1.

Table 1. The stress value of spherical inclusions around the wheel rim

	Straight condition	Curve condition	Turnout condition
Maximum equivalent stress value/MPa	226.009	302.822	248.195
Maximum principal stress value /MPa	228.446	317.324	255.223

The results show that the value of maximum equivalent stress and maximum principal stress is highest under curve conditions, and smallest under straight conditions. Under straight conditions, the wheel runs smoothly, compared with the other two conditions. Therefore curve conditions for wheel rims are relatively harsh operating conditions.

3.2. Compile the load spectrum

The general maximum axle load of a CRH5 type EMU guide wheel reaches 17 tons. The axle weight is 1.47 tons. Depending on statistics, the running mileage of a train is about 6×10^5 kilometers per year, and 350 load cycles occur per kilometer. Moreover, a maximum load occurs for each of the 10^6 cyclic loads. Tests by the Academy of Railway Sciences indicate that an eight level load spectrum can represent actual working conditions [12]. The literature [13] gives the eight level load ratio, the ratio of each load cycle, and lateral and vertical load spectra under curve conditions as shown in Table 2 and Table 3 respectively.

Table 2. Load ratio at all levels and its cycle ratio

Load level i	Load ratio F_i/F_l	Recycle ratio n_i/n_l
1	1	1
2	0.950	106
3	0.850	6834

4	0.725	174620
5	0.575	611563
6	0.425	201936
7	0.275	4927
8	0.125	13

Table 3. Vertical and lateral load spectrum

Load level i	Vertical load amplitude F_{Zi}/N	Lateral load amplitude F_{Y2i}/N	Frequency
1	115437.500	64645.00	105
2	109665.625	61412.75	11130
3	98121.875	54948.25	717570
4	83692.188	46867.63	1.8335×10^7
5	66376.563	37170.88	6.4214×10^7
6	49060.938	27474.13	2.1203×10^7
7	31745.313	17777.38	517335
8	14429.688	8080.63	1365

4. Confirming the critical plane of the danger point and calculating its stresses

Based on the McDiarmid model, the plane of the maximum shear stress amplitude at the danger point is a critical plane. The maximum shear stress amplitude is equivalent to the absolute value of the difference of the maximum shear stress and the minimum shear stress. The minimum shear stress in the plane of a load cycle is approximately equal to zero. So it is considered that the maximum shear stress amplitude plane can be determined directly by the critical plane of the maximum shear stress. The steps for determining the critical plane, shear stress and normal stress are as follows.

1) Finite element analysis is carried out on the wheel rim, and the stress nephogram is shown in Fig. 4. The stress tensor σ and the transformation matrix β for the danger point can be extracted by analyzing the stress nephogram. The relationship between the stress tensor and the transformation matrix is $\sigma^{'} = \beta\sigma\beta^{T}$, and the principal stress tensor $\sigma^{'}$ is

$$\sigma^{'} = \begin{bmatrix} \sigma_1 & 0 & 0 \\ 0 & \sigma_2 & 0 \\ 0 & 0 & \sigma_3 \end{bmatrix} \quad \sigma = \begin{bmatrix} \sigma_x & \tau_{xy} & \tau_{xz} \\ \tau_{yx} & \sigma_y & \tau_{yz} \\ \tau_{zx} & \tau_{zy} & \sigma_z \end{bmatrix} \quad \beta = \begin{bmatrix} l_1 & m_1 & n_1 \\ l_2 & m_2 & n_2 \\ l_3 & m_3 & n_3 \end{bmatrix}$$

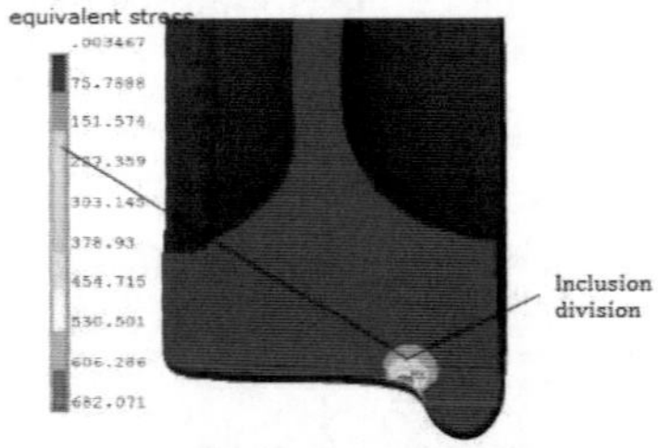

Figure. 4. The stress nephogram of the wheel rim with spherical inclusions

2) The maximum shear stress $\tau_{\max}$ at the danger point is in parallel with the intermediate principal stress σ_2, but which has 45 degrees between the maximum principal stress σ_1 and the minimum principal stress σ_3 respectively. Therefore, the direction cosine of the critical plane is $\begin{bmatrix} a^{'} & b^{'} & c^{'} \end{bmatrix}^{T} = \begin{bmatrix} \sqrt{2}/2 & 0 & -\sqrt{2}/2 \end{bmatrix}^{T}$ in the plane in which the principal stress is the reference coordinate. So the direction cosine $[a\,b\,c]^{T}$ of the critical plane in the global coordinate system is $[a\,b\,c]^{T} = [a'\,b'\,c']^{T}$. Then we can calculate the direction cosine of the danger point in the critical plane under the global coordinate system.

3) According to the Cauchy formula, the stress state on the plane of the global coordinate system of the wheel rim under each load level is calculated by $[T_1\,T_2\,T_3]^{T} = [\sigma][a\,b\,c]^{T}$. Therefore, the shear stress and normal stress on the critical plane can be obtained by (2).

$$\begin{cases} \sigma_n = T_1 a + T_2 b + T_3 c \\ \tau_n = \sqrt{T_1^2 + T_2^2 + T_3^2 - \sigma_n^2} \end{cases} \tag{2}$$

According to (2), the stress state at the danger point of the wheel rim under each load level can be transformed to the critical plane by coordinate transformation. The shear stress and normal stress on the critical plane under each load level is also obtained.

According to the above mentioned method, analyze the wheel rim 10 mm below the tread to confirm whether it contains spherical inclusions or not. The value of normal stress and shear stress at the critical plane of the wheel rim danger point is obtained under the eight level load spectrum as shown in Table 4.

Table 4. Normal stress spectrum and shear stress spectrum

Stress level i	No inclusion 10 mm below the tread		The spherical inclusion is located 10 mm below the tread		Frequency
	Normal stress value/MPa	Shear stress value/MPa	Normal stress value/MPa	Shear stress value/MPa	
1	159.9981	131.1945	182.4679	153.1566	105
2	137.5671	112.3342	174.4935	144.5594	11130
3	124.7819	101.6324	157.5493	129.3526	717570
4	112.3956	91.5843	141.7427	110.1936	1.8335×10^7
5	101.7859	79.0685	123.7724	98.109	6.4214×10^7
6	87.2134	61.2481	101.8344	76.1831	2.1203×10^7
7	62.0159	49.0032	82.0274	64.7765	517335
8	51.2379	34.7645	69.9137	41.8139	1365

5. Estimating the fatigue life of the wheel rim when containing spherical inclusions

Firstly, based on the shear stress and normal stress spectrum at the critical plane of the danger point, and combined with the McDiarmid model, the fatigue life of the wheel rim under each level of cyclic load is calculated. Then, the fatigue damage under each level of the cyclic life load is calculated. The total fatigue damage is obtained by linear cumulative calculation, and lastly the total fatigue life is obtained.

5.1. Fatigue life under each load level in curve conditions

Firstly, get the shear stress and normal stress of the wheel rim danger point. Then according to the McDiarmid criterion we estimate the fatigue life under each stress level. Lastly, the fatigue life of the wheel rim danger point (whether containing spherical inclusions or not) can be obtained for curve conditions. The results are expressed in Table 5.

Table 5. Fatigue life of danger points under all levels of stress

Stress level i	Fatigue life without inclusions(cycle time N)	Fatigue life with spherical inclusions(cycle time N)
1	1.567301×10^8	8.153729×10^7
2	2.376982×10^8	9.143913×10^7
3	4.461356×10^8	2.38399×10^8
4	5.19804×10^8	3.578407×10^8
5	6.99815×10^8	4.399726×10^8
6	8.724591×10^8	6.411495×10^8
7	1.058579×10^9	9.172023×10^8
8	1.40217×10^9	1.820413×10^9

5.2. Estimate the fatigue life under curve conditions

The so-called damage refers to the slight structural changes in the early stage of the fatigue process and the formation and extension of late cracks. One of the most widely used linear cumulative damage theories is the Miner rule. The rule stipulates that if the loading history of the specimen is made up of different stress levels $\sigma_1,\sigma_2,...,\sigma_l$. Then the fatigue life of each stress level is $N_1, N_2,...N_l$ and the number of cycles under each stress level is $n_1, n_2,...n_l$. The specimen is damaged when the damage reaches the value $D=\sum_{i=1}^{l} n_i/N_i = 1$.

Based on the cumulative fatigue damage theory, combined with fatigue life under cyclic loading at all levels, the total damage D of the wheel rim can be obtained under curve conditions.

The total damage for the current demand is a load spectrum. That is the total damage of the car in actual work for one year. At the same time, according to the formula $B=\frac{1}{D}=1/\sum_{i=1}^{l}\frac{n_i}{N_i}$ we can obtain the total number of load spectrum blocks until the wheel rim becomes invalid. Accordingly, the total operating life of the wheel is achieved.

The total damage *D* under the action of a load spectrum and load spectrum blocks B are obtained through the above method regardless of whether the wheel rim contains spherical inclusions or not as shown in Table 6.

Table 6. Total damage under the action of a load spectrum and load spectrum blocks

	No inclusions under curve conditions	Containing spherical inclusions under curve conditions
Total damage *D*	0.153479554	0.233955917
Load spectrum blocks *B*	6.51552584	4.274309506

From the above table, we can see that the total damage at the danger point under the action of a load spectrum is *D*=0.153 when the wheel rim without inclusions is in curve conditions. But the total damage is *D*=0.234 when there are spherical inclusions. Because the load spectrum used in this paper is the load spectrum for a train running for a whole year, the train running time is equal to the load spectrum block, which is calculated from the total damage *D*. The fatigue life of the train wheel rim without inclusions is 6.515 years when the is train running 300 thousand kilometers per year. The fatigue life of the train wheel rim is 4.274 years when it contains spherical inclusions. The fatigue life of the wheel rim is significantly shorter when spherical inclusions exist.

6 . Conclusions

The finite element method is used to analyze the equivalent stress value and the maximum principal stress value of CRH5 train wheel rims containing spherical inclusions in straight, curve and turnout conditions. The value of maximum equivalent stress and maximum principal stress is the largest under curve conditions, and smallest under straight conditions. Compared with straight and turnout conditions, the influence of curve conditions on the fatigue life of the wheel rim is relatively larger. Therefore, the fatigue life of the wheel rim with spherical inclusions in curve conditions is estimated using the McDiarmid criterion, and compared with the fatigue life of the wheel rim without inclusions. It is found that the fatigue life of the wheel rim is significantly shorter when spherical inclusion exists. The results of this analysis can provide a theoretical basis and technical support for further analysis of reliability and sensitivity and is also important for reliability analysis models of wheel shaft systems.

References

[1] Yao Weixing. Structural fatigue life analysis. Beijing: National Defence Industry Press, 2003

[2] Wang Wensheng, Zhao Lu. High cycle fatigue life prediction based on local stress strain method. Rail transit equipment and technology, 2015,1:50-51.

[3] Feng Lei. Effect of nonmetallic inclusions on local stress concentration in the material. Journal of Mechanical Engineering, 2013,49(08):41-48.

[4] Brown MW, Miller K J. A Theory for Fatigue Failure under Multiaxial Stress and Strain Conditions. Mechanical Engineers, 1973:745-755.

[5] McDiamid DL. A general criterion for high cycle multiaxial fatigue failure. Fatigue Fract Engng Mater Struct, 1991,14 (4):429-453.
[6] Li Shuling, Shi Qilong, Yang Jianwei. Fatigue life analysis of CRH3 type EMU wheel.Mechanical design and manufacturing,2011,(12):228-230.
[7] Ekberg A,Kabo E, Andersson H.An engineering model for prediction of rolling contact fatigue of railway wheels. Fatigue Fract Eng Master Struct 2002,25:899-909.
[8] Brant Stratman, Yongming Liu, Sankaran Mahadevan. Fatigue crack initi-ation life prediction of railroad wheels. International Journal of Fatigue, 2006,(28):747-756.
[9] Aleksander Karolcauk, Ewald Macha. A review of critical plane orientations in multiaxial fatigue failure criteria of metallic materials. International Journal of Fracture,2005,134:267-304.
[10] Liu Haiyan, Tan Wenjie. Multi axis high cycle fatigue life estimation of diesel engine crankshaft. Mechanical strength ,2013,35(6):844-849.
[11] Cai Li. Complex shape of inclusions on the effects of the EMU wheel rim crack. China Mechanical Engineering,2017,28(2):240-245.
[12] Wei Changzhu. Reliability calculation of fatigue life of double deck passenger car axle in 25.5m air conditioning system. Rail truck,1993,(11):8-13.
[13] Zhang Tingxiu. Fatigue life and reliability sensitivity analysis of the wheel rim of the EMU. Zhejiang: Zhejiang Sci-Tech University,2015.

Acknowledgment
Supported by the State Key Program of National Natural Science of China (Grant No.U1234207),the National Natural Science Foundation of China (Grant No.51475432), and the Zhejiang Provincial National Natural Science Foundation of China (Grant No.LZ13E050003).

IOP Conf. Series: Materials Science and Engineering **339** (2018) 012025 doi:10.1088/1757-899X/339/1/012025

Design of Ablation Test Device for Brick Coating of Gun

YAO shirui, CHEN yongcai, WANG fei, ZHAO jianxin

Artillery Engineering Department, Army Engineering University Shijiazhuang Campus, No. 97 Heping West Road ,shijiazhuang hebei, China

E-mail:ysryaoshirui@163.com

Abstract. As a result of the live ammunition test conditions, the barrel resistance of the barrel coating has high cost, time consuming, low efficiency and high test site requirements. This article designed a simple, convenient and efficient test device. Through the internal trajectory calculation by Matlab, the ablation environment produced by the ablation test device has achieved the expected effect, which is consistent with the working condition of the tube in the launching state, which can better reflect the ablation of the coating.

1. Test principle of ablation test device

The whole ablation test device mainly consists of three parts: gun barrel, new coating ablation test device and connecting device. The working principle is as follows: Before the test, first flat tube and unscrew the muzzle brake, will try to spin the embedded high pressure cylinder connected to the barrel. Loading pressure controlled bomb in high-pressure cylinder chamber, then in a certain amount of gunpowder, and then screwed sealing plug in high pressure cylinder is connected with the 24V electrode, DC. In experiments, the powder is ignited by the current and burned to produce a propellant gas with a certain pressure. When the powder gas pressure reaches a certain value, the pressure control was forced to play into high-pressure cylinder chamber and forward movement. At the same time, the gunpowder continued to burn and the pressure of gunpowder and gas increased. When the pressure bomb through the specimen, the propellant gas pressure reaches the maximum value, resulting in the best effect of using ablation treatment. After the pressure controlled projectile is released from the test piece, it continues to fly forward in the barrel until it hits the breech. The propellant gas ejected from the specimen expands in the tube, resulting in a sharp drop in temperature and pressure, eventually leaking from the gap between the breech and the tube in the atmosphere. Repeat these actions, the multiple ablation test of the test piece can be realized. At the end of the test, the test specimen is removed from the high pressure cylinder, and the ablation condition of the coating specimen can be studied by using relevant instruments to test the specimen.

2. Design of ablation test device

Based on the above test ideas and experimental principle, a test device for the ablation performance of a large caliber gun is designed. The device can be used for many ablative tests on the coated specimens. The inner environment generated by the ablation environment test device is similar to the firing conditions, the coating specimen after ablation can detect the ablation condition through various methods, so as to assess the prepared coating erosion resistance. In addition, a number of specimens can be placed at the same time during the ablation test to compare the ablative properties of the coated specimens.

Figure 1. The coating erosion test device

Figure 2. The gun barrel connector

Figure1. is a coating erosion test device machined by machining. The test device is screwed on the muzzle connecting device and connected with the gun barrel. Figure2. is a muzzle connection device, and Figure3. is the actual installation mode of the test device. The ablation test device is connected to the muzzle, the coating can be fixed ablation test device effectively, reduce the impact of combustion on the test device, but also can make the control from the test pressure of projectile impact of gun breech and stop, improve safety. On the other hand, the gas produced by the burning of the gunpowder can be stored in the barrel of the gun to dissipate slowly, and the smoke produced in the test can be slowed down. Furthermore, the noise produced in the test can be reduced.

Figure 3. The install pattern of erosion device

In order to explain the specific internal structure of the coating erosion test device intuitively, a three-dimensional model of the coating ablation test device is established by using the Pro-E three-dimensional modeling software, as shown in figure4.. Coating device ablation test in Figure4. connection device connected with the muzzle through threaded connection, the ignition device and the pressure cylinder also adopts threaded connection, the ablation test device through the connecting device and the muzzle of gun tube. The control pressure bomb is a copper material similar to the shape of a bullet. The chamber were used for powder grains after installing pressure controlled bomb and electric ignition device, the chamber for closed environment.

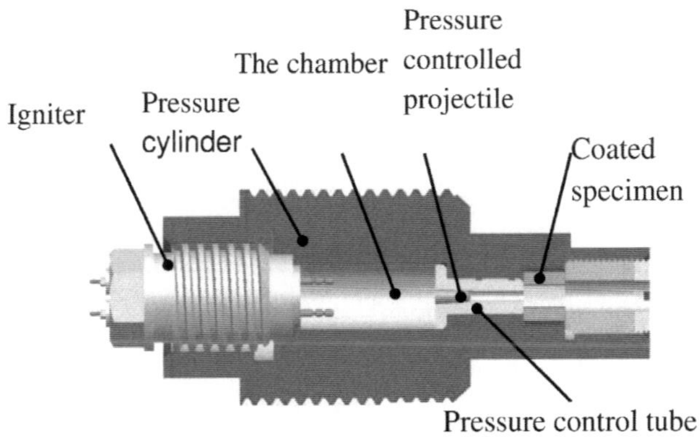

Figure 4. The 3D model of erosion device

The coating device can be placed up to one-time ablation test of 4 specimens, 4 specimens of the ablation test, installation can be convenient for ablation of a plurality of test pieces were analyzed to evaluate the coating erosion resistance. The coating specimens are installed in the retaining ring, and the standard coating specimen size is 11mm * 7mm * 9mm. Figure5. shows how the mounting rings and coated specimens will be fitted.

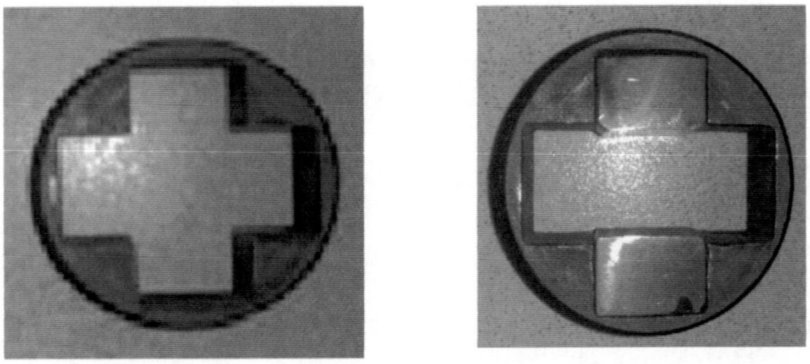

(a) Mounting ring (b) Installation of Coating Specimens

Figure 5. The install pattern of coating specimens

3. Internal ballistic calculation of ablation test equipment

The test process of a ablative test device can be approximated as an internal ballistic process. By solving the ablation test device of internal elastic coating problems can reflect the ballistic characteristics accurately and directly, the main parameters of internal ballistics are obtained and can directly reflect the powder gas pressure ablation test unit condition in the ablation process and coated specimens are in test. In addition, by comparing the interior ballistic parameters of the gun under the condition of live ammunition shooting, the simulation of ablation process of the ablation test device can be further described. During the calculation of the ballistic problem in the ablation test apparatus, the following basic assumptions are made [1]: (1) the combustion of propellant follows the law of geometrical combustion. (2) the particles are burned at average pressure and follow the law of burning speed. (3) the surface heat dissipation of the ablation device is indirectly corrected by reducing the powder force f or increasing the specific heat k. (4) use the coefficient φ to consider various secondary functions. (5) the propellant gas obeys the Nobel Abel equation of state. (6) the f, residual α and specific heat ratio k are constant in the whole combustion process.

3.1.basic equations of interior ballistics
In the internal ballistic process of the ablation device, the following basic interior ballistic equations [2]:

$$\left.\begin{aligned}
&\psi = \chi Z\left(1+\lambda Z+\mu Z^2\right)\\
&\frac{dZ}{dt}=\frac{p}{I_k}\\
&Sp=\varphi m_q\frac{dv}{dt}\\
&Sp(l_\psi+l)=fm_\omega\psi-\frac{\theta}{2}\varphi m_q v^2\\
&\frac{dl}{dt}=v
\end{aligned}\right\}\qquad(3\text{-}1)$$

Among: Ψ—The percentage of gunpowder burned out. Z—The percentage of gunpowder burned out. X、λ、μ—The shape characteristic quantity of propellant grain. P—Propellant gas pressure. I_k —Total pressure impulse. S —bore cross-sectional area. φ—Coefficient of resistance is also called secondary work coefficient. υ—Projectile velocity. l_Ψ—Free chamber volume shrinkage long.

3.2. Initial conditions of internal ballistic calculation for ablative test equipment
*3.2.1.charge parameters of ablative test equipment.*Charge quantity m_ω=10g .
Grain thickness 2e1=0.78×10-2 dm. Gunpowder f=950000 kg.dm/kg. Coefficient of surface burning χ1=0.72347956, λ1=0.3311631. Surface burning coefficient χs=1.4872194, λs=-0.3685706. Gas volume α=1.0 dm3/kg. Propellant density δ=1.6 kg/dm3 . Ignition charge ωB=10g. Effects of ignition f=300,000 kg.dm/kg.

3.2.2. Related internal ballistic parameters of ablation test equipment.(a) chamber volume.
According to the specific size, Figure6. Figure7. pressure cylinder pressure control pipe size and assembly relation between chamber volume can be calculated approximately V=7.84x10-2d m3.(b) cross-sectional area of a pressure control tube. The pressure control pipe cross-sectional area is similar to the internal ballistic gun bore transverse area from Figure7. pressure tube size can be calculated to control the pressure pipe cross-sectional area of =28.26 mm2.

3.2.3. The length of the projectile, the mass of the projectile and the pressure of the projectile.
The length of the projectile is the length of the control tube. As shown in Figure7., the stroke length of the projectile is =46mm. The projectile is made of brass, and the quality is 10g. The projectile is squeezed into the pressure reference gun and the barrel of the gun is squeezed into the pressure(30Mpa).

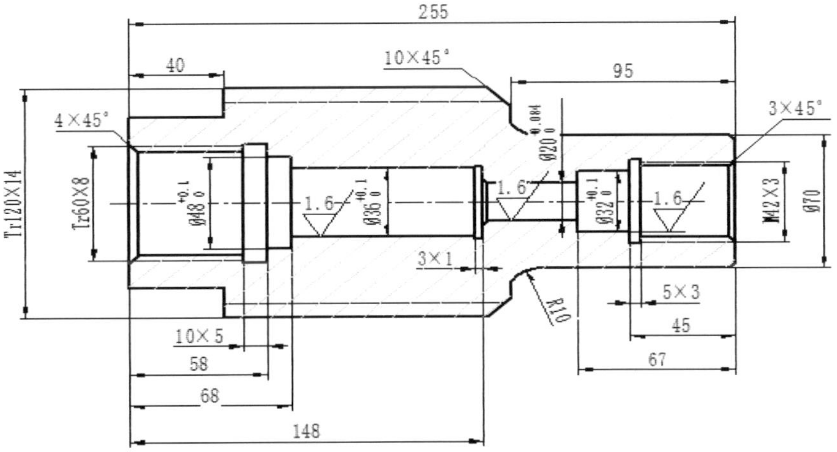

Figure 6. The size of pressure cylinder

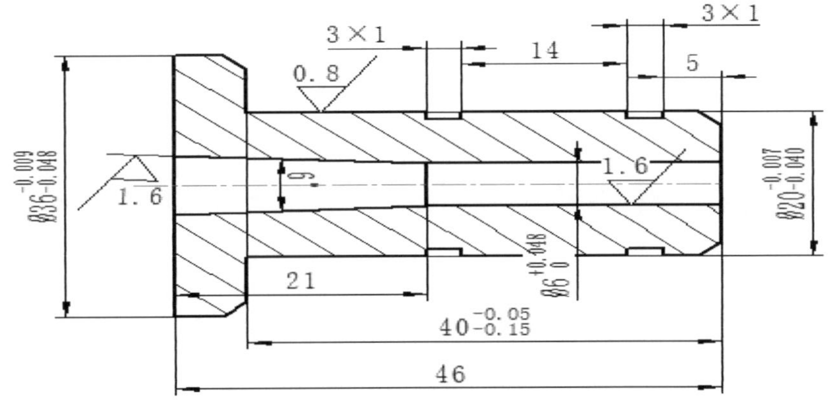

Figure 7. The size of pressure-control tube

3.3.Internal ballistic solution of ablation test device

Based on the known interior ballistic condition, the internal ballistic solution of the ablation test device can be obtained by using the internal ballistic solver. Because the test uses the powder as the porous propellant, therefore the gunpowder combustion has the enhancement stage and the reduction surface stage. The sign at which the end of the combustion is increased, i.e., Z=1. Gunpowder reduced surface combustion. At the end, Ψ=1. Table 1 is the internal ballistic table of the test device calculated, and Figure8. is the relevant internal ballistic parameter curve. From the results of internal ballistics calculation, the theoretical maximum pressure point position and the maximum point of pressure control projectile velocity are all located at the end face of the pressure control pipe. The maximum pressure at the bottom of the projectile is 186MPa, and the maximum velocity of the control projectile is 192.8m/s. It can be seen that the position of the coating specimen is the maximum pressure point position, and the maximum propellant gas pressure at the moment of the test piece is 186MPa. In addition, the coated specimen is also affected by the action of the powder gas and the mechanical wear of the powder residue. Working conditions of bore contrast during gun firing, the ablation ablation test device and the actual environmental emission is close to that of the coated specimens are close to ablation ablation and bore gun steel. According to the theoretical calculation results, the ablation environment produced by the ablation test device has achieved the desired results.

IOP Conf. Series: Materials Science and Engineering **339** (2018) 012025 doi:10.1088/1757-899X/339/1/012025

Table1. The internal trajectory of erosion

Time t/m	Propellant Burning Coefficient Z1	Ignition Coefficient of Propellant Z2	Length of Projectile Travel l/m	Projectile Velocity v/m*s^{-1}	Bottom Pressure pd/MPa
0.0000000	0.13855	0.03626	0.000	0.00	30.0000
0.0000300	0.14507	0.61428	0.000	6.05	158.2328
0.0000409	0.14910	1.00000	0.000	10.09	167.9846
0.0000409	0.14910	1.00000	0.000	10.09	167.9846
0.0000709	0.16038	1.00000	0.001	21.43	169.3717
0.0001009	0.17173	1.00000	0.001	32.87	170.7353
0.0001309	0.18317	1.00000	0.003	44.39	172.0740
0.0001609	0.19467	1.00000	0.004	56.01	173.3866
0.0001909	0.20626	1.00000	0.006	67.71	174.6716
0.0002209	0.21791	1.00000	0.008	79.50	175.9279
0.0002509	0.22963	1.00000	0.011	91.37	177.1541
0.0002809	0.24142	1.00000	0.014	103.33	178.3491
0.0003109	0.25328	1.00000	0.017	115.36	179.5117
0.0003409	0.26520	1.00000	0.021	127.47	180.6406
0.0003709	0.27719	1.00000	0.025	139.65	181.7348
0.0004309	0.30134	1.00000	0.034	164.24	183.8150
0.0004609	0.31350	1.00000	0.039	176.63	184.7990
0.0004909	0.32571	1.00000	0.044	189.09	185.7443
0.0004998	0.32936	1.00000	0.046	192.81	186.0179

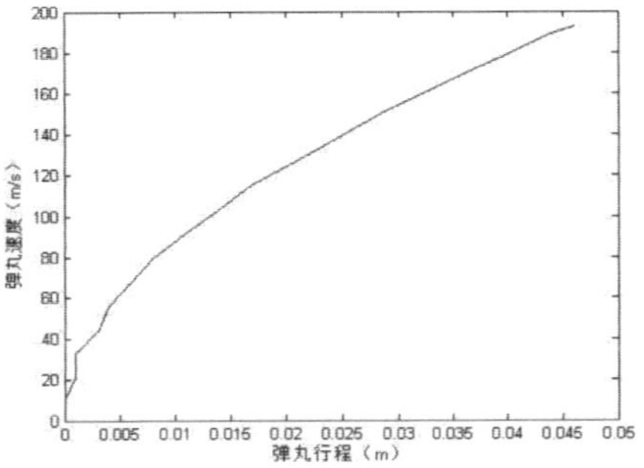

(a) $v \sim l$ curve

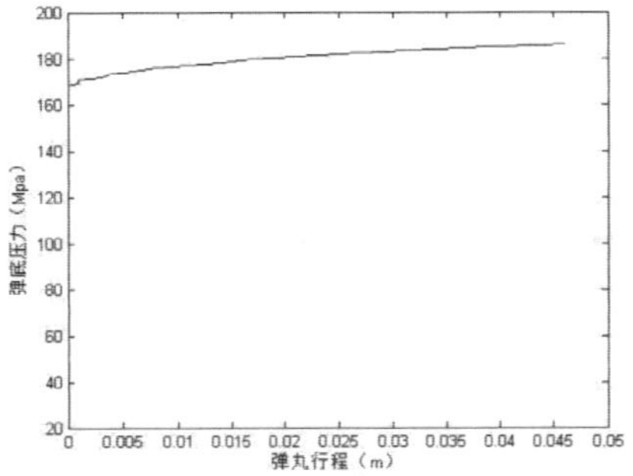

(b) $p \sim l$ curve

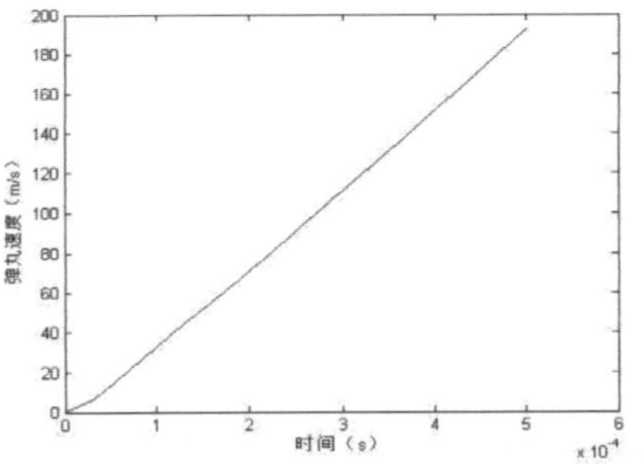

(c) $v \sim t$ curve

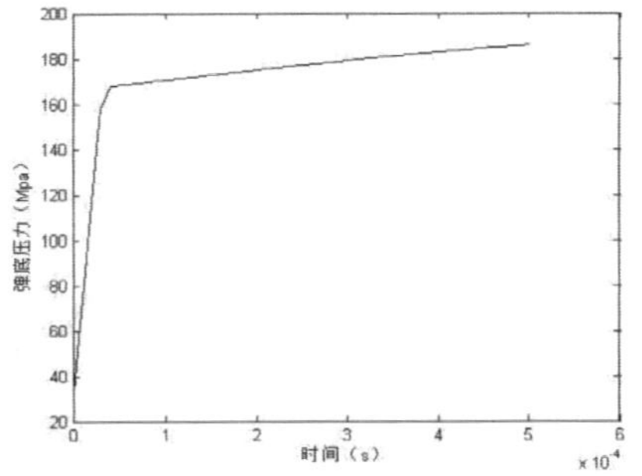

(d) $p \sim t$ curve

Figure 8. The curve graph of internal trajectory parameters

4.Conclusion

The design of a new type of coating erosion test device is described. The device can approximately simulate the bore condition when the gun is fired, and explain the working principle and test method of the ablation test device.The calculation of internal ballistic related problems of ablation test device, the internal ballistic solution of the test device is obtained, and the ablation condition of the ablation test device is basically consistent with the actual launching conditions.

Reference

[1] Wang Lianrong, Zhang Peicheng. Artillery Interior Trajectory Calculation Manual [M]. Beijing: National Defense Industry Press, 1987(6)pp37-39.
[2] Jin Zhiming, Yuan Yaxiong, Song Ming. Modern Interior Ballistics [M]. Beijing: Beijing Institute of Technology press, 1992(3)pp113-129.

A hybrid feature selection and health indicator construction scheme for delay-time-based degradation modelling of rolling element bearings

Bin Zhang, Congying Deng and Yi Zhang

School of Advanced Manufacturing Engineering, Chongqing University of Posts and Telecommunications, Chongqing 400065, China

Email: zhangbin@cqupt.edu.cn

Abstract. Rolling element bearings are mechanical components used frequently in most rotating machinery and they are also vulnerable links representing the main source of failures in such systems. Thus, health condition monitoring and fault diagnosis of rolling element bearings have long been studied to improve operational reliability and maintenance efficiency of rotatory machines. Over the past decade, prognosis that enables forewarning of failure and estimation of residual life attracted increasing attention. To accurately and efficiently predict failure of the rolling element bearing, the degradation requires to be well represented and modelled. For this purpose, degradation of the rolling element bearing is analysed with the delay-time-based model in this paper. Also, a hybrid feature selection and health indicator construction scheme is proposed for extraction of the bearing health relevant information from condition monitoring sensor data. Effectiveness of the presented approach is validated through case studies on rolling element bearing run-to-failure experiments.

1. Introduction

Rolling element bearings are widely used mechanical components in machinery and their malfunctions is one of the major causes for breakdowns in rotating machines. According to [1], abnormalities of the rolling element bearing account for almost 45–55% of these equipment failures. In practice, bearings usually undergo degraded processes from their normal states till final failures, and condition monitoring of the bearings were extensively studied in the past few decades [2, 3]. To implement more proactive and predictive maintenance paradigm, the last decade witnessed the developing of prognosis which enables early failure warning and prediction of residual life. Even one aircraft saved from turbine engine bearing failures with prognosis would pay its development cost [4]. Currently, diagnostic technologies for rolling element bearings are relatively well developed, but accurate and efficient prognosis of bearing failure is considerably more difficult.

Generally, prognostics approaches are grouped as:1) physics-based, 2) data-driven, and 3) hybrid. Physics-based techniques are important if accuracy is a critical factor and testing is restricted, since prognosis is predicted with mathematical models of physics failure mechanism [5]. However, only a few physics models are now available for bearing prognosis, especially for those used in sophisticated systems. Data-driven methods rely on models learned directly from historical data to predict future health. Statistical and artificial intelligence approaches are the mainly data-driven prognostics and their applications in bearing failure prognosis are gaining popularity [6]. To take advantage of both the physics-based and data-driven prognosis, hybrid approaches require the physics failure knowledge as well as the learned degradation models [7]. In many practical cases when it is easier to gather data than to build physics-of-failure models, data-driven prognosis is often the more applicable solution.

Extraction of health relevant information from the background knowledge of historical sensor data is the first important step for a data-driven prognosis. For health monitoring of motors, gears and bearings, sensor data often comes in the waveform of vibration or acoustic signals, etc. Thus, signal processing is applied to extract features. Traditionally, the features for bearing fault diagnosis (such as kurtosis, RMS and peak-to-peak value) are also used for its prognosis. However, due to reasons of not sensitive to incipient deterioration or lacking of consistent trend, these features are only effective for certain defect during certain service stage. To address this issue, construction of health indicators that relate to the bearing degradation is considered. In [8], empirical mode decomposition and self-organizing map (SOM) are used to calculate a confidence value on the bearing health state. With feature selection and weighting before fusing by SOM, Lei et al [9] constructed a health indicator of weighted minimum quantization error. Also, a recurrent neural network based health indicator was proposed in [10]. Before construction of a health index, features for the synthetization should be selected. Nevertheless, compared to the feature selection for diagnosis, much needs to be done on sensor selection for prognosis [4,11,12]. In this paper, a hybrid scheme for feature selection and health indicator construction is proposed to extract bearing health relevant information from sensory signals.

Degradation modelling for characterizing evolution of the degradation is another crucial task for a data-driven prognosis. If a proper model is established for the health indicator, then the model can be used for online prognosis or directly for maintenance decision-making. Stochastic process models and general path models are the two large classes used to model degradation [13].After the introduction to degradation literature by Lu et al [14], general path models have found widely applications [15]. On the other hand, Wiener process, Gamma process and inverse Gaussian (IG) process are the frequently used stochastic process models for degradation modelling [16]. Recently, in view that the failure of a bearing is generally a slow degradation process instead of a sudden breakdown, some researchers proposed to model the bearing degradation with two stages. Deterioration modelling is not focused on the whole life of the bearing, but on the second stage once an incipient fault is detected [17,18]. Actually, these degradation models fall in the delay-time concept which has been studied in industrial asset inspection and maintenance [19], and can be generalized to cases of more than two stages. For the modelling of the bearing degradation, delay-time-based framework is considered in this paper.

The remaining parts of the paper are organized as follows. Section 2 formulates bearing degradation in the delay-time concept. Section 3 presents a detailed description of the proposed hybrid scheme. Section 4 demonstrates the developed method with experimental case studies on rolling element bearing degradation. And Section 5 summaries the presented research and future work.

2. Delay-time-based modelling of bearing degradation

2.1The delay-time concept

The delay-time concept itself is simple to divide the failure process of an asset into two stages. The first stage is the normal operating stage from new to the point that an identifiable defect has been detected. The second stage defined as the failure delay time is from the point of defect identification to failure. It is the existence of such a failure delay time that provides the opportunity for carrying out maintenance to predictively and proactively remove or rectify the identified defects before final failure happens. With the durations of these two stages appropriately modelled, optimal inspection intervals can be identified to optimise an objective function of interest.

Originally developed by operational research researchers, delay-time-based models have attracted significant attention from mathematical modellers and found widely applications in the maintenance field [20, 21]. Relaxing the binary state description of one industrial item (i.e. normal and failure) to step closer to reality, assumption of two-stage failure has been generalized to three-stage process by Wang [22]. For its impacts on industrial plant maintenance from both theoretical and practical point of views, the delay-time concept is tailored to model degradation of the bearing in this paper.

2.2Modelling of bearing degradation

The failure of deployed bearings is not a sudden case, but generally a gradual process, taking place in one stage when the bearing is in normal working status without any identifiable fault and the irreversible degradation stage. To practice better operational management in the industrial field, the second stage may further be divided into two or more substages to account for slight, middle and severe bearing degradation, etc. This coincides well with the delay-time concept. Thus, degradation process *Y*(*t*) of the bearing can be modelled as

$$Y(t) = g(t, \boldsymbol{\Theta}) + \varepsilon, \quad t \geq t_D \tag{1}$$

where $g(\bullet)$ is the degradation path with parameter set $\boldsymbol{\Theta}$, t_D is a random variable relevant to the fault detected point, and ε is the error term normally distributed with zero mean and constant variance.

In the delay-time-based model for bearings degradation, modelling is only focused on the later stage triggered by the incipient fault identification. In contrast to the few two-stage degradation modelling for bearings, the degradation path in the proposed model is not assumed to follow one functional forms till failure but can be consisted of two or more segments demonstrating quite different patterns. Stochastic process and general path models can then be used to model the segments of the degradation stage upon change-point detection. For detection of the bearing incipient fault and the change point in the degradation stage, approaches as statistical process control can be applied. Also, the rich source of methodologies by the delay-time framework will facilitate the bearing degradation modelling.

3. The proposed hybrid scheme

3.1Flowchart of the hybrid scheme

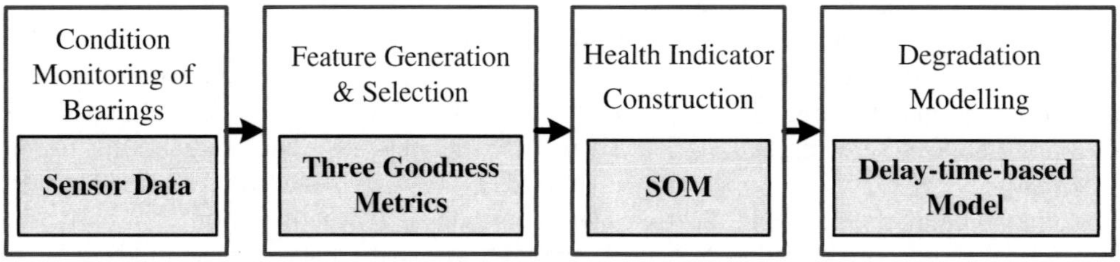

Figure 1. Flowchart of the proposed hybrid scheme.

The proposed scheme is given in figure 1. Bearing condition monitoring sensor data is pre-processed using signal processing techniques and a number of candidate features are thus generated. Then these generated features are evaluated by goodness metrics and informative features are selected. Fusing the selected features into one-dimensional health indicator that represents health condition of the bearing, SOM is used here. Finally, the bearing degradation is modelled with the delay-time concept into stages.

3.2Feature evaluation and selection

To mine features for bearing fault diagnosis and prognosis, signal processing has progressed from time or frequency techniques to advanced time-frequency or time-scale approaches. Unfortunately, appropriateness of the features often varies with nature of the failure, severity of the degradation and signal to noise ratio, etc. It is better to process the vibration signals with different methods simultaneously, and then evaluate the possible features based on metrics and select the optimal ones.

Focusing on bearing deterioration and failure prediction, statistical features of dimensional and non-dimensional ones as RMS and crest factor (CF), and wavelet packet node energy (WPNE) have been considered in this paper. Specifically, 29 features can be obtained upon processing each vibration waveform signal. To evaluate the effectiveness of an individual feature, the three goodness metrics proposed in our previous work [11] are revisited as Equation (2).

Among the three metrics, Corr measures the linearity between the feature and the usage time, Mon assess consistently increasing or decreasing trend of the feature, and Rob reflects the tolerance of the feature to outliers. Features with high scores of the three metrics should be selected and retained as optimal features, since these features have better correlations with the bearing lifetime and depict consistent trends with less vulnerability to interferences.

$$f(t_j)=f_T(t_j)+f_R(t_j),\quad Corr(F)=\frac{\left|K\sum_j f_T(t_j)t_j-\sum_j f_T(t_j)\sum_j t_j\right|}{\left(K\sum_j f_T(t_j)^2-\left(\sum_j f_T(t_j)\right)^2\right)^{1/2}\left(K\sum_j t_j^{\,2}-\left(\sum_j t_j\right)^2\right)^{1/2}} \tag{2}$$

$$Mon(F)=\frac{\left|\sum_j \delta\left(f_T(t_j)-f_T(t_{j-1})\right)-\sum_j \delta\left(f_T(t_{j-1})-f_T(t_j)\right)\right|}{K-1},\quad Rob(F)=\frac{1}{K}\sum_j \exp\left(-\left|\frac{f_R(t_j)}{f_T(t_j)}\right|\right)$$

where $f(t_j)$ is value of the feature F at the time t_j ($j = 1,2,\ldots,K$) with the trend of $f_T(t_j)$ and residual $f_R(t_j)$, and $\delta(\bullet)$ is simply the unit step function.

3.3Health indicator construction

SOM is an unsupervised neural network that provides a way of representing multi-dimensional features into a one or two-dimensional space [23]. Each neuron in the SOM network is represented by a weight vector and only data under normal operation of the bearings is needed in the training of the SOM. With the trained SOM, for each input feature vector **F** composed of the optimally selected features, there is a best matching unit (BMU) whose weight vector **m** is the closest to the input vector. And the distance (usually in the form of Euclidean distance) between the input feature vector and the weight vector of the BMU, which actually quantizes how far the current condition is from the normal operation state, is defined as minimum quantization error (MQE). Then based on the MQE, a health indicator (HI) y that represents the health state of the bearing is constructed as

$$y=\exp\left(-\frac{MQE}{c_0}\right),\quad \text{with } MQE=\left\|\mathbf{F}-\mathbf{m}_{BMU}\right\| \tag{3}$$

where $\mathbf{m}_{\mathrm{BMU}}$ is the weight vector of the BMU of the input **F**, and c_0 is a scale constant determined by the MQE under normal conditions.

This HI synthesized with SOM integrates the optimally selected multi-dimensional features. In fact, the proposed HI describes the degradation of one bearing well in the range of [0, 1], in that HI lessens with worsens of the bearing condition. With the proposed hybrid scheme for selection of degrading features and construction of health indicator, background health knowledge of the bearings can be constructed in the offline modelling to lay basis for the online prediction in a data-driven prognosis.

4. Experiment and analysis

4.1Description of the experiment

The bearing run-to-failure data sets used to validate the presented method are taken from the bearing accelerated life tests by the PRONOSTIA platform [24]. In the experiment, both horizontal and vertical vibration accelerations of the bearing housing were recorded with a sampling frequency of 25.6 kHz and the data length of 2.56 k, while the recording interval was 10 s. Failure of the bearing was claimed when the amplitude of the vibration signal overpassed 20 g. All the tested bearings were run till failure, thus each bearing could fail due to any of the possible failure modes like failure of balls, rings, cage or their combinations. Such data is very common in the real industrial applications and has been utilized for validation of bearing prognostics approaches [8,9,12]. More details about the experiments can be found in [24]. In this paper, the seven bearings tested under the operational condition (i.e.1800 rpm and 4000 N) are analysed.

The horizontal vibration signals for the whole lifetime of two tested bearings are illustrated in figure 2. Although they were from the same category and experimented under the same condition, the two bearings depict very different patterns with the usage time. Life durations of the seven bearings range from shorter than 3 h to longer than 7 h and demonstrate a high variability. This and the undocumented failure modes pose additional challenges in degradation modelling and prognosis.

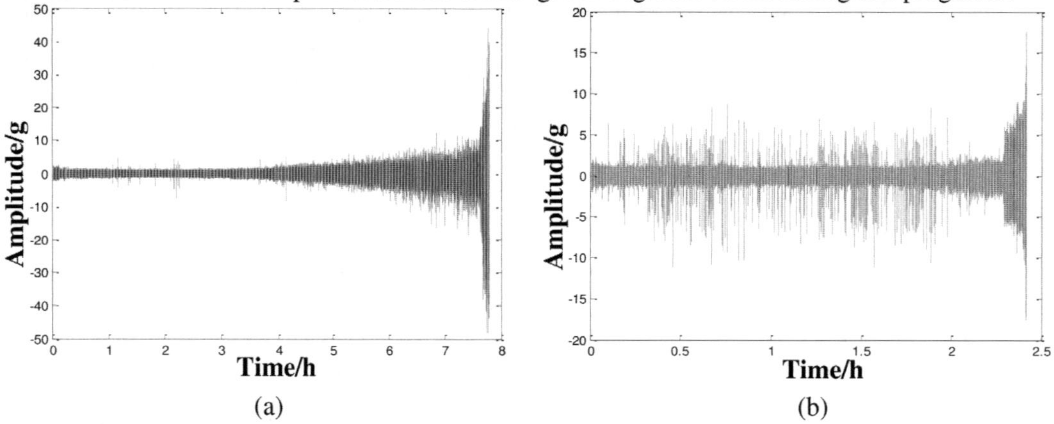

Figure 2. Horizontal vibration signals of two bearings: (a) bearing 1; (b) bearing 2.

4.2 Results and analysis

Based on the signal processing techniques in the proposed hybrid scheme, a total of 58 features are extracted for each bearing using both the horizontal and vertical vibration signals. Then the three goodness metrics of Corr, Mon and Rob are calculated for each kind of feature of the seven bearings. In the analysis, the feature is trended with the moving average method. And evaluation results of the tested bearings are further statistically analysed to sift the consistent features to secure appropriateness for all the seven bearings. After these steps, four features are optimally selected and two of them are shown in figure 3. The two features F1 and F2 respectively have an increasing and decreasing trend with consumption of the bearing lifetimes, and share consistency among the seven bearings.

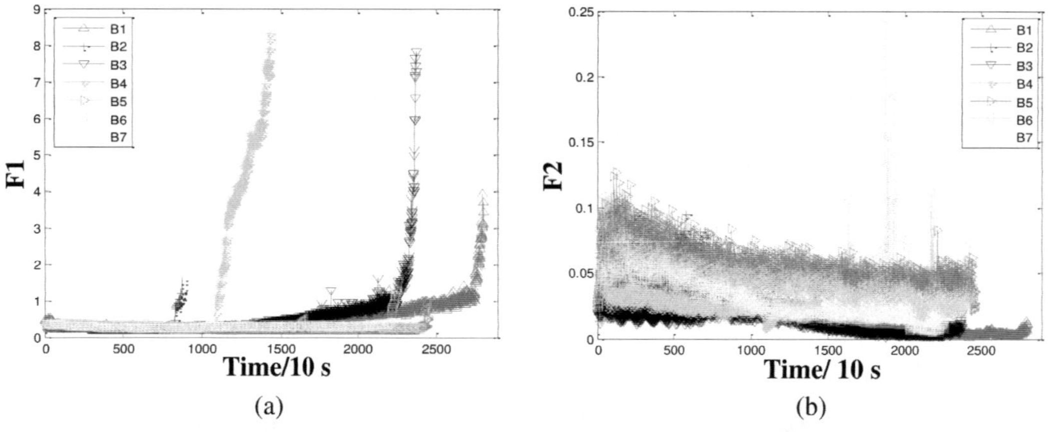

Figure 3. Two of the optimally selected features for the seven bearings: (a) F1; (b) F2.

With the feature vector of four optimally selected features as input, the HI of each bearing is constructed following the hybrid scheme presented in Section 3. Two stages of normal working before any identifiable fault and the degrading stage for all the seven bearings can be obviously observed from figure 4(a). However, mainly due to different failure modes, as also discussed in [12], the deteriorative paths diverge in the second stage and this stage can further be divided into two substages or segments that demonstrate quite different varying patterns. In the following, the proposed delay-time-based model is further analysed on the degradation of three bearings.

To analysis the failure process of the bearings, the second stage of deterioration can be modelled into two segments of degradation substage I and II with the delay-time-based model in this paper. For bearing 1 and 7, degradation substage I consumes most of their lifetimes. While for bearing 2, it operates in the normal condition with a long duration and then runs into breakdown quickly. Based on the constructed HI and delay-time modelling, sharp degradation can be detected long before the bearing fails. Actually, degradation substage II accounts no less than 1% of the bearing life for all the seven tested bearings. With the detectable defect identified and change points for substage partition detected by statistical methods, each substage of bearing degradation can be fitted with parametric or nonparametric approaches, and then bearing failure can be further prognosed.

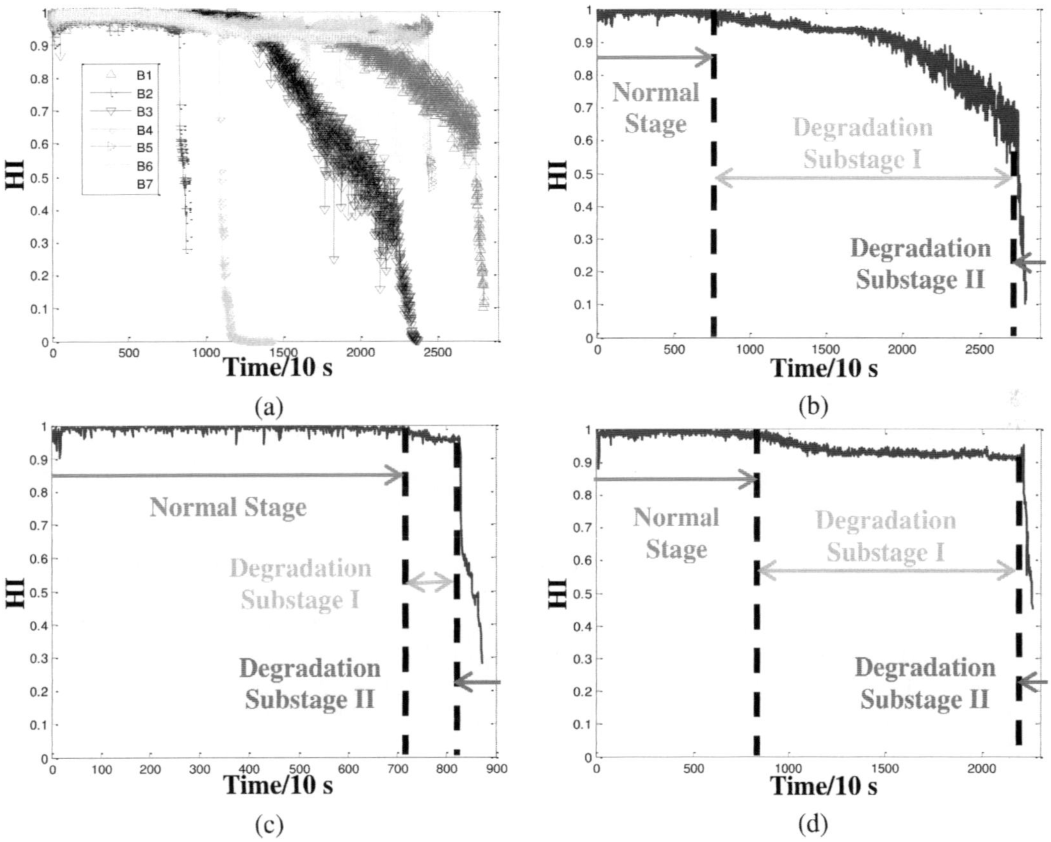

Figure 4. HI of the tested bearings: (a) HI of all seven bearings; (b) HI of bearing 1; (c) HI of bearing 2; (d) HI of bearing 7.

5. Conclusion

For accurate and efficient prognosis, degradation of rolling element bearings is modelled with delay-time-based model in this paper. Feature selection based on goodness metrics and health indicator construction with SOM are presented in a hybrid scheme. With the incipient fault identified, bearing degradation can be modelled by two or more substages using the delay-time framework. Case studies on bearing run-to-failure experiments validate effectiveness of the presented approach. To easy and enrich applicability of the proposed method, identification of the incipient fault as well as detection of the change points will be explored in the future.

References

[1] Rai A and Upadhyay S H 2016 A review on signal processing techniques utilized in the fault diagnosis of rolling element bearings *Tribol. Int.* **96** 289–306

[2] Randall R B and Antoni J 2011 Rolling element bearing diagnostics—a tutorial *Mech. Syst. Signal Process.* **25** 485–520
[3] El-Thalji I and Jantunen E 2015 A summary of fault modelling and predictive health monitoring of rolling element bearings *Mech. Syst. Signal Process.* **60** 252–72
[4] Camci F, Medjaher K, Zerhouni N and Nectoux P 2013 Feature evaluation for effective bearing prognostics *Qual. Reliab. Eng. Int.* **29** 477–86
[5] Cubillo A, Perinpanayagam S and Esperon-Miguez M 2016 A review of physics-based models in prognostics: Application to gears and bearings of rotating machinery *Adv. Mech. Eng.* **8** 1687814016664660
[6] Cerrada M, Sánchez R V, Li C, Pacheco F, Cabrera D, de Oliveira J V and Vásquez R E 2018 A review on data-driven fault severity assessment in rolling bearings *Mech. Syst. Signal Process.* **99** 169–96
[7] Xu D, Huang J E, Zhu Q, Chen X, Xu Y C and Wang S 2012 Residual fatigue life prediction of ball bearings based on Paris law and RMS *Chin. J. Mech. Eng.* **25** 320–27
[8] Hong S, Zhou Z, Zio E and Wang W B 2014 An adaptive method for health trend prediction of rotating bearings *Digit. Signal Process.* **35** 117–23
[9] Lei Y G, Li N P, Gontarz S, Lin J, Radkowski S and Dybala J 2016 A model-based method for remaining useful life prediction of machinery *IEEE Trans. Reliab.* **65** 1314–26
[10] Guo L, Li N P, Jia F, Lei Y G and Lin J 2017 A recurrent neural network based health indicator for remaining useful life prediction of bearings *Neurocomputing* **240** 98–109
[11] Zhang B, Zhang L J and Xu J W 2016 Degradation feature selection for remaining useful life prediction of rolling element bearings *Qual. Reliab. Eng. Int.* **32** 547–54
[12] Kundu P, Chopra S and Lad B K 2017 Multiple failure behaviors identification and remaining useful life prediction of ball bearings *J. Intell. Manuf.* https://doi.org/10.1007/s10845-017-1357-8
[13] Ye Z S and Xie M 2015 Stochastic modelling and analysis of degradation for highly reliable products *Appl.Stoch. Model. Bus.* **31** 16–32
[14] Lu C J and Meeker W O 1993 Using degradation measures to estimate a time-to-failure distribution *Technometrics* **35** 161–74
[15] Li N P, Lei Y G, Lin J and Ding S X 2015 An improved exponential model for predicting remaining useful life of rolling element bearings *IEEE Trans. Ind. Electron.* **62** 7762–73
[16] Wang Z Q, Hu C H and Fan H D 2017 Real-time remaining useful life prediction for a nonlinear degrading system in service: Application to bearing data *IEEE-ASME Trans. Mech.* DOI: 10.1109/TMECH.2017.2666199
[17] Wang D and Tsui K L 2017 Statistical modeling of bearing degradation signals *IEEE Trans. Reliab.* DOI: 10.1109/TR.2017.2739126
[18] Wang Y, Peng Y Z, Zi Y Y, Jin X H and Tsui K L 2016 A two-stage data-driven-based prognostic approach for bearing degradation problem *IEEE Trans. Ind. Inform.* **12** 924–32
[19] Christer A H and Wang W B 1995 A delay-time-based maintenance model of a multi-component system *IMA J.Manag. Math.* **6** 205–22
[20] Wang W B 2012 An overview of the recent advances in delay-time-based maintenance modelling *Reliab. Eng. Syst. Saf.* **106** 165–78
[21] Zhao F, Wang W B and Peng R 2015 Delay-time-based preventive maintenance modelling for a production plant: a case study in a steel mill *J. Oper. Res. Soc.* **66** 2015–24
[22] Wang W B 2011 An inspection model based on a three-stage failure process *Reliab. Eng. Syst. Saf.* **96** 838–48
[23] Qiu H, Lee J, Lin J and Yu G 2003 Robust performance degradation assessment methods for enhanced rolling element bearing prognostics *Adv. Eng. Inform.* **17** 127–40
[24] Nectoux P, Gouriveau R, Medjaher K, Ramasso E, Morello B, Zerhouni N and Varnier C 2012 PRONOSTIA: An experimental platform for bearings accelerated life test *Proc. IEEE Int. Conf. on Prognostics and Health Management* pp.1–8

Acknowledgments

This work was financially supported by the Science and Technology Research Program of Chongqing Municipal Education Commission (Grant Nos. KJ1704109 and KJ1704087), the Chongqing Research Program of Basic Research and Frontier Technology (Grant Nos. cstc2015jcyjBX0066 and cstc2017jcyjAX0005).

The research on construction and application of machining process knowledge base

Tan Zhao, Lihong Qiao, Yifan Qie and Kai Guo

School of Mechanical Engineering and Automation, Beihang University, 37 Xueyuan Road, Haidian District, Beijing 100191, China

E-mail: lhqiao@buaa.edu.cn

Abstract. In order to realize the application of knowledge in machining process design, from the perspective of knowledge in the application of computer aided process planning(CAPP), a hierarchical structure of knowledge classification is established according to the characteristics of mechanical engineering field. The expression of machining process knowledge is structured by means of production rules and the object-oriented methods. Three kinds of knowledge base models are constructed according to the representation of machining process knowledge. In this paper, the definition and classification of machining process knowledge, knowledge model, and the application flow of the process design based on the knowledge base are given, and the main steps of the design decision of the machine tool are carried out as an application by using the knowledge base.

1. Introduction

Process knowledge is an important intellectual capital of the manufacturing industry. It is a process-related knowledge that runs through the product's full life cycle in the product design, production preparation, manufacturing and operation and management activities of the enterprise[1]. In the face of complex and diverse knowledge of enterprises, the traditional process knowledge resource management has not adapted to the development of modern enterprises, which has affected the accumulation, communication, use and innovation of enterprise process knowledge[2]. In recent years, knowledge base construction technology developed rapidly, many scholars, experts have put forward new research results[3-5]. From the current construction and expression of process knowledge, the process knowledge in the knowledge base and knowledge expression model has no uniform standards, and the process of knowledge representation and knowledge model is the key to decide whether the knowledge can be shared and reused [6-8]. Based on the analysis of the knowledge of machining process, this paper classifies the knowledge of machining process and studies the expression of machining process knowledge in knowledge base. The process knowledge model, including the machining scheme, the machining processes, the manufacturing resource and the cutting parameter, is established, which is beneficial to the reuse and sharing of the machining knowledge in the enterprise.

2. Definition and Classification of Machining Process Knowledge

Machining process knowledge refers to all the process knowledge involved in the process of forming blanks into products[9]. The classification of machining process knowledge can be described from different levels. According to the product process, the machining process knowledge can be divided into process level knowledge, working procedure level knowledge and step level knowledge. The process level knowledge includes process route knowledge, processing method selection knowledge, mold and blank design knowledge and so on. The knowledge of working procedure level includes the selection knowledge of machine tools, the mode of workpiece clamping, step design knowledge, process inspection knowledge and other knowledge. The knowledge of step level includes machining margin determination, cutting parameter knowledge and tool selection knowledge, measuring tool usage knowledge, etc[10]. According to the application mode of knowledge in machining process design, machining process knowledge can be divided into three categories. They are process rule knowledge, manufacturing resource knowledge and cutting parameter knowledge. In this paper, the knowledge of machining process will be discussed according to the application pattern division.

Process rule knowledge is used to express the corresponding relationship between part feature information and process decision method. It can support intelligent process decision. This kind of knowledge includes the processing scheme knowledge, processing method knowledge and process constraint knowledge. Process constraint knowledge includes sequence knowledge in the processing phase, generally follow the first face after the hole, the first base after the other, the first rough after the essence of the principle.

Manufacturing resource knowledge refers to the selection of tools and equipment used in the process of manufacturing, including the selection of common manufacturing resources, such as machine tools, tools, clamps and measuring tools.

Cutting parameter knowledge refers to the relationship between cutting tools, parts design information, machining methods, machine tools and cutting parameters, which can be used to guide the knowledge of cutting parameter intelligent decision making.

3. Representation of machining process knowledge

Knowledge representation is the process of expressing knowledge as a data structure that computer can obtain and process. Common knowledge expression methods include heuristic rule knowledge representation, semantic network representation, predicate logic representation, object-oriented representation, artificial neural network representation and ontology representation, etc. Each kind of knowledge expression method has its superiority as well as the certain limitation. Different kind of knowledge representation method are used in this paper in view of the application characteristic for the three kinds of knowledge mentioned above, by which the complication of simple problems as well as the incomplete definition of complex problems could be avoided simultaneously.

An important part of the intelligent process decision is to determine the processing chain, namely, the machining scheme, for each machining feature of the parts. Each machining scheme contains one or more machining methods. The knowledge expression of process rule includes the expression knowledge for the selection of machining scheme and machining method, and the expression knowledge for clustering rules and sequence rules considered in process sequencing. The factors that affect the selection of machining methods include the workpiece material, the size and accuracy of the machining features, surface roughness, hardness, the structure of the part, and production lot sizes and so forth.

The heuristic representation of knowledge index is adopted to express machining scheme knowledge. The index structure are used as the IF entry while the knowledge content as the THEN entry. Taking the machining method selection of a hole feature as an example, the knowledge index instance can be expressed in the form of IF-THEN as follows:

IF "Feature type = HOLE" AND "Machining accuracy (IT) = 11" AND "Surface roughness (Ra)= 12.5" AND "Workpiece material = CAST IRON" AND "Heat treatment type = QUENCHING" AND "Batch = LARGE",

THEN "Machining method = DRILLING".

The characteristics of the wide variety of enterprise manufacturing resources are the difference of parameters that describes resource capabilities. Therefore, object-oriented representation are involved in the process of knowledge representation of manufacturing resources. The basic attributes of each layer are defined in terms of the hierarchical structure of the manufacturing resources. And as a result, the storage management of manufacturing resources could be facilitated efficiently. Taking machine tools for example, machine tools are divided into machining centers, CNC machine tools and other sub classes. Sub class can further be divided, such as horizontal machining centers. Each class layer has its corresponding attribute information. For instance, the knowledge expression form of machine tools is as follows:

Object class:: =<class>[numerical control machine tool]
Object attributes:: =<attribute>[name]
Object attributes:: =<attribute>[detailed description]
Object sub-class:: =<sub-class>[CNC milling planer]
Object sub- attributes:: =<sub- attribute>[machine tool version]
Object sub- attributes:: =<sub- attribute>[spindle speed]
Object sub- attributes:: =<sub- attribute>[power]

The selection of cutting parameters needs to meet the constraints of positioning ability, clamping ability, machining capacity and workpiece quality requirements that are provided by the process and the system. In the representation of cutting parameters knowledge, the heuristic representation of knowledge index is adopted. Five factors, including machine tool, cutting tool, feature type, workpiece material and process strategy, should be considered comprehensively.

4. Construction of knowledge model for machining process

The knowledge base eventually stores structured knowledge in the form of a database. However, the difference from the traditional database is that the knowledge base contains not only a large number of simple facts, but rules and process knowledge as well. In the model construction of process rules knowledge, according to the heuristic rule knowledge representation, the reference result of machining method and machining scheme knowledge is taken as the content of process decision. Figure 1 shows a typical knowledge model of machining scheme and machining method. This knowledge model includes heat treatment method, production batch, material, roughness, dimension precision, feature type and so on.

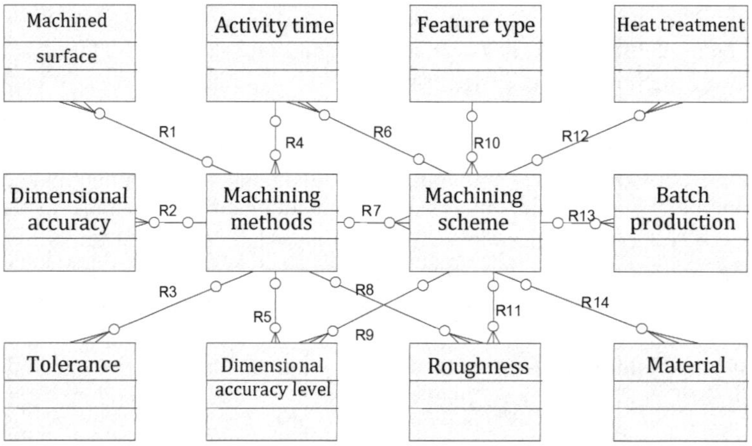

Figure 1. Knowledge model of machining scheme and machining method.

The knowledge model of manufacturing resource is built according to the object-oriented knowledge representation, as shown in figure 2. The manufacturing resource knowledge model describes the relationship between the specific types of manufacturing resources and the factors that influence manufacturing resource decisions as a bridge linking manufacturing resource types and manufacturing resource models.

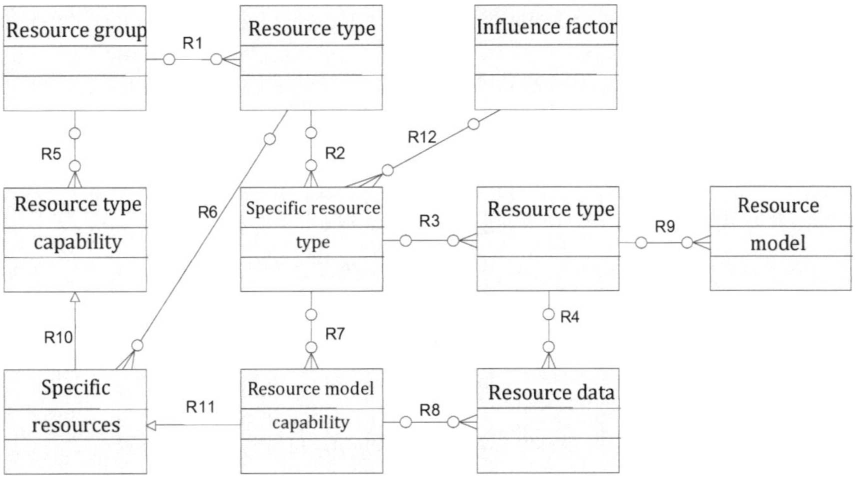

Figure 2. Knowledge model of manufacturing resource.

The knowledge model of cutting parameters mainly considers five factors: machine tool, cutter, feature, workpiece material and processing strategy. The cutting parameter knowledge model is shown in figure 3. Based on the hierarchical classification of manufacturing resource knowledge model, cutting parameters model establishes the knowledge of cutting parameters based on the tool type as the link. At the same time, the model describes the relationship among tools, parts design information, machining methods, machine tools and cutting parameters, which can be used for intelligent decision of cutting parameters.

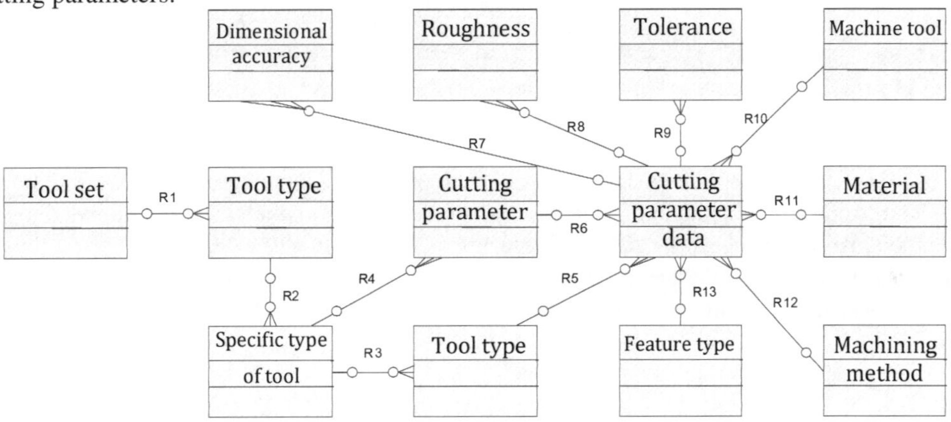

Figure 3. Knowledge model of cutting parameters.

5. Application of machining process knowledge base

This paper constructs the knowledge base as the starting point for the application, combined with machining process design system for rapid process planning. In the machining process design system, four knowledge bases are designed according to the knowledge model, which are machining scheme, machining method, manufacturing resource and cutting parameter. Taking the machining process design of the connecting parts as an example, the system interface and application flow are shown in Figure 4. And the decision-making process is as follows:

Step 1: 3D manufacturing feature information is extracted from the part. The feasible machining scheme set is generated and optimized to get the processing chain through querying the process rule knowledge base.

Step 2: The machining scheme is decomposed into a collection of machining methods. By making manufacturing resource knowledge and cutting parameter knowledge reasoning, the decision-making of machine tools, cutting tools and cutting parameters are carried out in turn, and the manufacturing resources and cutting parameters required by the machining method are obtained.

Step 3: Through the reasoning of the process rule knowledge base, the steps are optimized according to the process constraint rules. The process design results are generated.

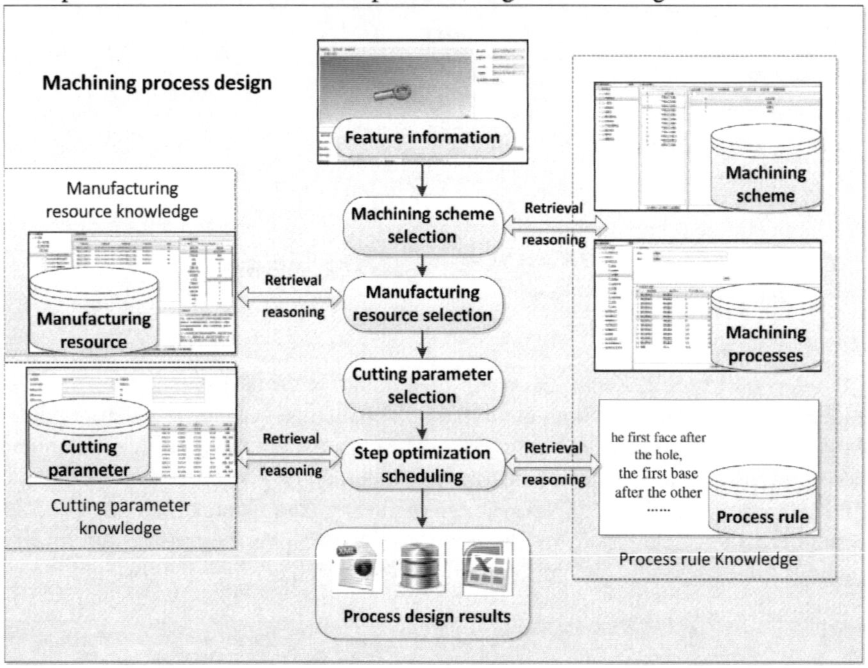

Figure 4. The process of machining process design based on knowledge base.

6. Conclusion

Based on the knowledge application, this paper analyzes the knowledge of machining process, establishes the knowledge model of machining process, and develops the knowledge base management system which meets the needs of enterprises. The knowledge base can support the rapid machining process design, facilitate the management of process data, adapt to the development of modern enterprises, and promote the accumulation, exchange, use and innovation of enterprise process knowledge.

References

[1] Lu H L and Chu Z B 2012 *Microprocessors.* **33** 90-92,95
[2] Peng X 2011 *Logist. Eng. Manage.* **33** 164-66
[3] Koyama K, Hara N and Ohsugi I J 2010 *Int. J. Adv. Intell. Para.* **2** 49-63
[4] Ken L D 2011 *J. Inf. Process. Manage.* **54** 243-53
[5] Niu F Q, Liu Y H and Gao Y 2009 *Agric. Network. Inf.* **2** 112-17
[6] Gao Y and Qiao L H 2014 *Aeronau. Manuf. Technol.* **16** 23-26
[7] Qiao L H and Dong W 2010 *Avi. Preci. Manuf. Technol.* **2** 45-49
[8] Gong Y Q and Lu M 2008 *Comput. Eng. Appl.* **26** 208-10
[9] Wu J R 2016 *Mod . Manuf. Technol. Equip.* **4** 122-24
[10] Guo C F 2011 *Coal. Technol.* **30** 3-5

Acknowledgments

This work is supported by research grant A0320131002. The authors would also like to thank Beijing Municipal Education Commission (Build a Project) for its support.

IOP Conf. Series: Materials Science and Engineering **339** (2018) 012028 doi:10.1088/1757-899X/339/1/012028

Non-contact control of the working condition of mechanical units of the steam compressor for desalination plant

A. I. Danilin, A.Zh. Chernyavsky, S.A. Danilin, V.V. Neverov, D.A. Voroh, E.V. Blagin

Samara National Research University, Samara, Russia

Email: evgenyblagin@gmail.com

Abstract. New methods and means for monitoring working condition of the rotating elements of steam compressor unit such as blade ring of the impeller and gears of multiplier are considered. Blade control is carried out by the signalling device of pre-emergency deformation of impeller blades. Control of the gears condition is carried out by apparatus system which allows to analyse change of the signal form caused by the gears wear. Influence of the wear types on the typical information parameters of the analysed signals is described. Technical characteristics of the devices and experimental research results are presented. Described control systems allow to detect deviations equal to 1-2% from initial condition. Application of such systems gives the opportunity to improve fault diagnosis and maintenance in 2-3 times.

1. Introduction

In modern desalination plants steam compressors are often used due to their high economic efficiency [1]. Steam compressor represents difficult mechanical unit which includes rotor with blade unit (impeller) and multiplier – assembly of gear mechanisms for conversion of rotary motion parameters. Such elements of dynamic systems must met enhanced requirements to manufacturing quality and the quality of the exploitation condition control. Today there are many publications directed at optimization of different compressors operation regimes. Some of them are researches of Samara University scientists [2,3], representing features of compressor working process optimization. Besides, suggested method of non-contact control of blades working conditions and turbomachinery elements fault diagnosis is successfully distributed on all blade elements of rotary machines – steam turbines [4], gas-turbine aircraft and ship engines [5-10] as well as different industrial fans. Active development of this research field in all industrial countries is connected with durability of most loaded units of energy plants, their lifespan and development of the criteria which determine exploitation of energy plants by their technical conditions.

2. Control of the working condition of the steam compressor impeller blade

Today, control of the working condition of the steam compressor impeller and shaft is executed by visual observation by the means of technical endoscope, vortex and magnet fault detectors. This is very hard work which requires professional skills and responsibility from technical staff. Despite the accepted actions statistics shows that there are emergency situations caused by shaft and impeller blades destruction (fig. 1).

In fact, impeller blade form deviations (gallings, blade bending) from calculated variant causes decrease of the steam compressor efficiency and, finally, causes steam compressor rotor restruction which restoration requires significant material and time costs. Thus, diagnosis of the fault condition of the steam compressor rotor is actual problem today.

Figure 1. Photo of the destroyed impeller blades

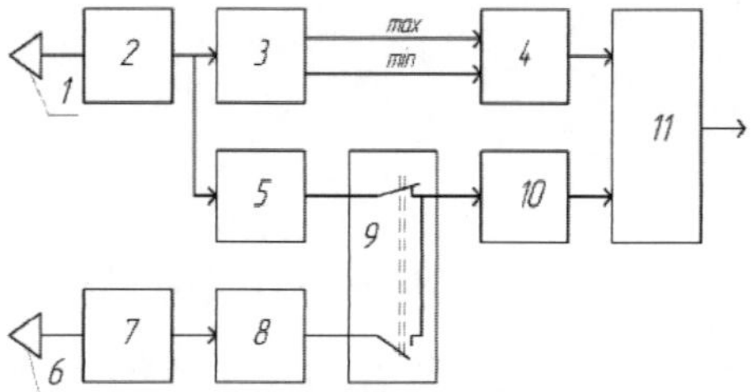

Figure 2. The block diagram of the signaling device for the pre-emergency state of the blade assembly of the steam compressor

As analysis of scientific literature shows that along developed methods and means of diagnosis and control of rotary units elements on exploitation steam generators most promising method is discrete-phase method (DPM), which allows to determine individual deformation condition of every impeller blade and change of the gaps between inner surface of the steam compressor shell and face part of the blade ring [11]. Acquired information is compared with geometric location of exact blade in the wheel in certain moments of time and corresponding interpretation of acquired values in the field of mechanical stresses and deformations.

Based on theoretical prerequisites of DPM, structural scheme of the signaling device of pre-emergency condition of the steam compressor impeller [12], presented on fig. 2. Structural scheme is based on principle of current time intervals (steps) measurement between faces of rotary blades, determination of the maximal and minimal values on every step and comparing this value with impeller-average step on every turn of the steam compressor rotor.

Corresponding to fig. 2., structural scheme is functioning as described below. Peripheral sensor 1, mounted in steam compressor shell above the blade face motion trajectory generates electric signal which is transformed by former 2 in rectangular impulse. Time intervals τi between rectangular impulses, corresponding to circular step between blades is converted to numerical code in block 3. Maximal and minimal values for every step are also determined in this block from N current rotor turns. Acquired extreme values of every step are transferred to block 4, where averaged value of τsi is determined by expression τsi =(τimax + τimin)/2 for every step.

Besides, rectangular impulses from former 2 are transferred to block 5 where time steps τi are summarized for N rotor turns and the average one for N turns is calculated as $T_s = \frac{1}{N}\sum_{j=1}^{N}\sum_{i=1}^{K}\tau_i$.

If design realization of steam compressor allows to mount sensor 6 of the rotor turn mark, its electric signals by former 7 are converted to rectangular signals whish are transferred to block 8 where average period of turning is determined as $T_s = \frac{1}{N}\sum_{j=1}^{N} T_j$.

Block of switches 9 allows to provide required operation regime of the device for both cases when turn mark is present or not. Acquired values of Ts in block 10 are divided by number of blade in the impeller, so average step along the impeller for N rotor turns is determined as τs= Ts /K.

Further, values of the τsi and τs, determined as described above, are compared between each other in comparison block 11. If τs differs from τsi on certain limiting value p, block 11 generates signal about fault occurrence in one or several blades.

On the basis of suggested algorithm and structural scheme signaling device of pre-emergency deformation of steam compressor blade unit (SCBU) is developed for steam compressors operating in desalination plants. External view and functional composition of SCBU is presented on figure 3. It consists of impulse sensor, mounted in steam compressor shell, measuring block with USB interface for connection to computer, patch cables with protection from noisy signal, special software.

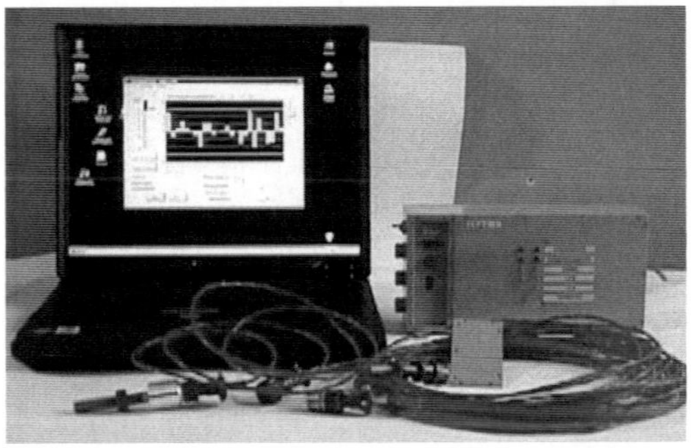

Figure 3. The appearance and functional composition of the SCBU

Signal of fault occurrence in one or several blades in form of alarm light and sound signals is transferred to a dispatcher console. If blade fault continues to develop, SCBU forms a signal for a system of automatic unit control for normal (or emergency) stop of the steam compressor.

There is a possibility to connect SCBU with personal computer via USB interface for control of the current condition of the blades, express-analysis of their operating ability, remembering their exploitation condition, creation of the technical passport of the blades. For this purpose special software was developed which allows to maintain on-line exploitation diagnosis of the blades. Besides, SCBU provides autonomous inner memory device which turns on after pre-emergency signal occurrence and writes dynamic of the blade fault development. Inner memory storage is energy independent and writes date, time and deformation condition for every blade of the impeller. SCBU is developed and realized on modern radio-technical element base which allows to obtain compact and reliable device with double reserving and current adaptation for any type of steam compressor and impeller with different number of blades.

Technical characteristics of SCBU:

1. Range of the steam compressor rotor rotation frequencies: 100 ... 35000 r/min.
2. Range of the operation temperatures for impulse sensors:

- option 1 -40 ... +150 °C; - option 2 -40 ... +1000 °C.

3. Range of the operation temperatures for measurement block -40 ... +50 °C.
4. Current supply for SCBU-1 is direct current with voltage +27 V.

Laboratory tests of the SCBU were carried out when fault in the form of blade face deviation was introduced in the one of the compressor blade. Fig. 4 shows SCBU operator program where deviations of every blade are shown including fault one up to 6 mm. Natural technological deviation of every other blade lies within the limits of ± 0,5 mm. On the base of the carried out experimental researches, recommendations for a choice of limitation values, corresponding to pre-emergency (2 mm) and emergency (2.5 mm) deviation of the steam compressor faces were developed.

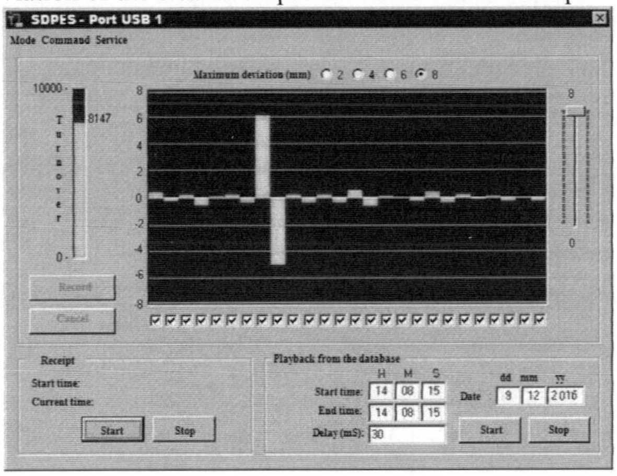

Figure 4. The screenshot of the operator's program of the SCBU with a defective blade

3. Control of the multiplier gear mechanism wear

Working lifespan of the gear mechanism determine the coefficient of the mechanism operation fail-safety which is in turn one of the critical elements of failure tree. Failure of the mechanical equipment due to wear or destruction of gear causes long fault time and requires large costs to restore its operating ability. Multipliers used in steam compressors of desalination plants, must met high requirements for reliability, precision of the gear mechanisms manufacturing and economic efficiency during exploitation. Such multipliers provide increase of the rotation frequency up to 35000 r/min. In the same time gears, used in given mechanism suffer colossal loads which causes their accelerated wear and frequent substitution. That's why control and diagnosis of the gears condition in multipliers during their operation, i.e. in exploitation regime, is a very actual problem.

Today, diagnosis of the faults, wear and integrity of the gears is usually carried out in static condition. In dynamic regime, spectral method of vibration signal analysis has found a wide use. These signals are received from the controlled mechanism. But spectral method of diagnosis has several serious disadvantages. Diagnosis of the signal envelope specter is quite difficult for physical explanation of gear teeth catching and it usually require good theoretical preparation of the vibration diagnostician or presence of the specialized computer expert system in his technical equipment [13].

Figure 5. Appearance of the experimental stand

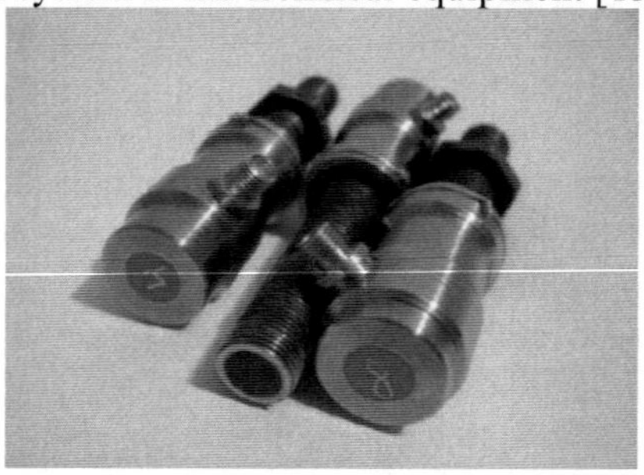

Figure 6. The design of the primary microwave converter

Non-contact radio wave method [14,15] was suggested for exploitation control and working condition diagnosis. This method is based on the on-line treatment of the signals, received after detection of the microwave radiation reflected from the gear teeth. Parameters of information signals are compared correspondingly with reference signal received at gear exploitation start. This method allows to obtain information directly about wearing degree of every exact tooth and fault occurrence in it in dynamic regime. Parameters of monitoring signals are determined by geometric parameters of the teeth and technological configuration of thee control zone. Microwave radiation can exist in oil media where gears operate. Advantages of this method are lack of high number of sensors and necessity of their precise adjustment. There is also lack of the necessity of frequent maintenance of sensor operating in quite harsh environment.

Several types of the tooth destruction are separated: tooth crash, spalling, teeth faces damaging, abrasive wear, crack occurrence, peeling and deep contact destruction of the material [13]. Several information parameters are separated during diagnosis from the signal of reflected monitoring signal converted to electric signal. Only teeth crack cannot be diagnosed by aforementioned method if they are not located on the researched area. Experimental unit was developed to carry out the tests and reference gear was manufactured with given tooth geometry. Upper part of the unit has an possibility to mount and regulate location of the modified wave emitter.

Results of theoretical and experimental researches show that received reflected signal after detection has a quasi-bell form of impulse [15]. Estimation of the steam compressor gear wear degree is based on information parameters which include signal amplitude, normalized duration of signal front, lack of signal. Every type of wear influences certain information parameters separated from reflected signal. For example, lack of signal means tooth crash. Teeth spalling influences signal amplitude and normalized signal duration. Abrasive teeth wear predetermines such information parameters as normalized duration of the signal, signal amplitude and normalized duration of the signal fronts.

Modified wave-guide emitter represents waveguide-coaxial transition and wave-guide with dielectric filling, for example, as quartz glass. Modified wave-guide emitter is screwed directly in the shell of multiplier system perpendicularly to pinion ring of controlled gear mechanism. Application of dielectric filling allows to decrease wave-guide dimension at constant wave length. Dielectric filling also helps to avoid oil splash inside the wave-guide and carbon deposit occurrence. Design of such primary converter is presented on figure 6. The proposed system for monitoring the working state of the gear wheels of energy-loaded multipliers can be used for diagnostics in operation in any dynamic modes. Ultimately, the proposed system allows to automate the control process, objectively evaluate and record its current operating state, and reduce the number of device preparations. The monitoring system generates a pre-emergency signal of the torque multiplier and thereby significantly reduces the probability of failure of a complex and responsible mechanical system and ensures operation of the torque multiplier of the steam compressor based on its actual technical state.

4. Conclusions

Suggested method allow to maintain non-contact control of working condition of some elements of the steam compressor for desalination plant. These elements include impeller of the compressor itself and gear mechanism for the multiplier. Both methods assume utilization of specially developed software. Described control systems allow to detect deviations equal to 1-2% from initial condition. Application of such systems gives the opportunity to improve fault diagnosis and maintenance in 2-3 times.

References

[1] Patent RU 2280011 Russian Federation

[2] Matveev V N, Baturin O V, Popov G M 2014. *Engineering Optimization IV - Proceedings of the 4th International Conference on Engineering Optimization* 821–826.

[3] Goryachkin E., Popov G., Baturin O., Kolmakova D. 2015 *ASME Turbo Expo* **2C** GT2015-43384

[4] Danilin A I, Adamov S I, Chernyavskii A Zh, Serpokrylov M I. 2007 *Power Technology and Engineering* **41(5)** pp. 295-301.

[5] Procházka P, Vaněk F 2012 *Proceedings of ISMA2012-USD2012* pp 3359-3371.

[6] Zielinski M, Ziller G 2005 *Proceedings of the International Symposium of Air Breathing Engines (ISABE '05).*

[7] von Flotow A, Mercadal M, Tappert P. 2000 *Proceedings of the 2000 IEEE Aerospace Conference* **6** pp 433–440.

[8] Carrington I B, Wright J R, Cooper J E, Dimitriadis G. 2001 *J. Aerospace Engineering* **215** pp 301-312.

[9] Zielinski M, Ziller G 2000 *Measurement Science and Technology* **11(7)** pp 847-856.

[10] Heath S, Imregun M. 1998 *J. Eng. Gas Turbines Power* **120(4)** pp 784-791.

[11] Danilin A I 2008 *Contactless measurement of the deformation parameters of the blades in the systems of monitoring and control of turbines* [in Russian] Samara University press p. 217

[12] Patent RU 2177145

[13] Rusov V I 2012 *Diagnostics of defects of rotating equipment on vibration signals* [in Russian] Perm university press p. 200

[14] Patent RU 2602488

[15] Danilin A I, Neverov V V 2016 *Proceedings of the SSC RAS* **18(4(6))** pp. 1265-1267

Acknowledgments

This work was carried out by lead performer of research and development effort with financial support of Russian Federation Ministry of Science and Education with the realization of Government decree №218 from 09.04.2010 according to contract about providing and using subsidy № 02.G25.31.0150 from 01.12.2015.

Numerical Simulation Analysis of High-precision Dispensing Needles for Solid-liquid Two-phase Grinding

Junye Li, Jinglei Hu, Binyu Wang, Liang Sheng, Xinming Zhang
College of Mechanical and Electric Engineering, Changchun University of Science and Technology, Changchun, China

E-mail: ljy@cust.edu.cn

Abstract. In order to investigate the effect of abrasive flow polishing surface variable diameter pipe parts, with high precision dispensing needles as the research object, the numerical simulation of the process of polishing high precision dispensing needle was carried out. Analysis of different volume fraction conditions, the distribution of the dynamic pressure and the turbulence viscosity of the abrasive flow field in the high precision dispensing needle, through comparative analysis, the effectiveness of the abrasive grain polishing high precision dispensing needle was studied, controlling the volume fraction of silicon carbide can change the viscosity characteristics of the abrasive flow during the polishing process, so that the polishing quality of the abrasive grains can be controlled.

1. Introduction

Abrasive Flow Machining (referred to as AFM), also known as extrusion grinding or extrusion honing, mainly used for polishing shaped holes and complex cavities and other traditional methods difficult to process the site, its principle is to help the pressure provided by the machine, the semi-solid processing medium containing high hardness abrasive grains is extruded through the inner and outer surfaces, edges and the channels of the workpiece, so as to achieve a certain processing purposes.

Abrasive grain processing technology is the last century developed by the United States, initially used in the aerospace field of complex geometric shape of the alloy workpiece deburring processing, after decades of development, has been applied in a number of areas. Domestic and foreign scholars related to its polishing method and polishing mechanism for in-depth study, And made a lot of results[1,2].

Sushil M et al. on the silicon carbide composite materials for abrasive grain polishing test study, the effects of processing parameters on the removal rate of the material, the change of surface roughness and surface morphology were observed, through the final experimental analysis, it is concluded that the squeezing force is the most significant factor affecting the roughness[3]. Wan S et al. Analyzed the effects of non-linear viscoelasticity and temperature on the viscosity of abrasive media, a non-elastic non-Newtonian model is used to obtain the shear dilution of the abrasive medium in the flow state, and the establishment of wall slip model, the wall shear rate function is used to calculate the relative motion between the abrasive medium and the wall[4]. Ding Jinfu et al. Analyzed the polishing mechanism of abrasive grain flow from the perspective of stress tensor, and according to Bowden theory, when the abrasive grains adhere to the surface of the workpiece, with the carrier flow forward, in the workpiece surface slip phenomenon, so that uneven surface of the workpiece plastic deformation; In addition, the abrasive particles in the radial force, will be pressed into the workpiece surface, produce plowing, so as to achieve the purpose of removing the material[5]. Zhao Jia et al. use the viscometer to get the relationship between the viscosity of the abrasive, based on the Rabinowicz single particle cutting model, the wear relationship of abrasive viscosity and temperature was obtained. The effects of abrasive viscosity on the metal wear were studied[6].

With the development of various dispensing processes, high-precision dispensing needle used in the LCD panel frame coating, vibration package, CCD sealant coating, BGA package coating range. Therefore, the accuracy of dispensing needles are getting higher and higher, high-precision dispensing needle surface quality directly affect the use of dispensing performance, when the high-precision dispensing needle hole on the inner wall of the existence of burr phenomenon, not smooth or residual pollutants, it is easy to cause dispensing uneven or lead to easy to plug the needle, so improve the surface quality of high-precision dispensing needles, it is very important to study the precision finishing of high precision dispensing needles[7-11].

2. Numerical Simulation of High Precision Dispensing Needles for Polishing Gravel

In this paper, high precision dispensing needle as the research object, The surface processing characteristics of the solid - liquid two - phase abrasive grains were numerically simulated. In the case of ensuring that the inlet pressure and the outlet pressure are constant, the effects of the volume fraction of silicon carbide particles on the dynamic pressure and turbulence viscosity of high precision dispensing needle channel were studied. As shown in Fig. 1 for the high-precision dispensing needle three-dimensional solid cross-section model, wherein the diameter of the small hole is 0.2 mm, big hole diameter of 2mm.

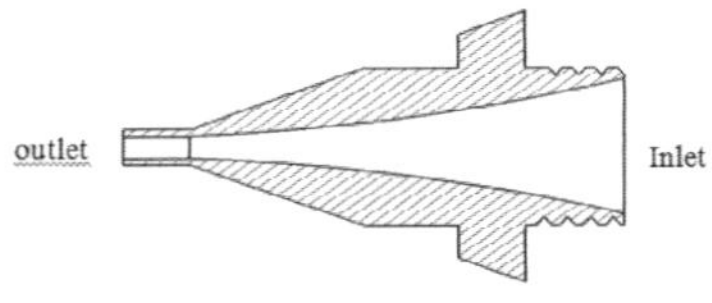

Figure 1. high-precision dispensing needle three-dimensional solid cross-section model

After the non-structured hexahedral mesh is divided into high precision dispensing needle model, using the import and export pressure boundary conditions for numerical simulation, according to the actual situation, set the inlet pressure to 6MPa, the volume fraction of silicon carbide is set to 0.1, 0.15, 0.2 and 0.25 for numerical simulation, the simulation results of the multi - physics coupling field under four volume fraction are obtained. Fig. 2 shows the dynamic pressure cloud image at different volume fractions.

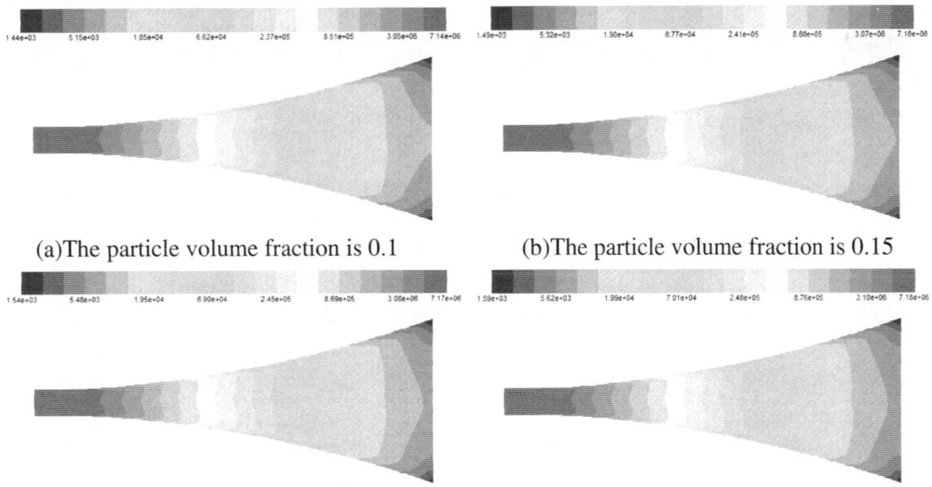

(a)The particle volume fraction is 0.1 (b)The particle volume fraction is 0.15

(c)The particle volume fraction is 0.2 (d)The particle volume fraction is 0.25

Figure 2. Dynamic pressure cloud diagram under different particle volume fraction

As can be seen from Fig. 2, from the entrance to the exit, dynamic pressure gradually increased, layered more obvious, the maximum pressure in the small hole, that is, abrasive flow on the hole grinding the most serious, polished the best.

Select the entry section as data 1 area, cross hole as data 2 area, small hole as data 3 area, the numerical values of the dynamic pressure values listed in the three data areas are shown in Table 1.

Table 1 different abrasive volume fraction of high precision dispensing needle dynamic pressure data distribution table

Particle volume fraction	Dynamic pressure (10^5MPa)		
	Data 1 area	Data 2 area	Data 3 area
0.1	0.184	8.54	71.4
0.15	0.190	8.60	71.6
0.2	0.195	8.69	71.7
0.25	0.199	8.76	71.8

Analysis of Table 1 data, can be derived from different abrasive volume fraction under the pressure changes: (1)First, the change of the dynamic pressure under the same particle volume fraction is analyzed, can be analyzed by the table to see the number of 3 areas> data 2 area> data 1 area, abrasive into the workpiece cavity, the minimum pressure at the abrasive inlet, with the inflow of abrasive, dynamic pressure gradually increased, dynamic pressure distribution of the distribution of more obvious, this is because the workpiece from the entrance to the hole of the cavity diameter is more smooth, no special mutation caliber, the pressure at the cross hole of the workpiece increases significantly, indicating that the abrasive at this time on the cross hole have a greater grinding effect, That is, the workpiece rounded; While the dynamic pressure in the small hole to reach the maximum, at this point the abrasive movement is more intense, the hole on the inner wall of the best polishing effect. (2) Second, the analysis of different abrasive volume fraction under the pressure of the situation, from the table analysis, with the abrasive grain volume fraction gradually increased, the whole workpiece flow in the dynamic pressure has increased, this is because the abrasive concentration increases, abrasive and the workpiece surface contact area increases, the grinding effect on the wall of the workpiece increases, thereby improving the abrasive flow on the workpiece surface polishing effect.

Under the same initial conditions, the numerical simulation of the turbulence viscosity in the grinding process of the workpiece under different volume fraction is carried out. The turbulence viscosity cloud images with different particle volume fraction are shown in Fig. 3.

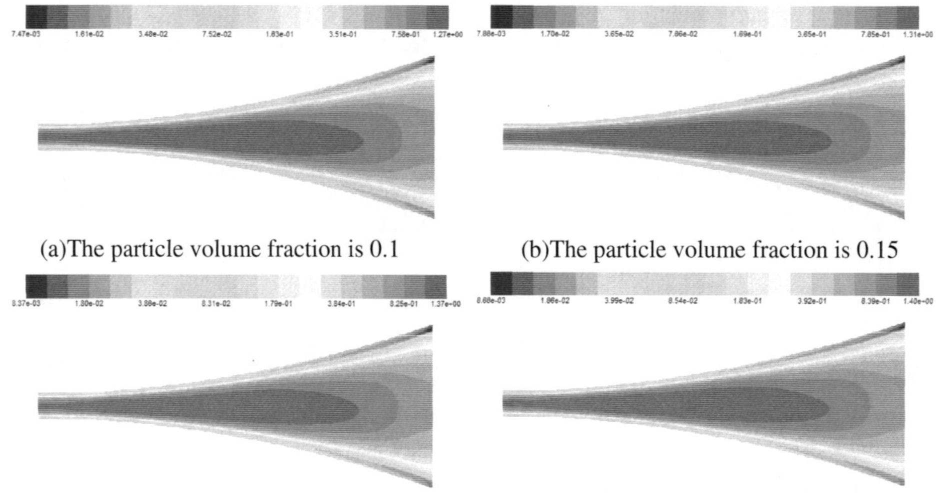

(a)The particle volume fraction is 0.1 (b)The particle volume fraction is 0.15

(c)The particle volume fraction is 0.2 (d)The particle volume fraction is 0.25

Figure 3. Cloud turbulence viscosities under different volume fraction

It can be seen from the turbulent viscosity cloud diagram under different volume fraction of Fig.3,due to the movement of viscoelastic silicon carbide abrasives in turbulent state, under the same inlet pressure boundary conditions, with the increase of the abrasive volume fraction, turbulence viscosity increases, due to the velocity gradient and viscoelasticity of the fluid, will inevitably lead to abrasive grain on the workpiece surface friction, so the abrasive particle volume fraction increases, the turbulence viscosity increases, thus improving the polishing effect of the workpiece.

IOP Conf. Series: Materials Science and Engineering **339** (2018) 012029 doi:10.1088/1757-899X/339/1/012029

3. Conclusion

Through the numerical analysis of the simulation results of the high precision dispensing needles, under certain pressure and abrasive volume fraction, the effect of volume fraction of silicon carbide on the processing effect was simulated by changing the effect of solid - liquid two - phase abrasive grains. The changes of dynamic pressure and turbulent viscosity in the process are analyzed, can get the greater the dynamic pressure, the greater the turbulence viscosity of the grinding effect has a significant impact, which can improve the polishing effect of abrasive grain flow.

References

[1] Shiming Ji, Chen Li, Dapeng Tan, et al. Study on Machinability of Softness Abrasive Flow Based on Preston Equation. Journal of Mechanical Engineering, Vol.47, Issue 17,2011, p.156-163.

[2] Yuanqiang Tan, Yi Li, Yong Sheng. On the Model and Pressure Simulation of Solid- fluid Two Phase Flow for Abrasive Flow Machining. China Mechanical Engineering, Vol.19, Issue 4,2008, p.439-441.

[3] Sushil M, Vinod K, Harmesh K. Experimental Investigation and Optimization of Process Parameters of Al/SiC MMCs Finished by Abrasive Flow Machining. Materials and Manufacturing Processes, Vol.30, Issue 7,2015, p.902-911.

[4] Wan S, Ang Y J, Sato T, et al. Process modeling and CFD simulation of two-way abrasive flow machining. International Journal of Advanced Manufacturing Technology, Vol.71, Issue 5-8,2014, p.1077-1086.

[5] Jinfu Ding, Runzhi Liu, Kehua Zhang, et al. Micro cutting mechanism of abrasive flow precision machining. Optics and Precision Engineering, Vol.22, Issue 12,2014, p.3324-3331.

[6] Jia Zhao, Liang Fang, Kun Sun, et al. Effect of suspension viscosity temperature characteristic on tribological properties of abrasive flow machining process, Vol.28, Issue 2,2008, p.173-177.

[7] Junye Li, Jinglei Hu, Xinming Zhang, et al. The research of polishing nozzle quality based on discrete element method. Journal of Measurements in Engineering. Vol.5, Issue 1,2017, p.29-39.

[8] Junye Li, Jinglei Hu, Ningning Su, et al. Numerical Analysis in Viscosity-Temperature Characteristics of Solid-Liquid Two-Phase Abrasive Flow Polishing. Journal of Measurements in Engineering. Vol.5, Issue 3,2017, p.115-124.

[9] Tzeng H J, Yan B H, Hsu R T, et al. Self-modulating abrasive medium and its application to abrasive flow machining for finishing micro channel surfaces. International Journal of Advanced Manufacturing Technology, Vol.32, Issue 11,2007, p.1163-1169.

[10] Li Junye, Liu Weina, Yang Lifeng, et al. Numerical simulation of the behavior of grape flow in the micro - hole of injector . Coal Mine Machinery, Vol.31, Issue 10,2010, p.56-58.

[11] Li Junye, Xu Ying, Yang Lifeng, et al. Experimental study on grinding flow of non - linear tube parts . China Mechanical Engineering, Vol.25, Issue 13,2014, p.1729-1733.

Acknowledgement

The authors would like to thank the national natural science foundation of china No. NSFC 51206011, Jilin province science and technology development program of Jilin province No.20160101270JC and No.20170204064GX, project of education department of jilin province NO.2016386.

IOP Conf. Series: Materials Science and Engineering **339** (2018) 012030 doi:10.1088/1757-899X/339/1/012030

The Light Plane Calibration Method of the Laser Welding Vision Monitoring System

B G Wang, M H Wu, W P Jia

Laboratory of Non-traditional Machining & Functional Material Preparing, College of Mechanical Engineering, Dalian University, Dalian 116622, People's Republic of China.

Email: 395839837@qq.com

Abstract. According to the aerospace and automobile industry, the sheet steels are the very important parts. In the recent years, laser welding technique had been used to weld the sheet steel part. The seam width between the two parts is usually less than 0.1mm. Because the error of the fixture fixed can't be eliminated, the welding parts quality can be greatly affected. In order to improve the welding quality, the line structured light is employed in the vision monitoring system to plan the welding path before welding. In order to improve the weld precision, the vision system is located on Z axis of the computer numerical control (CNC) tool. The planar pattern is placed on the X-Y plane of the CNC tool, and the structured light is projected on the planar pattern. The vision system stay at three different positions along the Z axis of the CNC tool, and the camera shoot the image of the planar pattern at every position. Using the calculated the sub-pixel center line of the structure light, the world coordinate of the center light line can be calculated. Thus, the structured light plane can be calculated by fitting the structured light line. Experiment result shows the effective of the proposed method.

1. Introduction

Because the laser welding has the advantages over the arc welding technique, such as the high welding speed, the smaller heat-affected zones, the formation of the high quality, laser welding technique has been researched and developed in the last decades[1, 2]. Nowadays the laser welding technique has been applied to weld the sheet steel parts in the aerospace, automobile industries. For improving the efficiency of the welding process, the robots have already been utilized in the laser welding process. Most of the robot used in the laser welding process are the "teach and playback", and lots of time should be spent in this kind of robot system to teach the welding path in advance.

Vision sensors have been employed as non-contact measurement equipment in the welding process because it can provide abundant information[3-7]. The vision information can be used to plan the welding path during welding and the welding path before welding. A number of studies have been done on controlling welding path by means of visual information. However, few researches are focused on the planning welding path before welding. Because of the seam width between two sheets steel parts are usually less than 0.1 mm, the error of fixture fixed will greatly affect the welding quality. In order to reduce the effect of the error of fixture fixed, the welding path planning before welding should be researched. In this paper, the method for calculating the light plane of the structured light used in the weld path planning before welding is studied.

2. Vision monitoring system

The structure of the vision monitoring system is shown in Figure 1. The vision monitoring system included a CCD camera, a band pass filter, and the low power laser. The low power laser and monochrome CCD camera are fixed on the laser head. The wavelength of low power laser is 450 nm. A spectral band-pass filter, which is sensitive to the wavelength from 440 nm to 460 nm, is installed in the front of the monochrome CCD camera for suppressing the light disturbance. The images acquired by the monochrome CCD camera are sent to the computer, where the feature of the structure light is performed. The resolution of the images are 1600×1200 pixels.

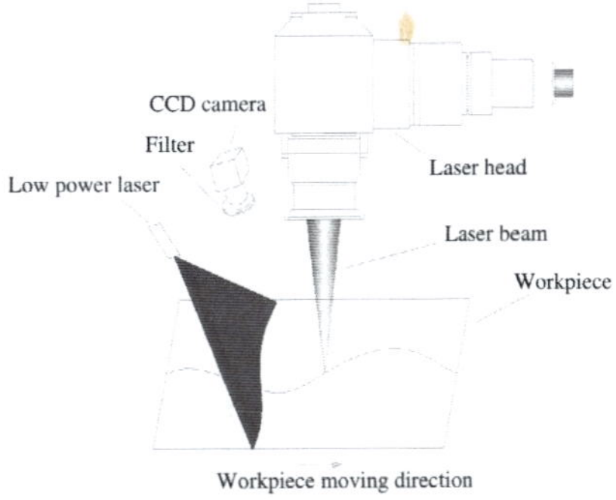

Figure 1. The vision monitoring system of laser welding

The vision monitoring system is setup on the laser head of three axis laser welding CNC machine. When calibrating the light plane of the line structured light, a planar checkerboard, which the size of each square in the grid is 20 mm, is located on the X-Y working platform. Furthermore, the planar checkerboard is vertical with the Z axis of the CNC machine. During the calibrating process, the laser head on the Z axis is control by the controller of the laser welding platform to move to three positions. The Z axis coordinates values of the three positions shown in the controller are 300 mm, 250 mm, and 200 mm respectively. At every position the monochrome CCD camera setup on the laser head shot two images of the planar checkerboard. The images of the planar checkerboard are shown in Figure 2, and the Z axis coordinates value of the CNC machine is 300mm. Thus, nine images be acquired. The Figure 2(c) is the image of the planar checkerboard shot by the monochrome CCD camera without band-pass filter in front of it. Therefore, the band pass filter which is setup in front of the monochrome CCD can eliminate the disturb light effectively.

(a) Image with structured light (b) Background (c) Image without the band-pass filter

Figure 2. The images of the planar checkerboard where the Z axis coordinate value of the CNC tool is 300 mm

3. Extraction of the light line center

It indicates in Figure 2(a) that the width of the light in the same image is not equal at different position, and in generally the light projected on the white grid is wider than the black grid. The structured light reflected by the black grid can be considered satisfying the Gaussian distribution. Therefore, Hessian matrix method can be used to acquiring the center line[8]. In order to reduce the influence of the outside light further, the background image Figure 2(b) is subtracted by the Figure 2(a). The image of the structured light is described by $f(x,y)$. The coefficient x and y is the horizontal ordinate and the longitudinal coordinates, respectively. The sub-pixel center line of the structured light can be calculated through the next three steps:

3.1. Calculate the Hessian matrix

The discrete two-dimensional Gaussian partial derivative kernels are given by

$$\begin{cases} g_{x,\sigma}(x_w,y_w)=g_\sigma(y_w)g_\sigma{}'(x_w) \\ g_{y,\sigma}(x_w,y_w)=g_\sigma{}'(y_w)g_\sigma(x_w) \\ g_{xx,\sigma}(x_w,y_w)=g_\sigma(y_w)g_\sigma{}''(x_w) \\ g_{xy,\sigma}(x_w,y_w)=g_\sigma{}'(y_w)g_\sigma{}'(x_w) \\ g_{yy,\sigma}(x_w,y_w)=g_\sigma{}''(y_w)g_\sigma(x_w) \end{cases} \tag{1}$$

Where, $\sigma \geq \frac{w}{\sqrt{3}}$; the w is the width of the structured light along the cross section;

$g_\sigma(x_w)=\frac{1}{\sqrt{2\pi}\sigma}e^{-\frac{x_w{}^2}{2\sigma^2}}$, $g_\sigma{}'(x_w)=\frac{-x_w}{\sqrt{2\pi}\sigma^3}e^{-\frac{x_w{}^2}{2\sigma^2}}$, $g_\sigma{}''(x_w)=\frac{x_w{}^2-\sigma^2}{\sqrt{2\pi}\sigma^5}e^{-\frac{x_w{}^2}{2\sigma^2}}$; $x_w \in [-w,w]$, $y_w \in [-w,w]$ 。

The line locally for each image point can be computed by convolving the image with the discrete two-dimensional Gaussian partial derivative kernels

$$\begin{cases} r_x=f(x,y)*g_{x,\sigma}(x_w,y_w) \\ r_y=f(x,y)*g_{y,\sigma}(x_w,y_w) \\ r_{xx}=f(x,y)*g_{xx,\sigma}(x_w,y_w) \\ r_{xy}=f(x,y)*g_{xy,\sigma}(x_w,y_w) \\ r_{yy}=f(x,y)*g_{yy,\sigma}(x_w,y_w) \end{cases} \tag{2}$$

Then the Hessian matrix is given by

$$\boldsymbol{H(x,y)}=\begin{pmatrix} r_{xx} & r_{xy} \\ r_{xy} & r_{yy} \end{pmatrix} \tag{3}$$

3.2. Calculating the direction (n_x,n_y)

The second directional derivative of $f(x,y)$ has two direction, which can be determined by computing the eigenvalues and eigenvectors of the Hessian matrix. The direction in which the maximum absolute value corresponding is perpendicular to the line, is given by (n_x,n_y).

3.3. The subpixel center of the light line

If $\|(n_x,n_y)\|=1$, then the point can be given by $(p_x,p_y)=(tn_x,tn_y)$. Moremove, the $(p_x,p_y)\in\left[-\frac{1}{2},\frac{1}{2}\right]\times\left[-\frac{1}{2},\frac{1}{2}\right]$ is required.

Where

$$t=-\frac{r_x n_x + r_y n_y}{r_{xx}n_x^2+2r_{xy}n_x n_y+r_{yy}n_y^2} \tag{4}$$

Figure 3 exhibits the subpixel center line of the structured light.

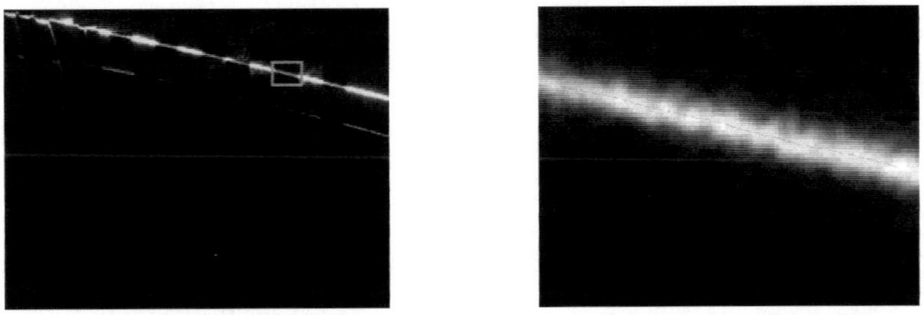

(a) Subpixel center line (b) Enlarged view

Figure 3. The subpixel center of the structure light line

4. Calculate of the structured light plane

The least-squares algorithmic technique can be used to fit the straight-line in the camera coordinate frame using the subpixel center points of the structure light line. When the fitted lines and the subpixel center line are shown on the Figure 2(c), the fitted straight line to the center points are drawn as Figure 4. The line l is the fitted line.

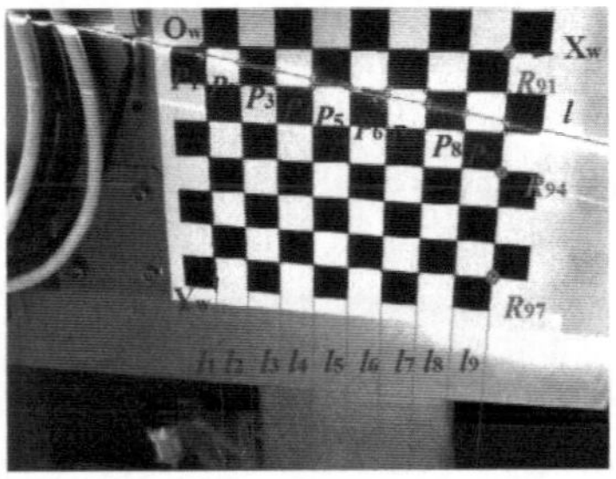

Figure 4. Fitting the straight line to the center points

Because the image corners coordinate can be extracted in the camera coordinate frame, the lines $l_{i,300}$ (i=1,2,…,9) are fitted in the camera coordinate frame using the least-squares algorithmic technique to the selected discrete set of image corner points. Where 300 indicate that the Z coordinate of the CNC machine is 300mm. The coordinate of points $P_{wi,300}$ (i=1,2,…,9) in the camera coordinate frame can be found by intersect the line l_{300} with the line $l_{i,300}$ (i=1,2,…,9). The $X_{w,300}Y_{w,300}$-plane of the world coordinate frame is aligned with the planar checkerboard. $O_{w,300}$ represents the origin point of the world coordinate frames. $Z_{w,300}$ is not displayed in the Figure 4, which is perpendicular to the planar checkerboard. In order to obtain the structure light plan, the world coordinate of the points $P_{i,300}$ should be calculated according to the cross ratio character. The cross ratio character is shown in Figure 5.

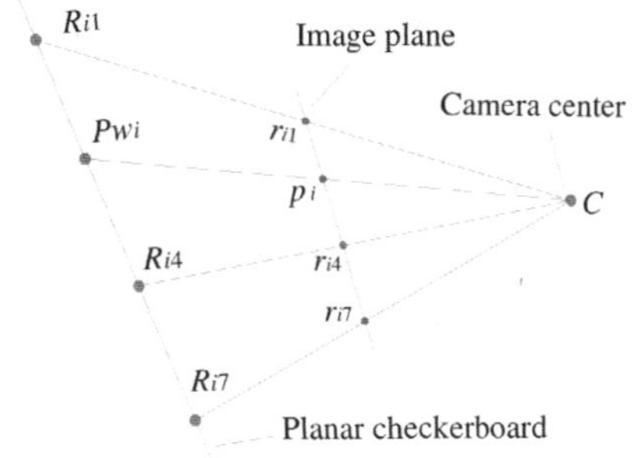

Figure 5. Cross ratios under projections

The cross ratio is the basic projective invariant of the 1D projective transformation. If four points are given, the cross ratio is defined by

$$Cross(r_{i1,300}, p_{i,300}, r_{i4,300}, r_{i7,300}) = \frac{|r_{i1,300}p_{i,300}||r_{i4,300}r_{i7,300}|}{|r_{i1,300}r_{i4,300}||p_{i,300}r_{i7,300}|} \tag{5}$$

Where

$$|r_{i1,300}p_{i,300}| = \det\begin{bmatrix} x_{i1,300} & x_{i,300} \\ y_{i1,300} & y_{i,300} \end{bmatrix} \tag{6}$$

Because the cross ratio is invariant to the projective coordinate frame, then

$$Cross(R_{i1}, P_{wi}, R_{i4}, R_{i7}) = Cross(r_{i1}, p_i, r_{i4}, r_{i7}) \tag{7}$$

The coordinate of point $\boldsymbol{P_{wi,300}}$ is represented by the vector $\boldsymbol{P_{wi,300}} = (X_{wi,300}, Y_{wi,300})^{\mathrm{T}}$. The coordinate value of $X_{wi,300}$ is decided by $20\times i$. Moreover, the coordinate of point $\boldsymbol{r_{i1,300}}, \boldsymbol{r_{i4,300}}, \boldsymbol{r_{i7,300}}$ in the image coordinate frame can be denoted by the vectors $\boldsymbol{r_{i1,300}} = (20\times i, 20)$, $\boldsymbol{r_{i4,300}} = (20\times i, 80)$, and $\boldsymbol{r_{i7,300}} = (20\times i, 140)$. Thus, the coordinate of $Y_{wi,300}$ can be solved. Furthermore, the coordinate of point $\boldsymbol{P'_{wi,300}}$ in the world coordinate frame is represented as $\boldsymbol{P'_{wi,300}} = (X_{wi,300}, Y_{wi,300}, Z_{wi,300})^{\mathrm{T}}$. The coordinate value of $Z_{wi,300}$ is decided by the **Z** axis value of CNC machine. When the Z axis value of the CNC machine are 200 mm, 250 mm, and 300 mm, the coordinate value of $Z_{wi,200}, Z_{wi,250}, Z_{wi,300}$ are 0 mm, 50 mm, and 100 mm, respectively.

The equations of the plane through $\boldsymbol{P'_{wi,k}}$ is $aX_w + bY_w + cZ_w = 1$, where $k = 200, 250, 300$. It has the matrix representation

$$\begin{bmatrix} X_{w1,300} & Y_{w1,300} & Z_{w1,300} \\ \mathrm{M} & \mathrm{M} & \mathrm{M} \\ X_{w9,200} & Y_{w9,200} & Z_{w9,200} \end{bmatrix}\begin{bmatrix} a \\ b \\ c \end{bmatrix} = \begin{bmatrix} 1 \\ 1 \\ 1 \end{bmatrix} \tag{8}$$

Then, the coefficient *a*, *b* and *c* can be calculated using the singular value decomposition method. The equation of the plane is $0.2288X_w - 0.0429Y_w + 0.0134Z_w = 1$.

5. Conclusions

Since the seam width of the workpiece in laser welding process is less than 0.1 mm, the accuracy of fixtures and workpiece will change the seam trajectory. Therefore, the welding trajectory should be planed for different workpiece. The recognition and center line extract methods of structured light line which is projected on the planar pattern are researched. Then, the structured light plane fitting approach is introduced. The experiment result shows the effective of the proposed method.

References

[1] Quintino L, Costa A, Miranda R, Yapp D, Kumar V and Kong C 2007 *Materials & Design.* **28** 1231-37.
[2] Fan D, Yu S R, Zhang J B and Katayama S 2003 *J. Lanzhou. Univ. Techno.* **29** 15-8.
[3] Luo Z J, Dai J S, Wang C Y, Wang F L, Tian Y L and Zhao M Y 2012 J. Mfg. Sys. **31** 2-7.
[4] Fang Z J, Xu D and Tan M. 2011 *IEEE/ASME.Trans.Mechatronics*. **16** 540-50.
[5] Graaf M D, Aarts R, Jonker B and Meijer J 2010 *Control. Eng. Pract.* 18 944-53.
[6] Colombo D, Colosimo B M and Previtali B 2013 Opt. Laser. Eng. **51** 34-46
[7] Gao X D, You D Y and Katayama S 2012 *Mechatronics.* **22** 370-80.
[8] Steger C. 1998 IEEE Trans. *Pattern. Anal. Mach. Intell.* **20** 113-125.

Acknowledgments

This paper is supported by the Doctoral Scientific Research Foundation of Liaoning Province (NO.20141121).

Research on torsional vibration modelling and control of printing cylinder based on particle swarm optimization

Y M Wang,W C Xu,S Q Wu,C W Chai,X Liu,S H Wang

Key Laboratory of Digitalization Printing Equipment,
Beijing Institute of Graphic Communication, Beijing 102600, China

Email: wangyiming@bigc.edu.cn

Abstract. The torsional oscillation is the dominant vibration form for the impression cylinder of printing machine (printing cylinder for short), directly restricting the printing speed up and reducing the quality of the prints. In order to reduce torsional vibration, the active control method for the printing cylinder is obtained. Taking the excitation force and moment from the cylinder gap and gripper teeth open & closing cam mechanism as variable parameters, authors establish the dynamic mathematical model of torsional vibration for the printing cylinder. The torsional active control method is based on Particle Swarm Optimization(PSO) algorithm to optimize input parameters for the serve motor. Furthermore, the input torque of the printing cylinder is optimized, and then compared with the numerical simulation results. The conclusions are that torsional vibration active control based on PSO is an availability method to the torsional vibration of printing cylinder.

1. Introduction

With the development of materials, electronics, control technology, active control has become a new way in the field of torsional vibration suppression in recent years. It has been noted(Parimal and Catalin,2013) that a comparative review of modelling and controlling torsional vibrations and experimentation using laboratory setups. Particle Swarm Optimization(PSO) is an artificial intelligence algorithm developed by authors based on the social behavior metaphor(Kennedy and Eberhart,1995). However, the problem of torsional vibration for impression cylinder of a complicated sheet fed offset printing machine has not yet been solved, so far no PSO application in this field.

Due to the torque on the rotary body and the angular displacement fluctuation changes with time, they lead to the printing cylinder system torsional vibration. The torsional vibration not only affects the printing cylinder in steady state process, but also affects the printing quality. Researchers had made a lot of effort in the printing machine dynamic testing and design.The author(Riese,2014) has studied the influence of torsional vibration to the printing machine drive motor on overprint, presenting a strategy for compensating existing deviations.

2. Printing cylinder torsional vibration phenomenon and characteristics

The paper studied the characteristics of transmission chain of a certain printing machine. And its printing cylinder structure schematic diagram is shown in Fig.1.

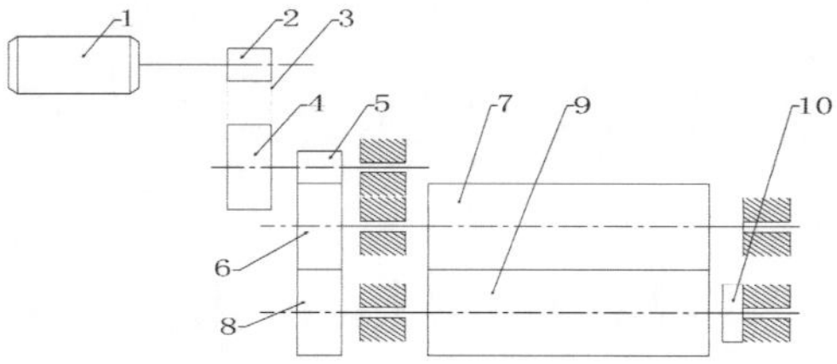

1- motor, 2, 4- timing belt pulleys, 3- timing belt, 5, 6, 8- helical gears, 7- impression cylinder, 10- open-closing cam of the printing cylinder, 9- blanket cylinder

Figure 1 Schematic diagram of the printing cylinder transmission chain

Parts of impression cylinder are too many so that we neglected the parts which have less influence on the torsion of impression cylinder, such as bolts. The structure schematic diagram of the printing cylinder transmission chain is shown in Fig.1.The motor providing control torque is the power source of the entire system.

3. Particle Swarm Optimization for printing cylinder

Particle Swarm Optimization（PSO）algorithm known as particle swarm optimization method is a branch of artificial intelligence, which imitates the social behavior of bird flocks searching for feed. Each particle stands for one set of possible solution, and all of them form a swarm. Swarm consists of m particles:

$$S_{warm} = \{x_{i1}, x_{i2}, \cdots, x_{iD}\} \tag{1}$$

In the standard PSO algorithm, the d-dimensional neighborhood function of the *i*-th particle at time *k* is calculated by:

$$v_{id}(t+1) = wv_{id}(t) + c_1r_1(p_{id}(t) - x_{id}(t)) + c_2r_2(p_{sd}(t) - x_{id}(t)) \tag{2}$$

$$x_{id}(t+1) = x_{id}(t) + v_{id}(t+1) \tag{3}$$

Combining F.(2) and (3), the optimal position $p_{id}^{(k)}$ of the individual history is the optimal position of the *i*-th particle in the previous *t* generation. The iterative formula is given by

$$p_{id}(t) = \begin{cases} p_{id}(t-1), f(p_{id}(t)) > f(p_{id}(t-1)) \\ x_{id}(t), \ etc \end{cases} \tag{4}$$

By F. (4), optimal position of the swarm is defined as:

$$p_{sd}(t) = \arg\min\{f(p_{id}(t)), i = 1,2,\cdots,\mathrm{m}\} \tag{5}$$

In the above Eqs., i=1,2,⋯,m.

In order to guarantee the stability of the algorithm in Eq.(5), when the neighborhood function generates new particle velocity, the constraint condition on speed is satisfied:

$$\left|v_{id}^{(k)}\right| \le V_{\max} \tag{6}$$

The meaning of each parameter in the F. (6) is shown in Tab.1.

Tab. 1 PSO algorithm basic parameters

	Inertia weighting factor	$V_{\max}$	Velocity vector restricts the constants
k	Number of iterations	c_1, c_2	Acceleration factor (cognitive, social coefficient)
$x_{id}^{(k)}$	Current position of the particle i	r_1, r_2	A random number between (0,1)
$v_{id}^{(k)}$	*i* Particle velocity	m	Population number of particles
$p_{id}^{(k)}$	*i* particle individual position optimal value	D	Search space dimension
$p_{sd}^{(k)}$	Population position optimal value	$f(\cdot)$	Fitness (target) function

Other parameters include the number of iterations, the maximum speed, and population size and so on. Where the swarm size depends on the complexity of the actual problem. For the general problem, the number of 30-40 particles is a good results, in this paper we select swarm size m = 50; the maximum velocity determines the maximum distance traveled by a particle in a seek cycle, is usually set to $V_{\max}=1$; The number of iterations is set to 60.

4. Realization of active controller simulation

For the parameters selection of the printing cylinder, the more optimization parameters, the better the optimization effect improved, while the difficulty of optimization parameters increases.

4.1 An optimal control model for torsional vibration of printing cylinder

In order to achieve control of torsional vibration for the printing cylinder, the whole system is regarded as a two order coupling with two conjugate poles, and the printing cylinder is simplified as two order system by modal reduction order.

Mathematical model is derived as follows:

$$J\ddot{\theta}+K\theta=M \tag{7}$$

The F.(7) is transformed into a state Eq. in which both sides are simultaneously left multiplied by J^{-1}, and rearranged

$$\ddot{\theta}=J^{-1}K-J^{-1}K\theta \tag{8}$$

The input torque and output position of the printing cylinder motor are taken as the optimization variables, and they can be expressed as:

$$X=\left[J_m,\theta_m\right] \tag{9}$$

If taking the weighted acceleration RMS values the surface of the printing cylinder of the objective function as the minimum of the target function by F.(8) and(9), it is expressed as:

$$\min E\left[X\right]=\min\left\{\delta=\left[\int_{0.5}^{80}v^2(E)\left|H(E)\right|_{z-q}^{2}G_q(E)df\right]^{\frac{1}{2}}\right\} \tag{10}$$

The objective function of the optimal design of the impression cylinder is finally determined as:

$$\min E\left[X\right]=v_1\delta_\theta^{\,2}+v_2\delta_J^{\,2} \tag{11}$$

4.2 Control simulation with Simulink

Through the acquisition of the signal of mechanical system, the response parameters are determined in the off-line module.

Select the *S*-function module in Simulink, and join the PSO *S*-function algorithm programmed in module, finally encapsulate them. Add the encapsulated module into the simulation model to get the simulation model based on PSO algorithm. Analysis of simulation result：After the simulation is completed, the optimized results are compared and analyzed. The contrast results are shown in Fig.2 (red line is parameter not optimized curve, blue line is parameter optimization curve).

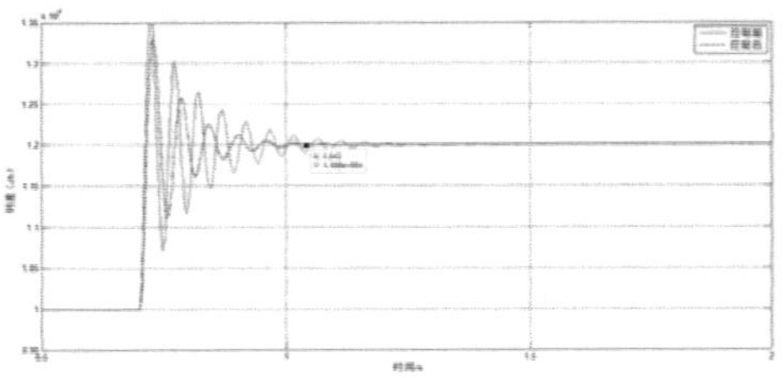

Figure 2. Simulation results of printing cylinder speed curve before and after active control

It is proved that the torsional vibration model was optimized by PSO algorithm, the torsional vibration of impression cylinder is controlled effectively, and the sensitivity of the input torque response is improved.

5. Experimental studies on torsional vibration test and active control

After system modeling and simulation of the printing cylinder, method and active control of torsional vibration of the printing machine are verified in printing cylinder experimental platform.

5.1 Experimental system design

The torsional vibration test platform is mainly composed of a four color sheetfed offset press, servo motor drive system, dynamic torque sensor and B&K data acquisition system.

1- servo driver; 2- dynamic torque sensor; 3- servo motor; 4- testing press; 5- B&K acquisition instrument; 6- digital tachometer; 7- software dongle; 8-pulse test system

Figure 3. Torsional vibration test platform of printing cylinder

The experimental platform is divided into driving unit, torsional vibration detection unit and load unit, the driving unit includes servo motor, bracket and elastic coupling. The test system used in the experimental platform is Denmark B&K PULSE vibration testing system. Test principle is that the vibration signals after the sensor acquisition, signal amplification and filtering processing, to achieve the signal processing, the processed data are input to the computer and displayed.

5.2 Torsional vibration test

The torsional vibration test platform is used to simulate the torsional vibration of the printing cylinder under the periodic excitation force, in which the excitation period of the gripper open-closing cam mechanism is determined according to the speed of the motor.

First, adjust the torque sensor to maintain a high degree of coaxial with the drive shaft and load shaft. Then servo motor drive, servo motor speed up to 10000 S/H, respectively, 12000 S/H, and acquisition speed of the system dynamic torque changes. The experiment measured 6 sets of experimental data, Fig.4 is the actual acquisition signal interface, where the ordinate represents the torque, and the abscissa represents the number of points, that is, sampling time.

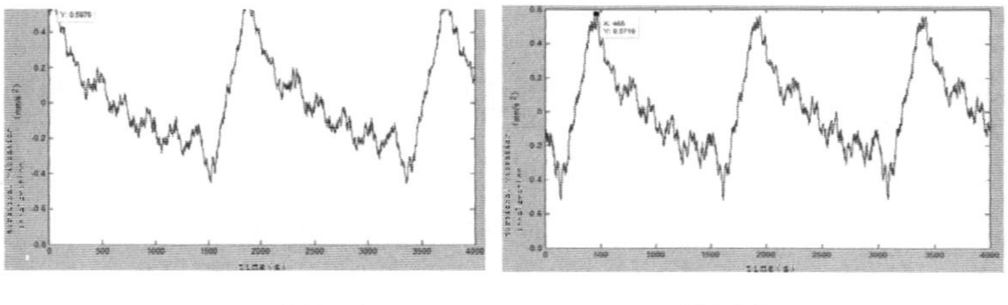

(a) 10000 S/H (b) 12000 S/H

Figure 4. Torsional vibration time domain diagram of printing cylinder

The torque the driving force to change the direction of the whole system to produce the torsional vibration phenomenon is obvious.

5.3 Torsional vibration control experiment

In the steady running of the printing stage, torsional vibration for printing cylinder measurement data is obtained and analyzed The torsional vibration controllable parameters are optimize and controlled with the built-in motion controller for servo drive system .

Increase the input torque in the angular position of the 6 peaks in a rotating period of the servo motor, and measure the torsional vibration of the cylinder again.

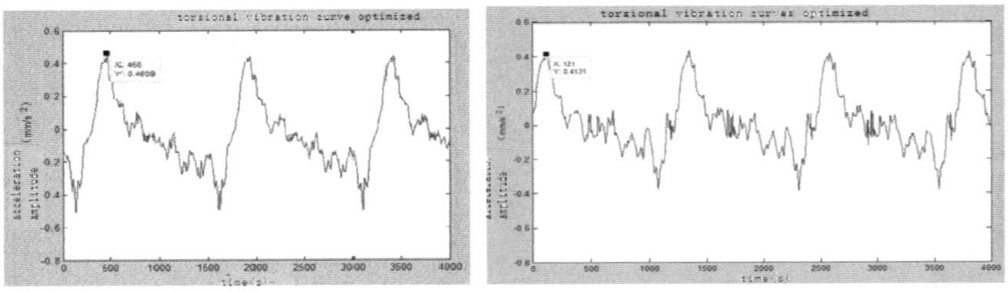

(a) 10000 S/H (b) 12000 S/H

Figure 5. Active vibration control time domain of printing cylinder

In combination with Fig.5, the torque variation curves of the roller transmission shaft before and after torsional vibration control are compared, shown in Tab.3.

Tab 3. Numerical simulation and experimental results of impression cylinder

Increasing speed rang(S/H)	Before control	After control	Amplitude cut off before and after control
	Maximum amplitude of acceleration	Maximum amplitude of acceleration	
8000 Speed up to 10000	0.5719	0.4609	19.4%
10000 Speed up to 12000	0.5386	0.4131	23.7%
average	0.5553	0.4370	21.6%

After application of the active control strategy with the servo motor, the torsional vibration of printing cylinder is significantly reduced at the teeth on-off cam mechanism action.

The results show that the experimental platform is suitable for measuring torsional vibration of impression cylinder at arbitrary speeds. And with the active control to the torsional vibration for printing machine, the response time of angular velocity and amplitude is improved to realize the optimal control effect of torsional vibration.

6. Conclusion

In view of the common torsional vibration problem of the printing press, the PSO optimization algorithm is used to optimize the input torque of the printing cylinder, which proves the effectiveness of the method in reducing the torsional vibration of the printing cylinder.

Experimental study on active control of the printing cylinder torsional vibration builds the experimental platform for motion and vibration control. The results show that the input torque of the printing machine cylinder optimized the fluctuation of time can effectively reduce the angular velocity and the amplitude of the printing cylinder.

The peak of the torsional amplitude of acceleration is cut down after application of the active control strategy with the servo motor. The conclusions are that torsional vibration active control based on PSO is an availability method to the torsional vibration of printing cylinder.

References

[1] Aubry E,Vaclabik I,Ragot P, et al(2006) Method for active compensation for transverse vibrations of a moving paper web. Experimental Techniques 30(1):47-50.

[2] Buck B, Knopf E,Schreiber S,et al(2005) Nichtlineare schwingung phanomene in bogenoffsetdruck maschinen. VDI Berichte 1887(1)：345-361.

[3] Cai LG, Xu B, et al(2014) Self tuning algorithm of input shaper parameters based on particle swarm optimization.Journal of electrical machinery and control 18 (10): 87-94.

[4] Hermanski M, Kohn K, Ostholt H(1995) An adaptive spectral compensation algorithm for avoiding flexural vibration of printing cylinders.Journal of Sound and Vibration 187(2):185-193.

[5] Kennedy J, Eberhart R(1995) Particle swarm optimization:Neural Networks. In: 1995 Proceedings International Conference on IEEE 1995：1942-1948.

[6] Parimal AP and Catalin T(2013) A comparative review of modelling and controlling torsional vibrations and experimentation using laboratory setups. Journal of Petroleum Science and Engineering. 112(11): 227–238.

[7] Riese M(2014) Method for compensating deviations in individual drives of a printing machine. Patent DE102014207445.

[8] Wang YM(2014)Printing machinery test technology. Beijing: Mechanical Industry Press.

Acknowledgments
This work was supported by Scientific Research Project of Beijing Educational Committee & Beijing Municipal Natural Science Foundation (KZ201510015016) and Collaborative Innovation Center of Green Printing & Publishing Technology (CGPT).

IOP Conf. Series: Materials Science and Engineering **339** (2018) 012032 doi:10.1088/1757-899X/339/1/012032

Parametric optimization of the MVC desalination plant with thermomechanical compressor

E.V. Blagin, V.V. Biryuk, M.Y. Anisimov, A.A. Shimanov, A.A. Gorshkalev

Heat Engineering and Heat Engines Department, Samara National Research University,
Samara, Russia

Email: evgenyblagin@gmail.com

Abstract. This article deals with parametric optimization of the Mechanical Vapour Compression (MVC) desalination plant with thermomechanical compressor. In this plants thermocompressor is used instead of commonly used centrifugal compressor. Influence of two main parameters was studied. These parameters are: inlet pressure and number of stages. Analysis shows that it is possible to achieve better plant performance in comparison with traditional MVC plant. But is required reducing the number of stages and utilization of low or high initial pressure with power consumption maximum at approximately 20-30 kPa.

1. Introduction

Lack of the fresh water became a problem in recent years. According to researches, in 2025 about half of the states will suffer fresh water deficit [1]. Thus, generation of the fresh water is one of the main tasks which must be solved to ensure ecological safety of the Earth population. There are several ways of fresh water generation. One of the most promising is seawater desalination or waste water treatment. There are several ways to desalinate seawater. They include reverse osmosis (RO) [2], electrodialysis (ED) [3], freezing [4] and distillation [5-7]. Distillation plants in turn can be divided on Multi-effect distillation plants (MED) [8], multi-stage flash distillation plants (MSF) [9], multi-effect distillation plants with thermal vapour compression (MED-TVC) [10] and plants with mechanical vapour compression (MVC) [11]. Every type of desalination plant has its own advantages and disadvantages. For example, MED plants can utilize waste heat of thermal power station, however, their efficiency is lower in comparison with RO plants. RO plants have a higher efficiency, however, they require often change of the membranes. ED plants are applicable on the water with relatively small salinity. Freezing plants are currently under development but in future they will allow using cold energy of liquid natural gas for desalination purposes. MVC plants have satisfactory energy consumption but they require careful design of the steam compressor which operation is crucial for plant.

Thermomechanical compressor is a device suited for gas compression by the heat adding in the closed volume [12]. From design point of view thermocompressor can be made with internal [13] or external regenerator [14] which can be represented in the form of annular gaps [15] and porous heat exchanger.

Thermomechanical compressor can be used in distillation plants as a substitution of mechanical compressor or as efficiency increase of other distillation plants. Principal possibility of thermomechanical compressor operation within desalination plant was justified in article [16].

IOP Conf. Series: Materials Science and Engineering **339** (2018) 012032 doi:10.1088/1757-899X/339/1/012032

2. Thermocompressor working principle

Working cycle of thermocompressor starts from high dead centre when both valves are closed and volume of the cold chamber is maximal (point 1). Piston starts to move downside pushing working fluid through regenerator to a hot chamber. Due to Amontons's Law of Pressure-Temperature, pressure is increasing because mean temperature of the working fluid inside the thermocompressor is increasing. When the pressure is high enough, output valve opens and working fluid with given pressure starts to be charged to a customer. When piston reaches low dead centre, output valve closes (point 3) and working fluid stops to be charged to a customer. Piston starts to move upside pushing working fluid back to a cold chamber. Pressure is decreasing due to decrease of the mean temperature inside the thermocompressor. When pressure drops low enough, input valve opens, charging new portion of working fluid inside the thermocompressor. Working cycle of thermocompressor is presented on figure 2.

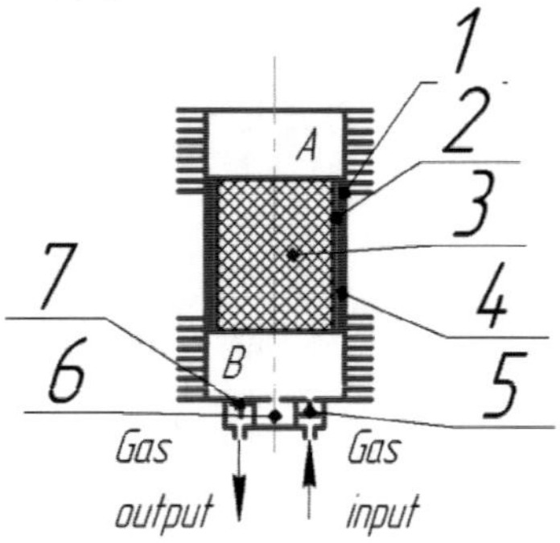

Figure 1. Principal scheme of the mechanical thermocompressor:
1 – thermocompressor shell; 2 – displacer-regenerator shell; 3 – regenerator; 4 – gap; 5 – input valve; 6 – rod; 7 – output valve; A – hot chamber; B – cold chamber.

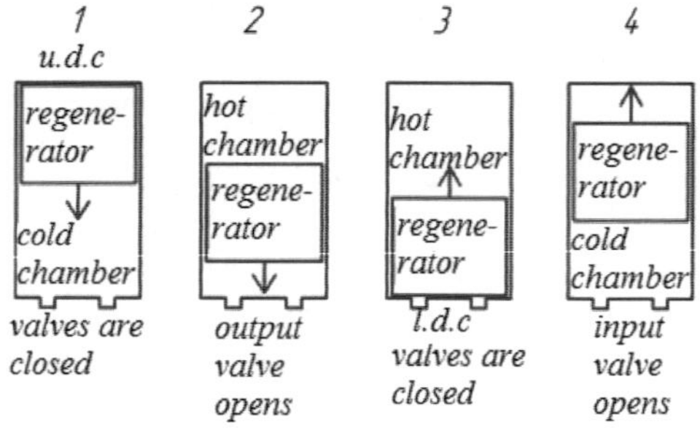

Figure 2. Working cycle of the thermocompressor.

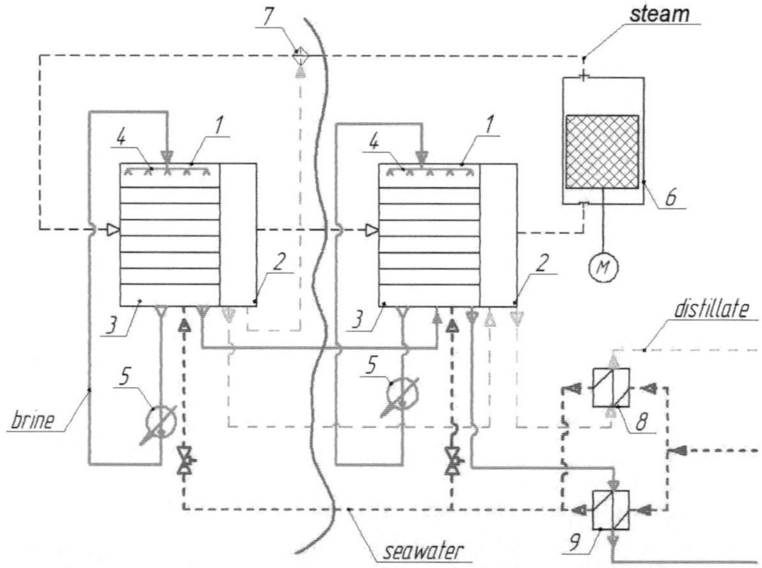

Figure 3. Desalination plant with thermomechanical compressor
1 – evaporator stage, 2 – distillate chamber, 3 – brine chamber, 4 – water sprayer, 5 – circulation pump 6 – thermocompressor, 7 – mixer heat exchanger, 8 – water-distillate heat exchanger, 9 – water-brine heat exchanger

Thus, for one working cycle thermocompressor rise the pressure of certain amount of gas Δm, which enters the TC through the input valve and leaves the TC from the output valve during one cycle.

3. The rmocompressor coupled work with distillation plants

The main feature of the MVC plant is requirement of only one energy source which is electricity. Electric energy is transformed in thermal energy and potential energy of pressure which are further used to execute consequential phase changes seawater → steam → distillate. Design scheme of the plant with thermocompressor is presented on figure 3.

In this case steam enters thermocompressor instead of mechanical compressor. During displacement through regenerator both its temperature and pressure are raised. Let's compare amount of energy necessary to raise the pressure in the same degree. Steam parameters of the last stage will be taken from the plant currently developed by Samara University together with "Metallist-Samara" OJSC. These parameters include: steam temperature 60 oC, steam pressure 20 kPa, required compression ratio is 1.6, steam flow rate is 0.111 kg/s.

Preliminary calculations show that at given configuration of desalination plant power consumption of the thermomechanical compressor is higher in comparison with traditional mechanical compressor. So main purpose of given research is to find such parameters and configuration of desalination plant where relation of power consumption of thermomechanical compressor to power consumption of mechanical compressor is minimal [16]. For further analysis let's investigate influence of different factors on thermomechanical compressor power consumption. Drinkable water output is kept constant as 2 m3/hour. Thermomechanical power consumption in ideal case is equal to:

$$N_{comp} = Q = G_{st} c_v (T_{out} - T_{in}) \tag{1}$$

Amontons's law states that in constant volume

$$\frac{p}{T} = const \tag{2}$$

Substituting (2) into (1) gives:

$$N_{comp} = Q = G_{st} c_v T_{in} (\frac{p_{out}}{p_{in}} - 1) \tag{3}$$

Dependency of saturation pressure on temperature is quite well described by empiric equation:

$$p_{sat} = 33.05 \cdot 10^{-5} (T - 273)^2 - 3.29 \cdot 10^{-2} (T - 273) + 0.993 \tag{4}$$

Previous researches show that optimal temperature difference in such evaporator-condenser is about 2 K [17]. So total temperature difference will be equal to 2n, where n is a number of stages. Considering that, thermomechanical power consumption will be equal to

$$N_{comp} = \frac{G_{wo}}{n} c_v T_{in} (\frac{33.05 \cdot 10^{-5} (T_{in} + 2n - 273)^2 - 3.29 \cdot 10^{-2} (T_{in} + 2n - 273) + 0.993}{33.05 \cdot 10^{-5} (T_{in} - 273)^2 - 3.29 \cdot 10^{-2} (T_{in} - 273) + 0.993} - 1) \tag{5}$$

Let's analyse dependency of consumed energy of thermomechanical compressor on two factors: initial pressure and number of stages. Initial pressure is connected with initial temperature by steady dependency because parameters on this point are taken from saturation line. So, initial pressure can be replaced by initial temperature as comparison criteria. To estimate dependency of power consumption on initial temperature it is necessary to find derivatives of the equation (5). Dependency of the isochoric heat capacity on the temperature and pressure was neglected for first calculation but must be taken into account in further calculations. Plot of the dN_{comp}/dT_{in} derivative is presented on figure 4. Similarly, derivative of power consumption by number of stages was also calculated and is presented on figure 5.

IOP Conf. Series: Materials Science and Engineering **339** (2018) 012032 doi:10.1088/1757-899X/339/1/012032

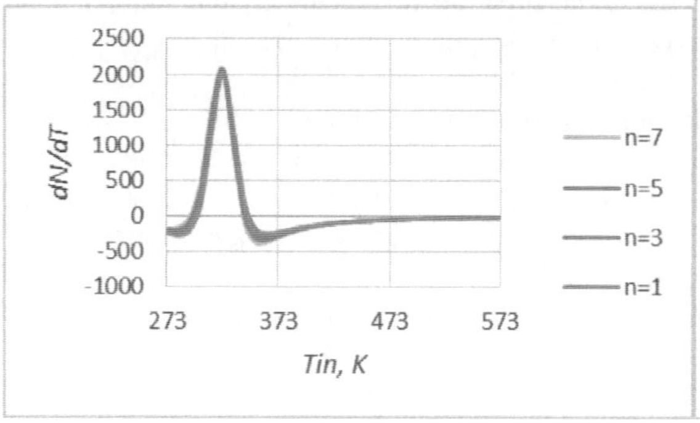

Figure 4. Dependency of the derivative dN_{comp}/dT_{in} on temperature and number of stages

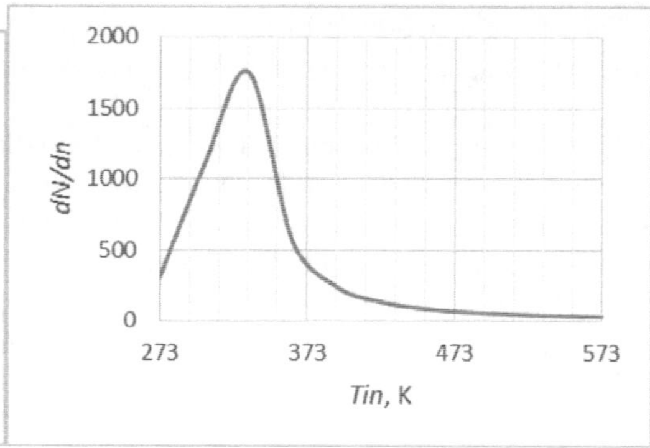

Figure 5. Dependency of the derivative dN_{comp}/dn on temperature

Analysis of the plot from figure 4 shows that derivative has a range of positive values between 287 and 346 K. So, conclusion can be made that maximal work of thermomechanical compressor will be at inlet temperature equal to 346 K, and it is highly recommended to avoid this temperature. Calculation of dNcomp/dn derivative shows that it almost does not depend on number of stages and depends only on temperature. Highest value of the derivative is achieved at temperature 333 K, so influence of the number of stage change will be maximal at this temperature. So, derivative analysis allow to make two main conclusions: inlet temperature must differ from 346 K and number of stages should be kept minimal, especially near inlet temperature equal to 333 K. This value actually is inlet steam temperature for the plant designed before, so direct replacement of the mechanical compressor to thermomechanical compressor gives high power consumption. Finally, power consumption dependency on inlet temperature and number of stages was determined and presented on figure 6. To compare optimized plant with thermomechanical compressor it is necessary to compare the values of their power consumptions. In ideal case power consumption of the mechanical compressor can be calculated as compression work in adiabatic process:

$$N_{comp} = G\frac{k}{k-1}RT_{evap}\left(p^{I}\right)\left[\left(\frac{p^{II}}{p^{I}}\right)^{\frac{k-1}{k}}-1\right], \tag{6}$$

where N_{comp} – power consumption, G – steam flow rate, k – specific heat ratio, R – specific gas constant, T_{evap} – evaporation temperature, p^{I} – gas pressure in the input of the compressor, p^{II} – gas pressure at the output of the compressor.

Plant with 5 stages and initial temperature equal to 333 K was considered as reference plant and all other plants were compared with it. Power consumption calculated by equation (5) was divided by power consumption, calculated by equation (6), resulted dependencies were plotted and presented at figure 7. It should be noted that inlet temperatures higher than 373 K require pressure higher than atmospheric one, so additional work must be spend on pressure increase of the water. The value of this work is relatively small but must be taken into account in further study. Figure 7 shows that it is possible to obtain better performance of the MVC plant with thermomechanical compressor than traditional MVC plant if appropriate optimization will be performed. However, several factors must be taken into account during careful investigation, such as deviation of real thermodynamic parameters of the steam from the equation (4), work spend on water pressurizing.

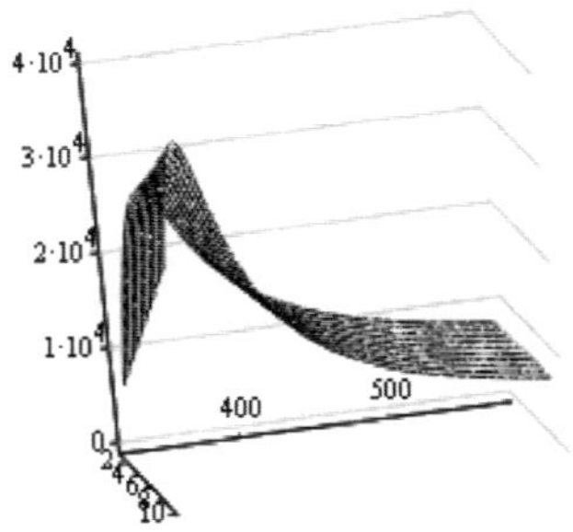

Figure 6. Surface of the power consumption dependence on temperature and stage number

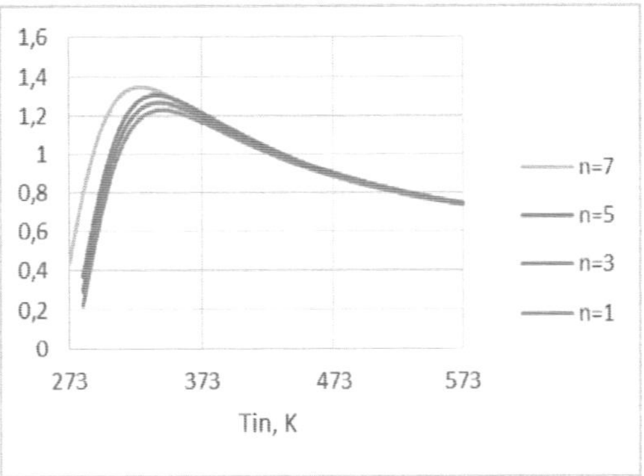

Figure 7. Power consumption ratio dependency on inlet temperature and stages

4. Conclusions

Parametric optimization of the MVC desalination plant with thermomechanical compressor shows that it is possible to achieve better plant performance in comparison with traditional MVC plant, however it is necessary to decrease number of stage to 1 and increase initial pressure. Decrease of the initial pressure to a certain point also helps to improve plant performance but low steam density at such pressure causes higher dimensions of the thermomechanical compressor and, thus, should be avoided.

References

[1] Deficiency of fresh water: problems and methods of solving http://thewallmagazine.ru/lack-of-fresh-water/
[2] Sarai Atab M, Smallbone A J and Roskilly A P 2016 *Desalination* **397** 174-184
[3] Largier T D, Wang D, Mueller J, & Cornelius C J 2017 *Journal of Membrane Science* **531** 103-110
[4] Cao W, Beggs C and Mujtaba I M 2014 *Desalination* **355** 22-32
[5] Maghsoudi K, Aliasghari M and Mehrpanahi A 2016 *Desalination and Water Treatment* **57**(38), 17707-17721
[6] Alasfour F N, Darwish M A and Bin Amer A O 2005 *Desalination* **174**(1) 39-61
[7] Zimerman Z 1994 *Desalination* **96** 51-58
[8] Fathalah K and Aly S E 1991 *Energy Conversion and Management* **31**(6) 529-544
[9] Darwish M A 1987 *Desalination* **63**(C) 143-161
[10] Maghsoudi K, Aliasghari M and Mehrpanahi A 2016 *Desalination and Water Treatment* **57**(38) 17707-17721
[11] Darwish M A, Jawad M A and Aly G S 1990 *Desalination* **78**(3) 313-326
[12] Bush V US Patent 2157229
[13] Edwards M J and Peterson R B 2007 *Proceedings of the Institution of Mechanical Engineers* **221**(4) 459-472.
[14] Kornhauser A A 1996 *Paper presented at the Proceedings of the Intersociety Energy Conversion Engineering Conference* **2** 1331-1336
[15] Lin W, Wu X, Yang J and Yang L 2013 *International Journal of Refrigeration* **36**(4), 1376-1387
[16] Blagin E V, Shimanov A A, Korneev S S, Uglanov D A 2017 *IOP Conference Series: Materials Science and Engineering (to be published)*
[17] Mu, X., Shen, S., Yang, Y., Liang, G., Chen, X., Zhang, J. Desalination and Water Treatment. **57** (21) 9993-10002.

Acknowledgments

This work was carried out by lead performer of research and development effort with financial support of Russian Federation Ministry of Science and Education with the realization of Government decree №218 according to contract about providing and using subsidy № 02.G25.31.0150 from 01.12.2015.

IOP Conf. Series: Materials Science and Engineering **339** (2018) 012033 doi:10.1088/1757-899X/339/1/012033

Failure Analysis on Tail Rotor Teeter Pivot Bolt on a Helicopter

WANG Qiang[1], DONG Zi-long[1]
[1] Civil Aviation Flight University of China of Xinjin branch , Qiaojin Up-street No.320 Xinjin County, China

Email: wang_qiang1984@163.com

Abstract. Tail rotor teeter pivot bolt of a helicopter fractured when in one flight. Failure analysis on the bolt was finished in laboratory. Macroscopic observation of the tailor rotor teeter pivot bolt, macro and microscopic inspection on the fracture surface of the bolt was carried out. Chemical components and metallurgical structure was also carried out. Experiment results showed that fracture mode of the tail rotor teeter pivot bolt is fatigue fracture. Fatigue area is over 80% of the total fracture surface, obvious fatigue band characteristics can be found at the fracture face. According to the results were analyzed from the macroscopic and microcosmic aspects, fracture reasons of the tail rotor teeter pivot bolt were analyzed in detail

1. Introduction

As the main rotor blades rotate, the fuselage will rotate the opposite direction if unopposed. Tail rotor system of helicopter as one of key components, not only provide anti-torque and achieve directional control, but also keep the helicopter's yaw stability, its working condition directly affects the safety and reliability of the helicopter[1]. Tail rotor teeter pivot bolt as a helicopter tail rotor system of the core force component, need to bear a greater bending moment and experiencing a lot of alternating load. Tail rotor flapping hinge bolt fracture will directly lead to the helicopter yaw control failure, even cause serious catastrophic accidents.

March 2016, a general aviation of domestic was carrying out flight training in field. The helicopter was climbing out when severe vibration prompted the commander to carry out an immediate autorotation landing in a field. The helicopter incurred severe damage during the landing. On inspection, The bolt attaching the tail rotor hub to the rotating fork assembly, about which the rotor is free to teeter, was found to have fractured.

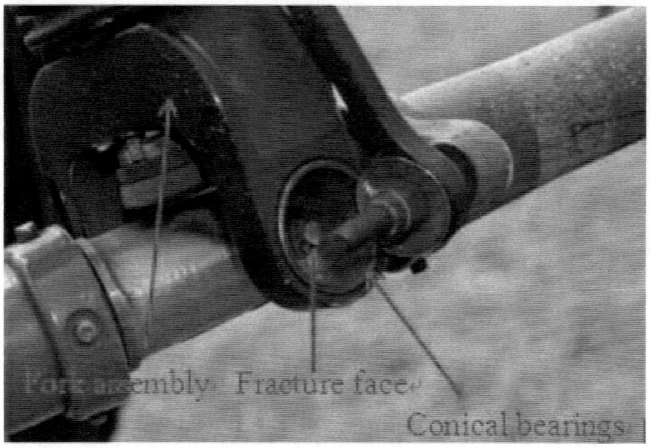

Figure 1. Fracture face of the tail rotor teeter pivot bolt

Examination of the fracture face of the teeter pivot bolt showed that it had failed in a plane very close to the exposed face of the threaded insert, allowing the dust shield and washers on one side of the hub to be lost (Figure.1). The thread insert is screwed into the hub forging. The fracture face incorporated a raised angular peak and trough at one edge with a ridge or step extending from the peak across the section separating the two halves of the fracture. Each half occupied a slightly different, though parallel, plane and the two were thus independent fracture surfaces. The paper on the fracture of the teeter pivot bolt were analyzed, the failure mode, analyzes the causes of failure[2].

2. Structural layout of tail rotor teeter pivot bolt

The tail rotor, mounted on the tail rotor transmission at the end of the tail-boom, counteracts main rotor torque and controls the heading of the helicopter. The tail rotor system consists of two variable-pitch blades mounted on a teetering delta-type hub assembly which is in turn mounted in a rotating fork. The fork incorporates two conical bearings and a bolt passes through the hub and both bearings. Teetering motion of the rotor is enabled by the presence of the bearings[3]. The bolt, in addition to passing through the bearings and the basic hub trunnion, passes through a shoe, within the hub, which locates the straps holding the two blades against centrifugal force. It also passes through an insert threaded into the hub. Shim washers of various thicknesses bear against the protruding inner races of the conical bearings and enable the pre-load to be adjusted. Two washers of larger diameter, acting as dust shields, are positioned on either side of the hub. structural layout of tail rotor hub components(Figure.2.)

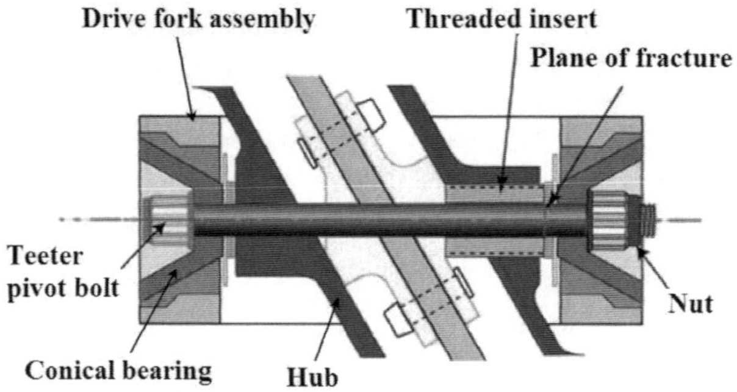

Figure 2. Structural layout of tail rotor hub components

The teeter pivot bolt is secured and tensioned by a nut which, when the assembly is correctly installed, ensures that the inner races, the nut, bolt, hub, shoe, threaded insert, shim washers and the dust shields all oscillate in unison relative to the fork as the rotor rotates during forward flight.

3. Analysis of tail rotor teeter pivot bolt

3.1. Appearance check

The whole morphology of fractured tail rotor teeter pivot bolt is shown figure 3. No obvious bending deformation was found in this section of the bolt. But, at ends of the surface visible obvious signs of corrosion, wear characteristic is not obvious. The fracture section is located on the left side of the bolt. The fracture face is located at the edge of the threaded end bearing, the corrosion feature of the fracture surface of bolt is serious.

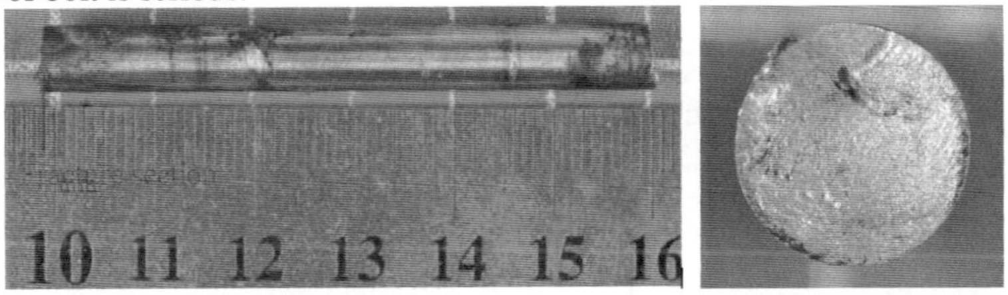

(a) Morphology (b) Fracture section

Figure3. Morphology and fracture section of teeter pivot bolt

According to the fracture profile of the bolt. Fracture section height slightly difference, the underside side is relatively flat, upper side is rough. It can be seen from the above figure 3, left and right sides of the lower radiation ridge feature near the surface of the origin of the position, indicates that its source fracture is located in this location.

3.2. Microscopic observation of fracture

Scanning Electron Microscope (SEM) examination of the bolt found that fatigue cracking had sourced at the below side surface of the fracture section, no obvious defect on the source location. Fatigue growth from the source area to the left and right sides of the fracture surface, fatigue strip fine, fatigue expansion full. Besides, according to the fracture feature, it can be seen that the cyclic stress of fatigue fracture is small and more fatigue expansion cycles (Figure.4).

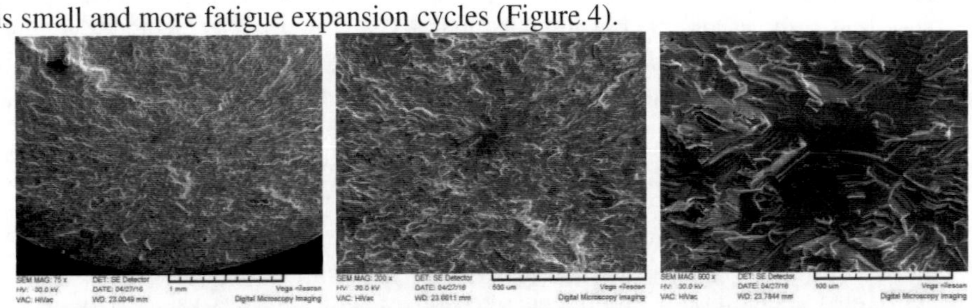

(a) Low magnification (b) Middle magnification (c) High magnification

Figure 4. Microscopic feature of fracture source section

a In initial period b In middle period c In late period

Figure 5. Microscopic characteristics on the left fatigue expansion of bolt

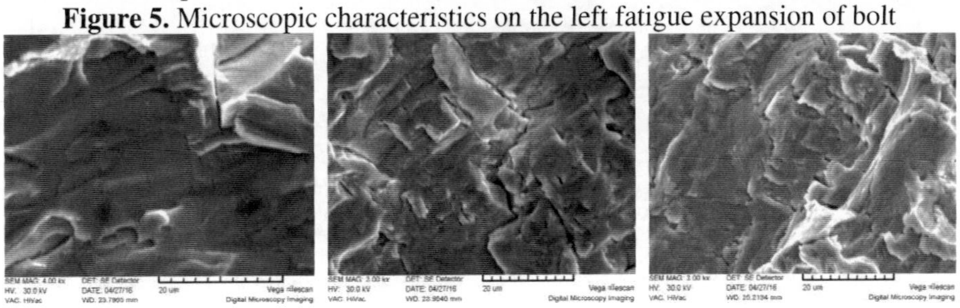

a In initial period b In middle period c In late period

Figure 6. Microscopic characteristics on the right fatigue expansion of bolt

The microscopic characteristics of fatigue expansion at the left and right sides, from the front, middle and late stages of the fatigue source of the bolt fracture, the fatigue characteristics of fracture are obvious, fatigue strip fine, fatigue expansion full (Figure.5and Figure.6). The fatigue had the characteristics of torsional shear stressing applied in two distinct directions. However, the torsional evidence was, not orientated about an axis coincident with that of the bolt. Therefore, it can be concluded that the fracture property of the bolt is fatigue fracture.

3.3. Material analysis of teeter pivot bolt
Energy spectrum analysis of the tail teeter pivot bolt, shown that the material for Ni-Cr alloy, contains Ni、Cr、Fe and other elements.

4. Analysis and discussion

4.1. Fracture property of bolt
The fracture face of tail rotor teeter pivot bolt at the thread root, originated in the near surface position of the bolt fatigue fracture, Obvious fatigue band characteristics can be found at each location of the fracture face, fatigue area is over 80% of the total fracture surface[4]. Therefore, it can be concluded that the fracture property of the bolt was high cycle fatigue fracture.

4.2. Analysis of causes
Reference to the maintenance manual of the aircraft revealed that a requirement exists for the preload in the teeter pivot bolt to be ascertained at regular intervals. An alternative method of tightening the nut, involving achieving a measured stretch in the bolt, was also available when installing bolt. The tightness of the teeter pivot bolt in its housing found during sectioning of the hub could have been caused by an unfavourable combination of manufacturing tolerances. The bolt was not, however, measurably different in diameter from another sample bolt measured. It was not possible to measure accurately the internal diameter of the threaded insert or the shoe due to the necessarily destructive method used to separate them.

Considering the bolt in the work not only bear the alternate load of the tail rotor, but also subjected to torsion of tail rotor transmission component, combined with the analysis of the bolt fracture characteristic, analysis that causes of the bolt fatigue fracture has the following 3 possible:

(1) The tail rotor fork bolt failed in fatigue after a period of operation with no pre-load present in the bolt/nut assembly. The lack of pre-load was probably a consequence of failure to establish the correct tension in the bolt at the time of last assembly. Failure to establish the correct tension may have resulted from a tight fit of the bolt within the threaded insert, disguising the fact that the bolt head was not correctly seated as torque was applied.

(2) The bolt is not fitted with appropriate thickness of shims when it was installed. The bolt in installation, failed to effectively measure fork assembly inner diameter, outer diameter and the teflon dust thickness of tail rotor hub, The fork assembly cannot eliminate the loose of conical bearing ends and the axial clearance of bolt, which lead to the normal flapping motion of the conical bearing to be blocked.

(3) Contamination of the bore of the threaded insert. Contamination of the bore of the threaded insert with teflon dust may have caused the tight fit of the bolt within the threaded insert, which leads to fatigue torsion of the bolt during installation.

It seems reasonable to analysis that the bolt failed as a result of a fatigue mechanism which, although not fully explained, nonetheless resulted from operation with a bolt which was not carrying the correct end-load consistently throughout its length. The installation irregularity responsible for this condition was most probably excessive friction of the bolt within the bore, shoe or threaded insert in the hub, experienced at the time of assembly, resulting in the bolt failing to seat fully.

5. Conclusion

The fracture face originated in the near surface position of the bolt, no obvious defect on the source location the fracture extends from the source area to the left and right sides, fatigue banding closely. The fracture of tail rotor teeter pivot bolt is caused by fatigue fracture, high cycle fatigue fracture due to high original stress and fatigue fracture development is more adequate. It shows that the fatigue crack has experienced a long time process from the initiation to the final fracture.

6. References

[1] Wang Qiang. Failure Analysis on Tail Rotor Teeter Hinge Bolt Fracture for a Certain Type of Helicopter[J], Aviation Maintenance & Engineering,2017,5:98-100.
[2] LI Chun-guang, SHU Ping, MA Xiao-ming, etal. Failure Analysis on Tailor Rotor Rod Components of a Helicopter[J], Failure Analysis and Prevention ,2013,8 (6):346-349.
[3] Sikorsky S-300CB Model 269C-1 Helicopter Maintenance Instructions[M/CD]. Sikorsky Aircraft Corp. Stratford, CT 06614, Revised: 30 Nov 2015.
[4] Zhang Dong, ZHONG Pei-dao, TAO Chun-hu, etal. Failure Analysis[M], Bei Jing: National Defence Industry Press ,2004,229-231.

IOP Conf. Series: Materials Science and Engineering **339** (2018) 012034 doi:10.1088/1757-899X/339/1/012034

Large - scale Rectangular Ruler Automated Verification Device

Hao Chen, Luping Chang, Minjian Xing, Xie Xie

Shanghai Institute of Measurement and Testing Technology.

No. 1500, Zhang Heng Rd. Pu Dong New District, 201203 Shanghai, China

Email: xingminjian@126.com

Abstract. This paper introduces a large-scale rectangular ruler automated verification device, which consists of photoelectric autocollimator and self-designed mechanical drive car and data automatic acquisition system. The design of mechanical structure part of the device refer to optical axis design, drive part, fixture device and wheel design. The design of control system of the device refer to hardware design and software design, and the hardware mainly uses singlechip system, and the software design is the process of the photoelectric autocollimator and the automatic data acquisition process. This devices can automated achieve vertical measurement data. The reliability of the device is verified by experimental comparison. The conclusion meets the requirement of the right angle test procedure.

1. Introduction

Rectangle ruler vertical detection method consists of optical gap method, amount of block comparison method and rectangular ruler to play the table method. The most commonly used method is optical gap method. The principle of it is to observe the actual gap through the visible gap to determine the size of the gap, the test results usually impacted by different visual testing personnel. Ruler measuring range in China is generally less than 600mm at present, and only little individual manufacturers can customize 1000mm. The detection object is generally 0 or 0 below of the rectangular ruler.

This device is designed to detect the vertical degree of large rectangular ruler, complete mechanical and electrical integration equipment, and establish an extended measurement system to form a complete product. The function of the verification device is as follows:

(1) to complete the automatic acquisition, motion control and wireless transceiver function;

(2) to achieve accurate position judgments and different span size adjustment function;

(3) to achieve automatically save the function during the measurement process.

2. The structure of the device design

Since the size of the large-sized rectangular ruler is large, the volume of the autoradiograph must be small enough to measure in the course. Therefore, horizontal measurement is chosen in the detection process. The rectangular ruler is flattened. The width of the two sides of the ruler is different, so the shorter side is held with a thimble. It should ensure that the rectangular ruler is stable with no shaking.

2.1. Optical axis design

The design of measurement axis uses high-precision optical axis. The optical axis is used to fit to the measured ruler. It will be fixed through the base in the body. The optical axis has no race movement, but axial movement, to ensure that the measurement axis and the side is line contact during the acquisition process. This design can ensure that the optical axis in the detection has not unnecessary radial round beating.

2.2. Drive part of the design

The transmission part is mainly driven by the stepping motor, the stepping motor is connected with a bevel gear, the other bevel gear is connected with the axle and the double bevel gear is in the engaged state, and the axle ends are respectively connected with the two driving wheels to form the transmission part.

The measurement part is mainly composed of measuring shaft. The measuring shaft and the drive wheel and auxiliary wheel are located on the same side. Since the measurement axis needs to be closely aligned with the side, and the verification device needs to move forward, the measurement of shaft thickness in the design process must be higher than the thickness of the four wheels, so as to ensure that the process of walking does not affect the measurement axis to collect data.

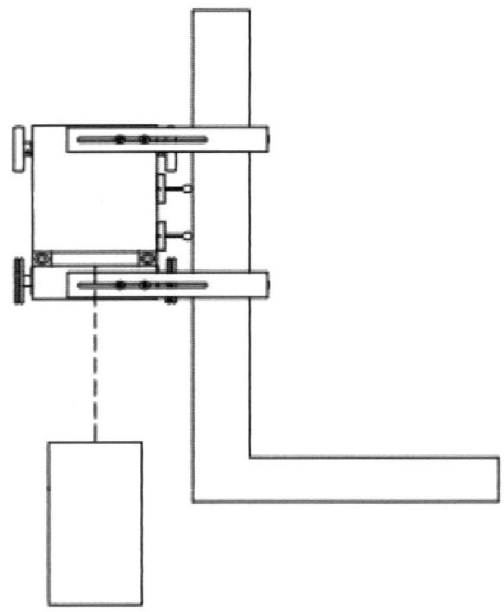

Figure 1. System connection diagram

2.3. Fixture device

The fixture part is an important device to determine whether the measured data is true or not. The setting of the clamping force needs to take into account the frictional force generated by the clamping block and the rectangular ruler.

If the friction is too large, the motor torque is not enough large to drive the entire structure forward. On the other hand, if the clamping force is too small, the measured plane will not be able to fit closely with the measuring shaft. This may lead to the wheel move left and right in the process and in separation with the ruler measured. The U-shaped connecting strip is fixed on the vehicle body, and the U-shaped groove is easy to move forward and backward. In order to facilitate the measurement of different types of rectangular rulers, the clamping device makes the measuring shaft of the automatic acquisition instrument always fit to the side of the rectangular ruler. The other of the U-shaped connecting strip is connected with a clamping block. The clamping block is connected by a spring washer to the U-shaped connecting strip. According to the different width of the rectangular rulers, U-shaped plate connects to the spring and the spring makes the rectangular rulers fit to the measurement axis through clamping block to gets the measurement data stable and reliable.

2.4. Wheel design

IOP Conf. Series: Materials Science and Engineering **339** (2018) 012034 doi:10.1088/1757-899X/339/1/012034

The left and right swing of the automatic acquisition device must be existed in the process. Therefore, the conventional wheel will lead to the distortion of the motion trajectory and increase the absolute clearance between the measuring axis and the measured surface.

The design uses the embedded wheel instead of the drive wheel, and the arc wheel instead of the auxiliary wheel, in order to meet the regulatory role, and appropriate to increase the friction of the wheel and the plane to ensure its stable movement.

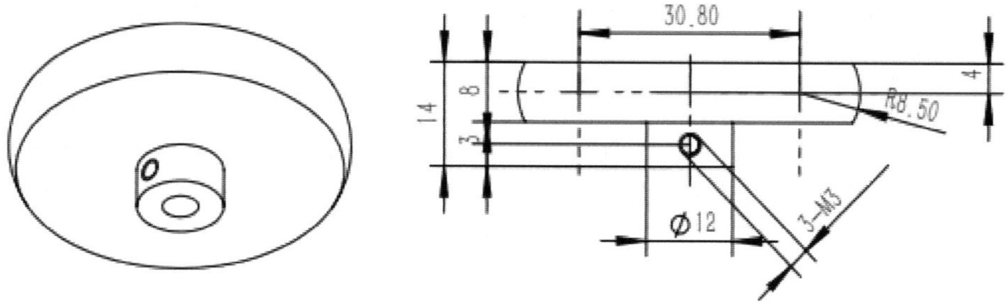

Figure 2. Radius wheel structure diagram

3. The control system design

According to the measurement principle and the overall program, the work flow in the concrete practical work is as shown in Figure 3.

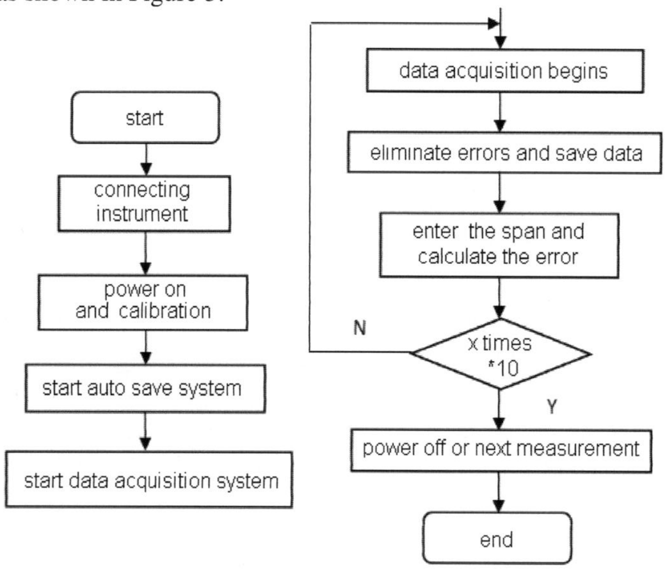

Figure 3. Total work flow chart

As shown in Figure 3, the center of the system is the process of the photoelectric autocollimator and the automatic data acquisition process. The number of data collected should be delineated according to the length of the measured gage.

Since the photoelectric self-collimator is set on the measured side of the ground through a dedicated tripod rack, the general measurement needs to be measured in a small vibration range of the room in the measurement process in order to avoid noisy environment and large vibration of the floor. Taking into account the impact of thermal deformation, the environment condition should be adjust to constant temperature and humidity state to avoid interference from other factors. In general, each measurement needs to be done ten times before the results be calculated.

As shown in Figure 4, the main hardware modules of the measurement system and the working relationship are shown. The core part of the system is STC89C52 microcontroller. It completes data collection, automatic data storage, traffic control, and the automatic acquisition instrument control.

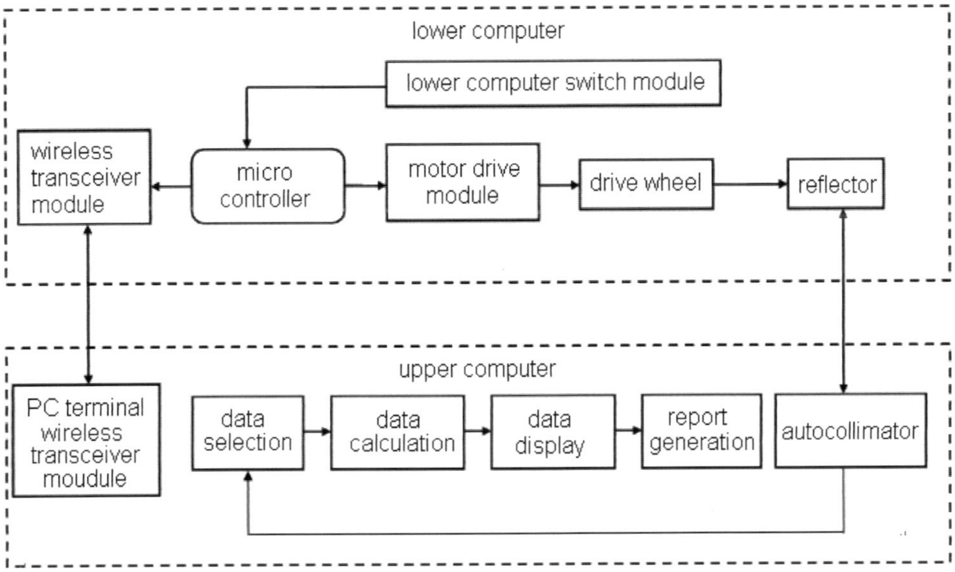

Figure 4. Hardware module and its working relationship diagram

4. Experimental results

The working environment during the inspection process is: constant temperature 19.96 °, humidity 41.6 °.

In the course of the experiment, 10 repeat experiments were carried out for the same rectangular scale, and the verticality was analyzed for each group of data. Then, the composite test was carried out. The following is a group of 10 data, as shown in Table 1.

Table 1. Rectangular ruler experimental data

Group \ No.	1	2	3	4	5	6	7	8	9	10
1	20.72	20.86	20.96	19.91	20.09	21.03	20.06	19.86	20.41	21.29
2	-35.23	-37.33	-39.01	-34.98	-35.77	-36.69	-35.02	-34.21	-35.63	-36.55
3	2.79	3.09	2.45	2.11	2.80	2.54	3.12	1.99	3.32	2.98
4	8.04	7.89	8.27	9.05	7.94	8.86	8.97	8.04	7.68	9.37
5	-3.85	-3.90	-2.92	-4.12	-4.67	-3.54	-3.05	-4.85	-4.19	-2.85
6	-37.54	-35.67	-35.74	-37.89	-36.48	-37.96	-35.25	-36.77	-34.99	-37.83
7	1.09	0.77	0.24	1.13	-0.03	2.01	1.73	1.60	0.89	1.84
8	57.43	57.88	56.13	58.49	56.62	57.71	58.05	56.36	57.57	56.24
9	26.69	25.17	26.98	27.22	26.74	25.72	28.02	25.85	26.46	27.63
10	3.71	4.43	3.29	3.11	4.83	5.07	4.86	3.99	4.58	3.81
11	73.08	73.77	74.21	72.63	73.81	74.26	72.87	73.33	72.90	74.59

12	-0.75	-0.92	-2.63	-1.86	-0.99	-2.07	-1.71	-1.42	-2.22	-1.45
13	-7.90	-8.59	-8.26	-9.03	-7.72	-8.68	-8.11	-7.35	-9.14	-7.99
14	67.90	65.90	65.78	66.25	67.16	65.93	67.47	66.64	65.95	67.16
15	21.90	23.95	22.89	23.23	21.39	24.05	23.56	22.72	23.41	22.37
16	7.90	7.11	6.95	6.79	8.13	7.74	7.48	8.25	7.56	6.90
17	-11.92	-13.00	-12.83	-12.54	-11.11	-10.98	-11.58	-10.18	-12.32	-13.12
18	-54.39	-54.99	-53.26	-52.75	-55.06	-53.94	-52.97	-54.01	-53.94	-52.99
Verticality	21.16	22.27	21.74	21.09	22.06	21.17	21.36	21.86	22.19	21.55

The ruler is $1000mm \times 630mm$, data is reached through the automated verification device. The collected values are relative values, and the results are analyzed using the minimum inclusive area method. The data analysis chart is shown in Figure 5.

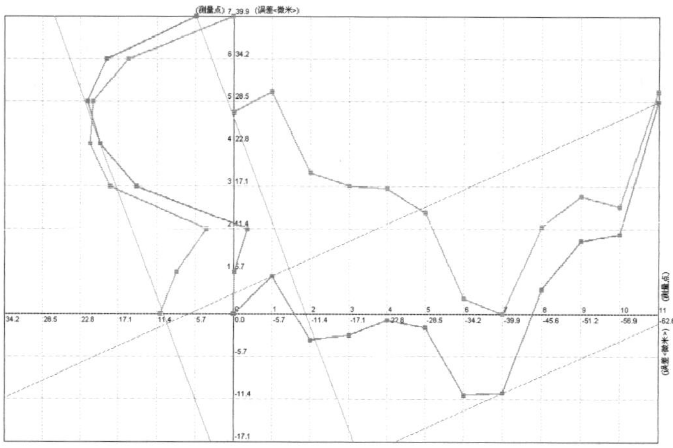

Figure 5. Error diagram with minimum area method of experimental data

In this paper, the experimental rectangular ruler is the calibration rectangular ruler, the calibration report shows the measured value is 23um which is close to the experimental value. So this experiment has the reliability and the system is feasible.

5. Conclusion

In summary, the device described in this paper can greatly reduce the positioning error of large rectangular ruler measurement. It improves the measurement accuracy and meets the 《JJG7-2004 rectangular test procedures》 technical requirements. The optical path of the optoelectronic self-collimator can reach the longest distance up to 10m, so the device can detect the rectangular ruler side length up to 8m theoretically. With the increase in the length of the edge, the difficulty of straightening work of the optical path of will increase, so the use training of optical self-collimator is needed.

References

[1] Wang Tao. Portable device for measuring verticality of right angle ruler [J]. Brand and standardization, 2011
[2] Zhao Guoxin. Uncertainty evaluation of verticality measurement results of wide seat right angle ruler [J]. China metrology, 2013
[3] Li Zhifei. Design of automatic verification device for electronic scales based on Labview[J]. Journal of China Jiliang University, 2012
[4] Gu Li. Development and application of automatic verification device for multimeter [J]. Measurement technology, 2015

[5] Wang Hua. Evaluation of uncertainty in measurement of wide seat square [J]. Industrial measurement, 2016
[6] Qiao Li. Research on straightness measuring device of knife edge ruler [J]. Measurement technique, 2015

IOP Conf. Series: Materials Science and Engineering **339** (2018) 012035 doi:10.1088/1757-899X/339/1/012035

Optimization of the High-speed On-off Valve of an Automatic Transmission

ZHAO Li-mei[1, 2], WU Huai-chao[2, 4], ZHAO Lei[2], LONG Yun-xiang[3], LI Guo-qiao[3] and TANG Shi-hao[1]
[1] Key Laboratory of Advanced Manufacturing Technology of Ministry of Education, Guizhou University, China
[2]College of Mechanical Engineering, Guizhou University, Guiyang , China
[3]Guizhou Winstar Hydraulic Transmission Machinery Co., Ltd, Zunyi, China

Email: zlm0226@163.com

Abstract. The response time of the high-speed on-off solenoid valve has a great influence on the performance of the automatic transmission. In order to reduce the response time of the high-speed on-off valve, the simulation model of the valve was built by use of AMESim and Ansoft Maxwell softwares. To reduce the response time, an objective function based on ITAE criterion was built and the Genetic Algorithms was used to optimize five parameters including circle number, working air gap, et al. The comparison between experiment and simulation shows that the model is verified. After optimization, the response time of the valve is reduced by 38.16%, the valve can meet the demands of the automatic transmission well. The results can provide theoretical reference for the improvement of automatic transmission performance.

1. Introduction

High-speed on-off valve is a new type of digital control hydraulic element [1]. In automatic transmission, high speed on-off valves play a lead role in the hydraulic buffer control of the vehicle shift clutch and directly determine the automatic control performance of planetary gear systems in the hydraulic transmission. At present, the main problem in the hydraulic control system is that the fast response performance of the high speed on-off valve is not reach the standard required since its design techniquete is bad. Because of this problem, control capability and the performance of the hydraulic control system in automatic transmission are greatly influenced.

In recent years, optimization of solenoid valves has been paid more and more attention [2-7]. Zhang Sheng-chang from Shanghai Jiaotong University has proposed a high speed solenoid valve performance prediction model and iteration algorithm, established hybrid optimization design method of high speed solenoid valve [3,4]. Dai Dong from Nanjing Sci- Tech University optimized the electromagnet parameters of the parallel spool type high pressure pneumatic solenoid valve and shortened the response time by multi – population genetic algorithm [5]. The finite element simulation method of structural dynamic design of ε type high-speed on-off valve electromagnet has brought forward by Xiang Zhong from Zhejiang University [6]. Tao Run has optimized the main parameters that influence the solenoid valve response time of the pressure valve and pressure reducing valve of the ABS braking system by genetic algorithm and shortened the response time [7]. But the above research works are mainly optimizations of the structural parameters of the electromagnet and do not involve the circle number, the working stroke parameter optimization. And the optimization of high-speed on-off valve of the automatic transmission is rarely reported.

In this work, the joint simulation model was built for the high-speed on-off valve in a certain automatic transmission. To reduce the response time, an objective function based on ITAE criterion was built and the Genetic Algorithms was used to optimize the key factors related to the dynamic response characteristic including circle number. So this work will offer technical support and theoretical reference for the development of a high-performance high-speed on-off valve.

2. Constitution and principle

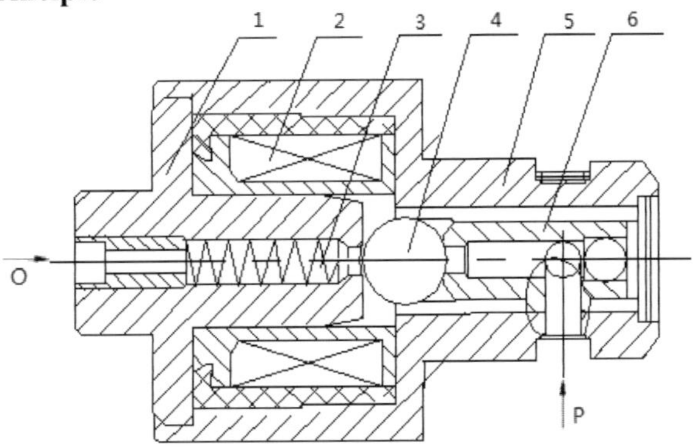

Figure 1. Schematic diagram of high-speed on-off valve
1 pole shoe 2 coil 3 spring 4 ball 5 valve body 6 valve seat

The structure of a certain high-speed on-off valve is shown in Figure 1. It is mainly composed of a pole shoe, coil spring, ball, a valve body and a valve seat. The high-speed on-off valve is a two position – three way and normally closed valve. Electromagnetic force generated when the coil is energized, the steel ball overcomes the spring force and moves to the left under the electromagnetic force and hydraulic pressure, so the left oil drain port is blocked, the oil from the P port flows out of the outlet by the ball, and then flows to other valves or equipments. When the coil power is out, the electromagnetic force disappears, the spring restoring force moves the ball to the right, so the ball againsts the seat, oil drain port opens, and hydraulic oil accumulated in the gas gap flows back to the tank through the drain port.

3. Modeling and verification of joint simulation model

Ansoft Maxwell and AMEsim software have a strong complementarity [8-11]. In this work, Ansoft Maxwell and AMEsim softwares were combined to establish a dynamic performance simulation model in this work.

3.1. Ansoft Maxwell simulation model

The simulation model of the high speed on-off valve is established using Ansoft Maxwell software as shown in Figure 2 Valve body, pole shoe, coil shell and ball, etc. magnetic materials were modeled. The loop material was set to copper, alve body, pole shoe and ball were set to Steel, and air model was set to air. The air gap length of high speed on-off valve is 0.6mm. With external circuit functions of Maxwell, the excitation of the winding was realized using field circuit coupling finite element calculation. High level voltage is 12V, the low level voltage is 0V, power on time is 20ms, and pulse voltage cycle is 40ms.

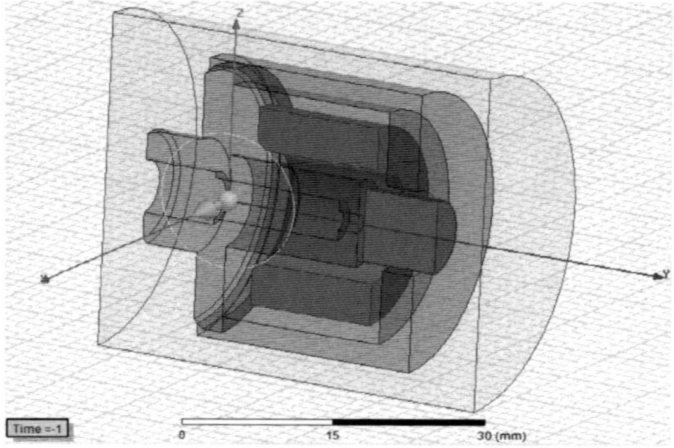

Figure 2. Ansoft Maxwell simulation model of high speed on-off valve

3.2. United simulation model

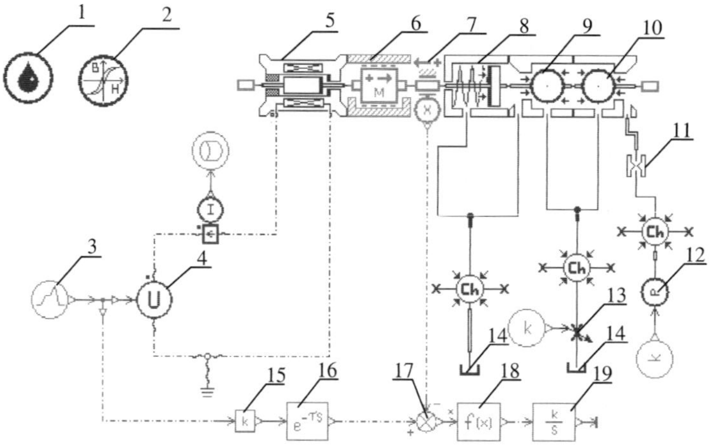

Figure 3. Dynamic performance united simulation and optimization model

The united simulation model of high speed on-off valve dynamic performance is shown as Figure 3. Each submodel in the simulation model was selected according to the specific structure of the high speed on-off valve and working principle [12]. In Figure 3, the submodel 1, 2 are used to define the characteristics of hydraulic oil and magnetic material properties. In electromagnetic model, the control signal model 3 and voltage source model 4 are used to simulate the control voltage signal. EMLT40 submodel 5 is used to simulate electromagnet. The ball quality is simulated by mass model 6, displacement sensor model 7 is used to simulate displacement sensor, spring module 8 is used to simulate spring and the ball valve sub model 9 is used to simulate spool and valve seat. According to the working principle of the high speed on-off valve, when the coil is powered, the ball moves to the left, block the oil drain port, so add a ball valve submodel 10 in the simulation model. Damping hole model 11 is used to simulate damping hole. Pressure source model 12 is used to simulate fuel supply pressure. Variable orifice model 13 is used to simulate load pressure. Tank model 14 is used to simulate tank. The electromagnet simulation results of Maxwell Ansoft software were catalogued in data tables that can be called by AMEsim software. Then, read the data tables into the simulation model by electromagnetic module EMLT40. So Maxwell Ansoft and AMEsim Softwares united simulation was realized.

4. Optimization of High-speed On-off Valve

4.1Mathematical model of optimal design

The selectivity and practicability of ITAE criterion are relatively good. ITAE criteria can reflect the comprehensive control performance of the system [12]. In order to achieve quick response requirement, the response time of the high-speed on-off valve is an important index in the dynamic design and is related to the error of transient displacement response. The objective function based on the ITAE criterion is the integral of the time between the actual output and the desired output. The objectiv function can take into account the high speed on-off valve response time and transient error, requires short response time and small integral of the transient error. Therefore, the objective function of the optimization has established as equation (14):

$$J = \int_0^{\infty} t\left|e(t)\right|dt \qquad (14)$$

$$e(t) = S(t) - S_0(t) \qquad (15)$$

Where, $e\ (t)$ is the error between the actual displacement and the desired displacement of the ball when step drive voltage inputting system, $s(t)$ is the actual displacement of the ball; $S_0(t)$ is the desired displacement of the ball.

In the simulation model, $S_0(t)$ is calculated by the driving voltage, proporional element 15 and delay link 16, $S(t)$ is measured by displacement sensors, the displacement error $e(t)$ is obtained by comparison point 17.

From the above analysis, the optimization objective of the problem is to minimize the $J(x)$. Circle number, spring preload, spring stiffness, the quality of the ball, working air gap five parameters are optimized variables. High speed on-off valve optimization design mathematical model is as follows:

$$\begin{aligned} &\min J(X) \\ &s.t.\quad X = \left(N, F_{k0}, K, m, \delta\right)^T \\ &\qquad\ X \in \Omega \\ &\qquad\ X_{\min} \le X \le X_{\max} \end{aligned}$$

Where, X is the optimization variables; N is the circle number; F_{k0} is spring preload; K is spring stiffness; m is the quality of the ball; δ is working air gap; Ω is design domain; x_{min} and x_{max} are lower bound and upper bound of optimization variables respectively. Considering the structure of the valve and actual working conditions of the automatic transmission, the constraint conditions including the circle number, spring preload, spring stiffness, the ball quality, working air gap are provided respectively as follows:

700 turn$\le N \le$1300 turn、27 N$\le F_{k0} \le$33 N、12 N/mm$\le K \le$18 N/mm、0.006 kg$\le m \le$0.014 kg、0.5mm$\le \delta \le$0.7mm.

4.2 Optimization result analysis

AMESim genetic algorithm module was used to optimize. The population quantity of genetic algorithm was set to100, the crossover probability was set to 80%, the variation rate was set to 10%, and the number of generations was set to 50. Parameters before and after optimization are compared in Table 1. In order to facilitate processing, the optimized parameters are modified. The analysis results show that the correction of parameters of high-speed on-off valve has little impact on response time. The comparisons of displacement response curves before and after optimization are shown as Figure 5. The opening response time is reduced from the original 8.2ms to 6.2ms, and closing response time is reduced from the original 12.5ms to 6.6ms. The total response time is decreased by 38.16% and meets the requirements of fast response performance. As shown in Figure 6., the maximum electromagnetic force is reduced from 39N to 37N, it doesn't fall much. The electromagnetic force is still able to meet the requirements of the work.

Table 1. Comparison of parameters before and after optimization.

Parameters	circle number / turn	spring preload /N	spring stiffness N/mm	the quality of the ball /g	Working air gap /mm	opening response time /ms	Closing response time /ms
Original parameter	1000.00	30.00	15.00	10.00	0.60	8.20	12.50
Optimized parameters	703.28	29.94	16.00	7.61	0.64	6.20	6.50

IOP Conf. Series: Materials Science and Engineering **339** (2018) 012035 doi:10.1088/1757-899X/339/1/012035

Modified parameters	703.00	30.00	16.00	7.60	0.64	6.20	6.60

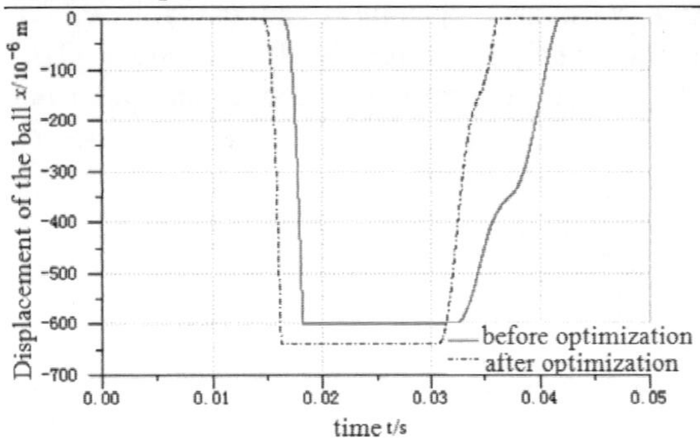

Figure 5. Comparison of displacement response curves before and after optimization

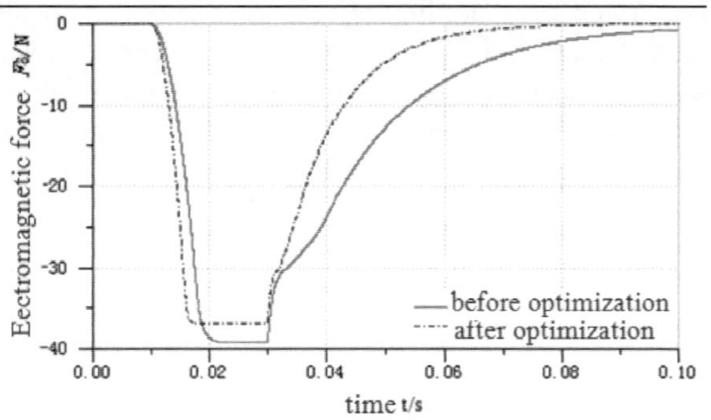

Figure 6. Comparison of electromagnetic force curves before and after optimization

5. Conclusion

A joint dynamic performance simulation model of the high-speed on-off valve was built by use of AMESim and Ansoft Maxwell software. The model has realized the coupling of electromechanical, liquid and magnetic. The experimental datas verified the reliability of the model.

To shorten the response time, the ITAE Criterion was used to create the optimization objective function and the genetic algorithm was used to find the optimal solution. The main factors affecting performance of the high speed on-off valve has optimized. Through the optimization, the opening response time is reduced from the original 8.2ms to 6.2ms, and closing response time is reduced from the original 12.5ms to 6.6ms. The total response time is decreased by 38.16%.

References

[1] Man Jun, Ding Fan and Li Yong Dynamic characteristics of high-pressure long-stoke high-speed on-off solenoid 2010 *J. China Coal Society* **35** 871-5

[2] Yuan Hong-bin, Zhang Min-qing and Sun Yan-tang AMESim-based simulation analysis of dynamic characteristics of direct-acting solenoid valve 2011 *J. Rocket Propulsion* **33** 30-5

[3] Zhang Sheng-chang, Xu Yang-zeng and Zhong Ting-xiu Extra-high Speed Digital Valve StudyBased on New Performance Predicting Methods 2001 *J. Hyd. Pneum. &Seals* **21** 1-3

[4] Zhang Sheng-chang, Zhong Ting-xiu and Xu Yang-zeng A New Algorithm for Coupled Solutions Systems in Dynamic Simulation of Digital Valve 2002 *J. Hyd. Pneum. &Seals* **22** 16-8

[5] Dong Dai, Li Xiao-ning and Guo Zhong-hua Optimal design of parallel-spool high-pressure solenoid valve 2014 *J. Transactions of the Chinese Society of Agricultural Eng.* **30** 43-50

[6] Xiang Zhong and Tao Guo-liang 2010 Research on the Key Technologies of Pneumatic High-Speed On-off Vavle (Hangzhou:Zhejiang University)

[7] Tao Run, Zhang Hong and Fu Dechun Simulation of ABS hydraulic system and optimization of solenoid valve 2010 *J. Transactions of the Chinese Society of Agricultural Eng.* **26** 135-9

[8] Su Ming and Chan Lun-jun 2011 Study on control characteristics and method of high speed on-off valve solenoid (Guiyang: Guizhou University)

[9] Wang Yangbin, Xu Bing and Liu Yingjie Simulation Analysis of the Dynamic Characteristics of the Solenoid Based on Ansoft and AMESim 2008 *J. Machine tools and hydraulic* **36** 104-8

[10] Zhang Tingyu and Zhang Guoxian Performance Analysis and Investigation to High Speed Digital Valve 2006 *J. Machine tools and hydraulic* **9** 139-42

[11] Kong Xiaowu and Li Shizhen Dynamic performance of high speed solenoid valve with parallel coils 2014 *Chinese J. Mech. Eng.* **27** 816-21

[12] Paul D. Walker, Bo Zhu and Nong Zhang Nonlinear Modeling and Analysis of Direct Acting Solenoid Valves for Clutch Control 2014 *J. Dynamic Systems, Meas., and Ctrl.* **136** 051023

Acknowledgements
This project is supported by training plan for high-level innovative talent in Guizhou Province (Grant No.Q.K.H.P.T.R.C[2016]5659), the science and technology major project of Guizhou Province in China (Grant No.Q.K.H.Z.D.Z.X.Z [2013]6015) and the introduction of talent fund of Guizhou University in China (Grant No. G.D.R.J.H.Z[2014]41).

IOP Conf. Series: Materials Science and Engineering **339** (2018) 012036 doi:10.1088/1757-899X/339/1/012036

Study on the influence of parameters of medical drill on bone drilling temperature

Xianchun XU[1,2], Yahui Hu[1,2,3], Jingwang Han[1,2], Lin Yue[1,2] and Wangbiao Jiang[1,2]

1.Tianjin Key Laboratory of Design and Intelligent Control of Advanced Mechatronical System, Tianjin, 300384, China;
2.Tianjin University of Technology, 300384, China.

Email: huyahuilg@126.com

Abstract. During surgical interventions, the temperature generated during cortical bone drilling can affect the activity of bone material, which may lead to necrosis. In this paper, with the purpose of reducing the temperature during cortical bone drilling, the influence of the parameters of medical drill were analyzed. The finite element model of the drilling process was established based on the parametric design of the dril. The relationship between the drill bit diameter, the point angle, and the helix angle to the drilling temperature was studied by the center composite experiment. The results showed that the drilling temperature is increased with the increase of drill diameter, vertex angle and helix angle in the range of certain research.

1. Introduction

During surgical operations, drilling in the bone with medical twist drills is extensively carried out before some operations. Drilling force, temperature and quality are the important factors during the drilling process that could influence the integrity of the bone-impact interface[1].

Drilling temperature, which arises from shear deformations and the friction among the drill, bone debris and the bone,may lead to cell or bone necrosis[2-5]or cause ring sequestra around drilled hole[6]. There are studies in literatures about the registration of temperature during bone drilling that indicates thermal necrosis in cortical bone when this reached a temperature of 47℃for 60s[7-8]. Other authors showed that temperature above 55℃for a period longer than 30s can cause great irreversible lesions in bone tissue[7-9].It is known that the heat generated in these processes is associated to the drilling parameters,such as rotational speed,feed rate,geometry of the drill bit,drill force,hole depth and the bone density[10].

The complex relationships among the geometry of the drill bit, drilling conditions, mechanism of the chip removal and the properties of the bone presents a great challenge in developing a mathematical model for the calculation of the heat generated during bone drilling[11]. Some studies showed that small angles drill has some advantages in drilling.The maximum axial force and temperature created by the drill with small angles is less than the ordinary standard twist drill, the damage area is smaller and the processing quality is better[12-13].This paper intends to present an approach to developing a mathematical relationship between the temperature produced during cortical

IOP Conf. Series: Materials Science and Engineering **339** (2018) 012036 doi:10.1088/1757-899X/339/1/012036

bone drilling and the geometric parameters of medical drill such as diameter, point angle and helix angle by using response surface methodology.

2. FE model of drilling and experiments

2.1 Parametric design of the drill

The geometry and the parameters of the drill bit is shown in figure.1

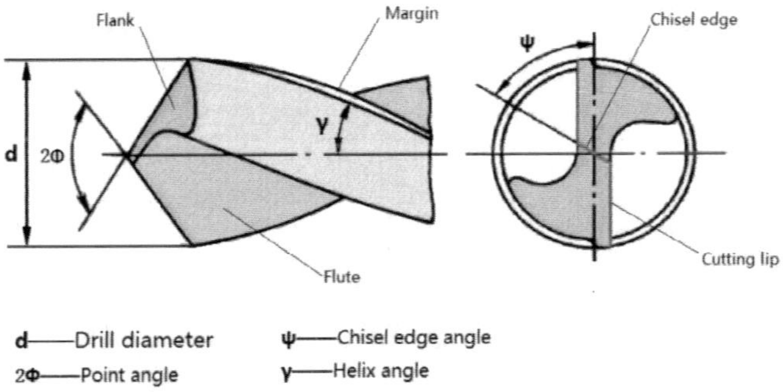

Figure 1. The geometry and the parameters of the drill bit

A 3D solid model with parametric design of the drill was established according to the mathematical model of its blade by UG\Open Grip.Some parameters to create the drill model we need were input from the human-computer interactive interface, shown as figure 2(a). Three dill models of diffierent parameters created by this method, shown as figure 2(b).

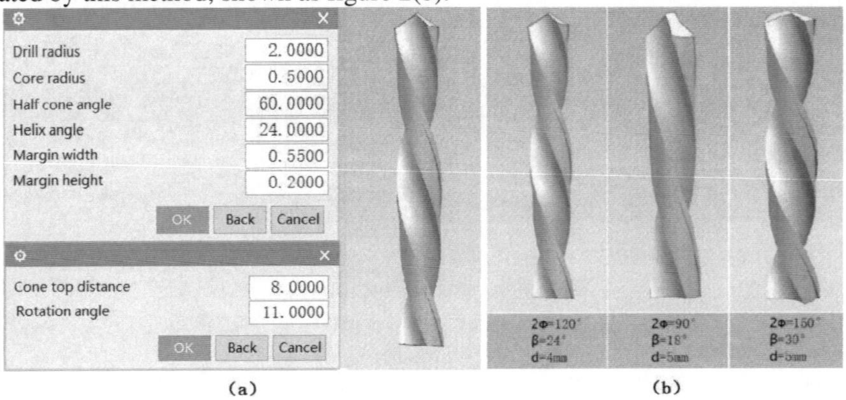

Figure 2. Models created by parametric design

Then the solid model of drill was saved as(.x_t) and transformed it into ABAQUS13.1. The model of cortical bone was replaced by the disk with a diameter of 10mm and a thickness of 4 mm[14].

2.2 Material properties

The material for the stainless steel medical twist drill is the 4Cr13.The cortical bone was modeled as transversely isotropic elasto-plastic rate dependent material to simplify our research[15].The material properties of the two parts are shown in Table 1.

Table 1. Material properties

Material	Cortical bone	Drill
Density/（$t \cdot mm^{-3}$）	2.00×10^{-9}	7.84×10^{-9}
Young's modulus/ MPa	20 000	210 000
Poisson's ratio	0.36	0.24
Plastic / MPa	106.99	585
Specific Heat /（$mJ \cdot t^{-1} \cdot ℃^{-1}$）	1.26×10^{9}	4.90×10^{8}

Conductivity / （Watt • m^{-1} • ℃$^{-1}$）	0.56	47.7

2.3 Boundary condition and mesh

In the boundary conditions setting,the degrees of freedom on the sides of the cortical bone are fixed at all.The drill-bit was modeled as a rigid body (reference point is RP-1)because the elastic stiffness of the stainless steel twist drill-bit used is in the range of 200-220 GPa as compared to 20 GPa for the cortical bone;this reduces the computational cost involved in the highly resource consuming drilling simulations[16].And the drill-bit was constrained to rotate only about its own longitudinal axis with a specified speed and feed vertically downwards into the work piece as shown in figure3(a).

The material properties and the unit types were assigned to each parts.The dill-bit was modeled with Tet unit, the unit type is C3D4T and meshed with 23318 elements; The cutting area where the bone is in contact with the drill bit was modelled with Wedge unit, the unit type is C3D6T and meshed with 6688 elements;Other areas on the cortical bone was modeled with C3D8RT and meshed with 29440 elements.The meshing is shown in the figure 3(b). Contact between the dill and cortical bone was defined by the surface to surface contact algorithm available in Abaqus/explicit.And the frictional contact between the dill and cortical bone was modelled with a constant coefficient of friction of 0.7[17].

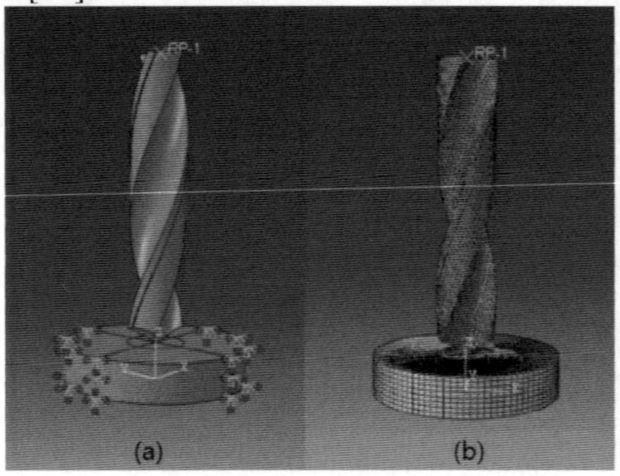

Figure 3. Boundary conditions and mesh

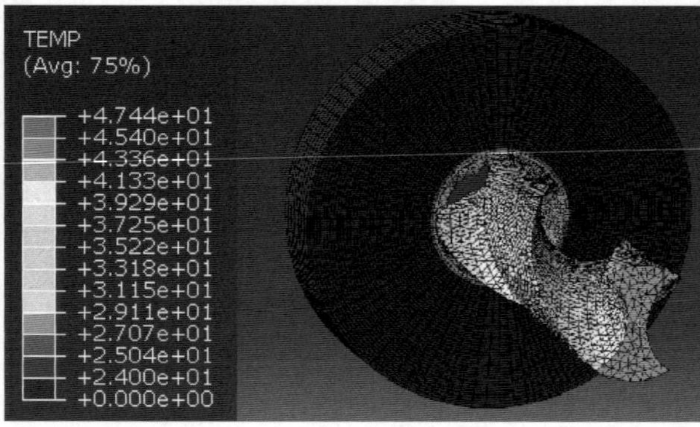

Figure 4. Cortical bone drilling temperature distribution

2.4 Experimental verification

In this simulation, the stainless steel 4Cr13 drill was used. And it had a diameter of 4mm, a point angle of 120°and a helix angle of 24°. Simulation was carried out at a speed of 1000rpm and a feed rate of 70mm/min. The initial temperature was set at 24℃. Running the model and the result was gotten as shown in figure 4. The maximum temperature of the drilling area after drilling is 47.44°.

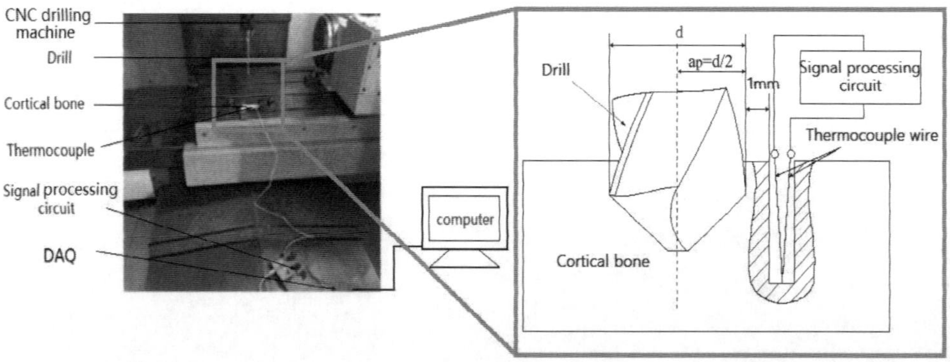

Figure 5. Measurement system of bone drilling temperature

The experiments were carried out on the YCM-V65A Vertical Machining Center, the parameters of the drill are the same as the simulation model. The measuring system consists of thermocouple sensor, signal processing circuit, DAQ card and a computer. The experimental setup is shown as figure 5.

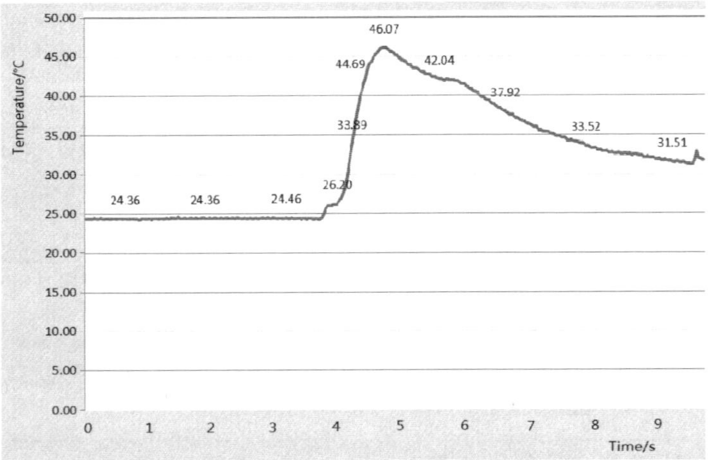

Figure 6. Curve of drilling temperature

When the experiment was carried out at a speed of 1000rpm and a feed rate of 70mm/min,the maximum temperature of the cortical bone drilling zone measured by thermocouple method is 46.07℃. And figure 6. shows that the drilling temperature firstly increased rapidly to the maximun value and then decreased moderately.

Figure 7. is the comparisons between the experimental results and the FE simulation results. In the first set of the comparisons, we used the same feed rate(70mm/min) and drilling in the cortical bone specimens were carried out at drilling speed of 800rpm, 1000rpm, and 1200rpm. In the second set of the comparisons, we used the same speed(1000rpm) and the feed rates were taken as 50mm/min, 60mm/min, and 70mm/min. The similar results were gotten from FE modeling and experimental tests. So the simulation model based on ABAQUS is credible.

Figure 7. Comparisons of simulation results and experiment results

3. Results and Discussion

3.1 Experiment design and results

IOP Conf. Series: Materials Science and Engineering **339** (2018) 012036 doi:10.1088/1757-899X/339/1/012036

RSM is an aggregation of mathematical and statistical tools that are efficacious for the modeling and analysis of process in which a response is influenced by several variables and the objective is to optimize that response[18-19].In this research,the influence of drill diameter, point angle and helix angle on drilling temperature were studied. We set up the center composite experiment by the software Design-Expert.The drilling process dynamics model was established by using ABAQUS finite element software to get the prediction model of drilling temperature,which provides theoretical guidance for the optimization of medical drill bits. The experimental factor levels were given in Table 2.

Table 2. Factors and levels for drilling model

Experiment factor / level	-1.6818	-1	0	1	-1.6818
Diameter/d（mm）	2.32	3	4	5	5.68
Vertex/2Φ(°)	69.55	90	120	150	170.45
Helix /β(°)	13.91	18	24	30	34.09

According to the central composite experiment, the experimental scheme of Table 3. was obtained. The drilling were carried out at rotational speed of 1000rpm and feed rate of 70mm/min. All simulations were performed at room temperature(24℃). The ABAQUS was used to calculated the schemes one by one, and the results were shown in the Table 3.

Table 3. Experimental conditions and results

Std	Run	Diameter d（mm ）	Point angle 2Φ(°)	Helix angle β(°)	Temperature (℃)
6	1	5.00	150.00	18.00	46.00
4	2	3.00	150.00	30.00	42.31
15	3	4.00	120.00	24.00	47.44
8	4	5.00	150.00	30.00	50.78
19	5	4.00	120.00	24.00	47.44
12	6	4.00	120.00	34.09	42.56
13	7	2.32	120.00	24.00	36.30
20	8	4.00	120.00	24.00	47.44
2	9	3.00	150.00	18.00	42.57
5	10	5.00	90.00	18.00	42.00
11	11	4.00	120.00	13.91	43.03
10	12	4.00	170.45	24.00	49.53
3	13	3.00	90.00	30.00	37.74
18	14	4.00	120.00	24.00	47.44
14	15	5.68	120.00	24.00	45.93
17	16	4.00	120.00	24.00	47.44
7	17	5.00	90.00	30.00	46.48
1	18	3.00	90.00	18.00	35.69
9	19	4.00	69.55	24.00	39.22
16	20	4.00	120.00	24.00	47.44

3.2 Response surface analysis

The results of Table 3 were input to the experimental design software(Design-Expert 8.0.6)for analysis.The quadratic regression equation of drilling temperature is shown as(1).

$$T = -26.49 + 0.42\Phi + 1.65\beta + 17.63d - 1.39\times10^{-3}\Psi\cdot\beta - 1.31\times10^{-2}\Psi\cdot d + 0.15\beta\cdot d - 1.03\times10^{-3}\Psi^2 - 4.12\times10^{-2}\beta^2 - 2.08d^2 \quad (1)$$

Where: Ψ (2Φ) —— Point angle (°).

The ANOVA results for the response surface model were given in Table 4. F is the significance, P is the confidence level. And the P value of the model is <0.0001 which means that the factor is significant. In this case A,C,A^2,B^2 and C^2 are significant model factors.

Table 4. AVOVA table for quadratic model

Source	Sum of Squares	df	Mean Square	F-value	P-value (Prob>F)	
Model	344.96	9	38.33	34.18	<0.0001	significant
A-Point angle	100.73	1	100.73	89.82	<0.0001	
B-Helix angle	7.71	1	7.71	6.87	0.0255	
C-Diameter	136.31	1	136.31	121.55	<0.0001	
AB	0.51	1	0.51	0.45	0.5174	
AC	1.24	1	1.24	1.11	0.3177	
BC	6.98	1	6.98	6.22	0.0318	
A^2	12.38	1	12.38	11.04	0.0077	
B^2	31.80	1	31.80	28.36	0.0003	
C^2	62.32	1	62.32	55.57	<0.0001	

The plot of the residuals against the predicted temperature is shown in figure 8. The figure shows that the residual points distributed near a straight line. So the model proposed predicts of the temperature with reasonable accuracy and can be effctively used for the prediction of temperature during bone drilling.The above mathematical model can be used to predict the values of the drilling temperature within the limits of the factors studied.

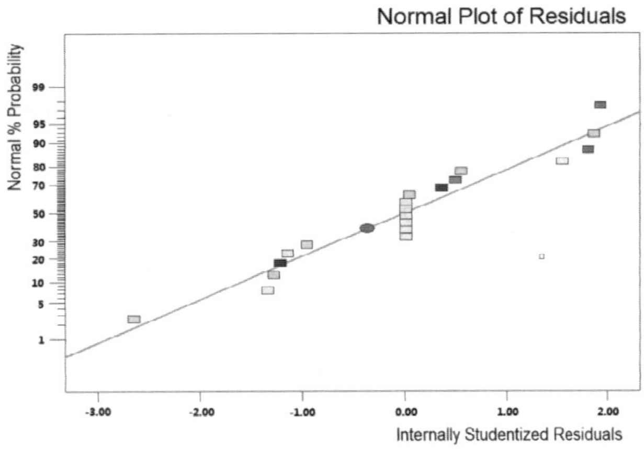

Figure 8. Curve of drilling temperature

The response surface plots obtained by the modeling of temperature in cortical bone drilling are shown in figure 9. It can be seen from the figure that the drilling temperature increases with the increase of the drill diameter, the point angle and the helix angle within the limits of the factors studied. The drill-bit diameter and the point angle have a great influence on the drilling temperature, and the helix angle has little effect on the drilling temperature.

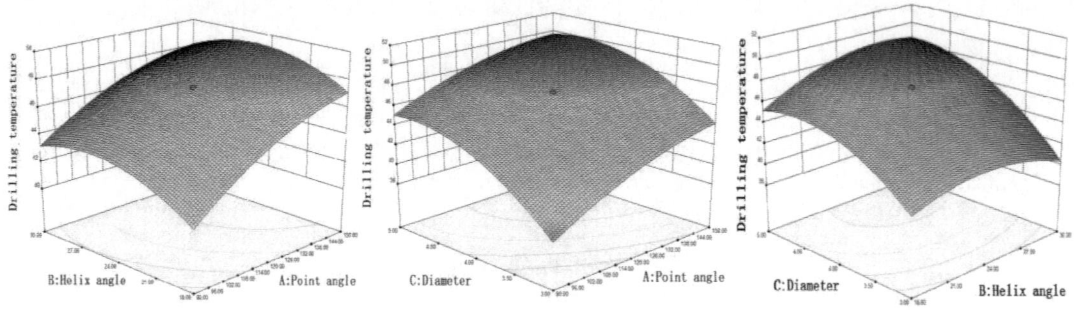

Figure 9. Response surface plots of drilling temperature during bone drilling as a function of point angle, helix angle and diameter.

4. Conclusions

In this paper,with the aim of decreasing the temperature during cortical bone drilling, a surgical drill bit has been optimized, the influence of the drill diameter, the point angle and the helix angle on the drilling temperature were analyzed.The following observations can be made in this research:

•The finite element model of drilling on cortical bone was developed using a FE software Abaqus/Explicit based on the three-dimensional solid model of the drill with parametric design.The parameters that can be controlled are diameter, point angle and helix angle.

•The finite element model predicted cortical bone drilling temperature with reasonable accuracy when compared to the experimental results.

•The contribution of each factor and their interaction on the temperature produced during cortical bone drilling was evaluated by using RSM. It was observed that the drilling temperature increased with an increased in diameter, point angle and helix angle.The drill diameter has the highest influence on drilling temperature, followed by the point angle and the helix angle has a insignificant influence on drilling temperature.

Reference

[1] Augustin G, Zigman T, Davila S, Udilljak T, Staroveski T, Brezak D, Babic S 2012 *Clinical Biomechanics* **27** 313-325

[2] Fuchsberger 1988 *Unfallchirurgie* 14(4) 173-183

[3] Bachus K N, Rondian M T, Hutchinson D T 2000 *Medical Engineering and Physics* **22** 685-691

[4] Kreith F 2000 *Boca Raton: CRC Press LLC*

[5] Augustin G, Davila S, Udilhak T, Vedrina D S, Bagatin D 2009 *Archives of Orthopaedic and Trauma Surgery* 129(5) 703-709

[6] Davidson S R H, James D F 2003 *Journal of Biomechanical Engineering* **125** 305-314

[7] Tu Y K, Chen L W, Ciou J S, Hsiao C K, Chen Y C 2013 *J Med Biol Eng* **33** 269-274

[8] Eriksson AR, Albrektsson T 1983 *J Prosthet Dent* **50** 101-107

[9] Hillery M T, Shuaib I 1999 *J Mater Process Technol* **92-93** 302-308

[10] Sezek S, Aksakal B, Karaca F 2012 *Comp Mater Sci* **60** 13-18

[11] Rupesh K P, S.S. Panda 2013 *Journal of clinical orthopaedics and trauma* **4** 15-30

[12] Yang Y X, Wang C Y, Qin Z, Song Y X, Chen H Y 2010 *Machinery Design & Manufacture* **11** 118-120

[13] Wang M H, Sun G Q, Sun Y 2012 *Manufacturing Technology & Machine Tool* **12** 182-185

[14] He L, Hu Y H, Sun J B, Yang C L 2015 *Tools and technology* **49** 25-28

[15] Reilly D T, Burstein A H 1975 *Biomech Eng Aug* **3** 245-252

[16] Waqas A L, Kaddour B M, Ian A 2015 *Journal of mechanical behavior of biomedical materials* **42** 32-42

[17] Davidson S R H, James D F 2003 J *Biomech Eng* **125** 305

[18] Myers R H, Montgomery D C 1995 *Wily, New York*

IOP Conf. Series: Materials Science and Engineering **339** (2018) 012036 doi:10.1088/1757-899X/339/1/012036

[19] Kilicap E 2010 *Int J Adv Man Technol* **49** 911-923

Acknowledgments
This study was funded by the National College Students' Innovative Entrepreneurial Training Program (Grant No.201610060026) and the Tianjin Natural Science Foundation of China (Grant No.15JCZDJC32800)

The influence of installation angle of GGIs on full-tensor gravity gradient measurement

Hongwei Wei and Meiping Wu *

College of Mechatronics and Automation, National University of Defense Technology, Changsha 410073, China

Email: meipingwu@263.net

Abstract. Gravity gradient plays an important role in many disciplines as a fundamental signal to reflect the information of the earth. Full-tensor gravity gradient measurement (FGGM) is an effective way to obtain the gravity gradient signal. In this paper, the installation mode of GGIs in FGGM is studied. It is expected that the accuracy of FGGM will be improved by optimizing the installation mode of GGIs. In addition, we analysed the relationship between GGIs' installation angle and FGGM by establishing the measurement model of FGGM. Then the following conclusions was proved that there was no relationship between GGIs' installation angle and the measurement result. This conclusion showed that there was no optimal angle for the GGIs' installation in FGGM, and the installation angle only need to satisfy the relationship shown in the conclusion section of this paper. Finally, this conclusion was demonstrated by computer simulations.

1. Introduction

The gravitational gradient signal, as a high-precision signal for the change of the particle density in geological body, plays an important role in many disciplines [1-4]. Many measuring instruments have been developed for the measurement of gravity gradients [1, 5]. The earliest gravity gradiometer was torsion balance, developed by the Hungarian physicist Eötvös. This measurement device has poor precision and low efficiency, so it is quickly replaced by more efficient and high-precision measurement equipment.

One of the high-precision gravity gradient measurement equipment is rotating-accelerometer-based gravity gradiometer, which is widely used in mineral exploration. The core component of the gradiometer is the gravity gradient instrument (GGI) developed by Lockheed Martin. An independent component and a combined component of the gravitational gradient tensor can be measured by one GGI[6]. In order to achieve all the measurements of gravity gradient components, three GGIs are required. With these GGIs, three independent components and three combined components can be obtained.

As a kind of conservative field, the second derivative of gravitational potential satisfies Poisson equation. Therefore, there are only five independent components in the gravity gradient tensor. This means that there is a redundancy in the full-tensor gravity gradient measurement (FGGM). However, in the actual FGGM using Lockheed Martin's full tensor gravity gradiometer (FTG) [7], there was no redundancy. The installation mode of GGIs in FTG is shown in figure 1: The rotational axes of the three GGIs, shown in figure 1, are orthogonal to each other. The angle between the projections of the axis of the axes on the x-y plane of the platform coordinate system is 120°, and the angle between the axes and the x-y plane is 35°.

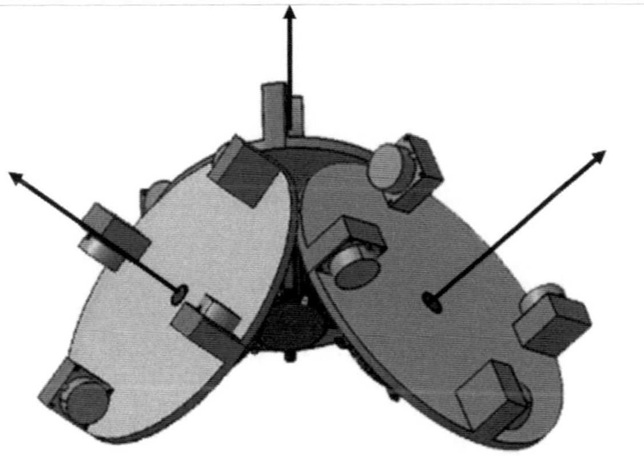

Figure 1. Installation mode of GGIs in FTG. The rotational axes of the three GGIs are orthogonal to each other.

In this article, the installation mode of GGIs will be analyzed. Furthermore, the reason for the loss of redundant information in FGGM is also explained. Finally, in order to improve the measurement accuracy, the relationship between the GGIs installation angle and FGGM is given in this paper.

2. The principle of GGI

The structure of GGI is shown in figure 2. There are four accelerometers symmetry mounted on a disk. The direction of the accelerometer sensitive axis is parallel to the tangent of the disk. The sensitivity of the two accelerometers is opposite. The radius of the disk is r, and the disk is rotating at constant angular velocity ω.

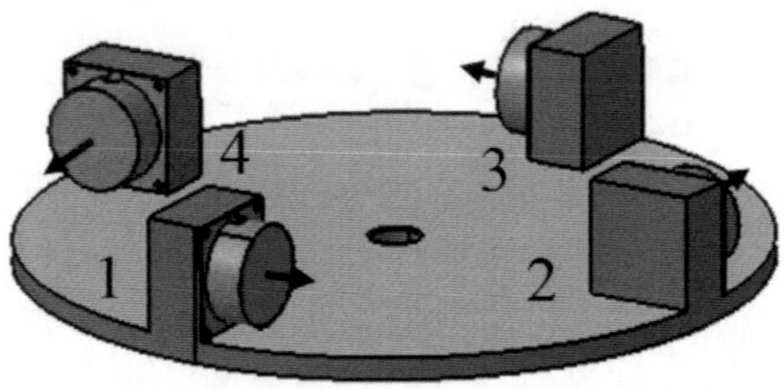

Figure 2. The structure of GGI. It consists of four accelerometers, the arrows represent the direction of the sensitive axis of the accelerometers.

The output value of the accelerometers on the turntable are denoted by f_i, where i is the number of the accelerometers.

The output signals of the four accelerometers are combined as follows:

$$U = f_1 + f_3 - f_2 - f_4 \tag{1}$$

Considering the direction of the accelerometer sensitive axis for accelerometer 1 and 3, 2, and 4 are opposite. The acceleration signal generated by the carrier motion in the combined signal U will be cancelled, thus the common-mode rejection for carrier motion acceleration was achieved. Finally, the relationship between the combined signal and the gravity gradient components is obtained as follows:

$$\begin{aligned} U &= 4r\left(\Gamma_{xy} + N_1\right)\cos\left(2\omega t\right) \\ &\quad - 2r\left(\Gamma_{xx} - \Gamma_{yy} + N_2\right)\sin\left(2\omega t\right) \end{aligned} \tag{2}$$

where Γ_{xy}, Γ_{xx}, and Γ_{yy} are the gravitational gradient components; N_1 and N_2 respectively represent the disturbance signals that are coupled together with the measurement signal Γ_{xy} and $\Gamma_{xx} - \Gamma_{yy}$, which can be calculated by the observation of the movement of the carrier.

Equation (2) shows that the gravity gradient signal is modulated to twice the rotation speed. The independent component Γ_{xy} and the combined component $\Gamma_{xx}-\Gamma_{yy}$ can be achieved by demodulating the combined signal U at the double frequency.

3. Relationship between GGIs installation angle and FGGM

In order to determine the relationship between GGIs installation angle and FGGM, the following co-ordinate frames as shown in figure 3 are used:

The Platform frame (*p*-frame) with its origin at the location of the stable platform, and axes aligned with the directions of north, east, and the local vertical (down).

The GGI frame (gi-frame, $i \in \{1,2,3\}$) with its origin at the center of the i-th GGI; the x-axis points to the starting position of the disc, and the z-axis points to the rotation axis of the disc, and the y-axis is in the plane of the disc.

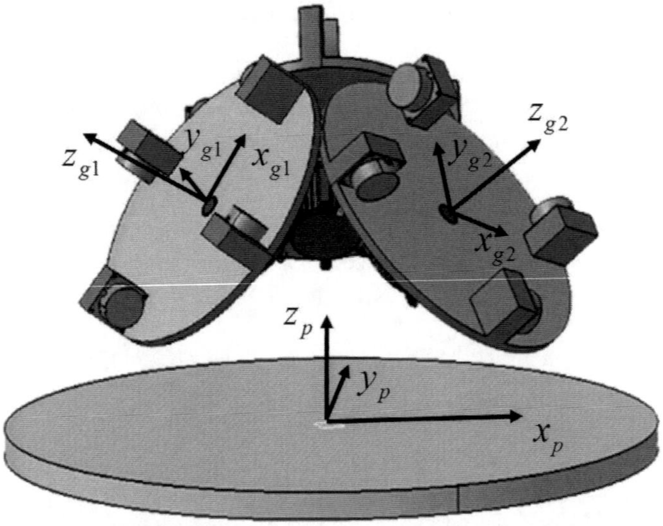

Figure 3. The diagram of the platform frame (*p*-frame) and the GGI frame (gi-frame).

In figure 3, the GGIs are placed on a stable platform, which tracks the local geographic frame.

Furthermore, it is necessary to clarify the transformation relation of gravity gradient tensor in different coordinate systems. According to the study of Zhao L, et al. [8], the transformation relationship between the gravity gradient tensor in different coordinate systems is as follows:

$$\mathbf{\Gamma}^{gi} = C_p^{gi}\mathbf{\Gamma}^p C_{gi}^p \tag{3}$$

where $\mathbf{\Gamma}^{gi}$ and $\mathbf{\Gamma}^p$ are the representation of gravity gradient tensor in the gi-frame and the *p*-frame, respectively; C_p^{gi} is the direction cosine matrix from *p*-frame to gi-frame, and C_{gi}^p is the transpose of C_p^{gi}.

Then, substituting the direction cosine matrix C_p^{gi} and gravity gradient tensor expressed in the *p*-frame $\mathbf{\Gamma}^p$ into equation (3) yields the gravity gradient components $\Gamma_{xx}^{gi}-\Gamma_{yy}^{gi}$ and Γ_{xy}^{gi} which are measured by the *i*-th GGI:

$$\begin{cases} \Gamma_{xx}^{gi}-\Gamma_{yy}^{gi} = \left(\left(c_{11}^i\right)^2-\left(c_{21}^i\right)^2\right)\Gamma_{xx}^p + 2\left(c_{12}^i c_{11}^i - c_{21}^i c_{22}^i\right)\Gamma_{xy}^p + 2\left(c_{11}^i c_{13}^i - c_{21}^i c_{23}^i\right)\Gamma_{xz}^p + \left(\left(c_{12}^i\right)^2-\left(c_{22}^i\right)^2\right)\Gamma_{yy}^p \\ \qquad + 2\left(c_{12}^i c_{13}^i - c_{22}^i c_{23}^i\right)\Gamma_{yz}^p + \left(\left(c_{13}^i\right)^2-\left(c_{23}^i\right)^2\right)\Gamma_{zz}^p \\ \Gamma_{xy}^{gi} = c_{11}^i c_{21}^i \Gamma_{xx}^p + \left(c_{21}^i c_{12}^i + c_{22}^i c_{11}^i\right)\Gamma_{xy}^p + \left(c_{21}^i c_{13}^i + c_{23}^i c_{11}^i\right)\Gamma_{xz}^p + c_{22}^i c_{12}^i \Gamma_{yy}^p + \left(c_{22}^i c_{13}^i + c_{23}^i c_{12}^i\right)\Gamma_{yz}^p + c_{13}^i c_{23}^i \Gamma_{zz}^p \end{cases} \tag{4}$$

where:

$$
C_p^{gi} = \begin{bmatrix} c_{11}^i & c_{12}^i & c_{13}^i \\ c_{21}^i & c_{22}^i & c_{23}^i \\ c_{31}^i & c_{32}^i & c_{33}^i \end{bmatrix}
$$

$$
\begin{cases} c_{11}^i = \cos\theta_i \cos\psi_i, \; c_{12}^i = \sin\phi_i \sin\theta_i \cos\psi_i - \cos\phi_i \sin\psi_i, \; c_{32}^i = \sin\phi_i \cos\theta_i \\ c_{21}^i = \cos\theta_i \sin\psi_i, \; c_{22}^i = \cos\phi_i \cos\psi_i + \sin\phi_i \sin\theta_i \sin\psi_i, \; c_{33}^i = \cos\phi_i \cos\theta_i \\ c_{31}^i = -\sin\theta_i, \; c_{23}^i = \cos\phi_i \sin\theta_i \sin\psi_i - \sin\phi_i \cos\psi_i, c_{13}^i = \sin\phi_i \sin\psi_i + \cos\phi_i \sin\theta_i \cos\psi_i \end{cases} \tag{5}
$$

For the three GGIs in figure 2, the relationship between the measured value and $\mathbf{\Gamma}^p$ is as follows:

$$
G = MP \tag{6}
$$

in which

$$
G = \left(\Gamma_{xx}^{g1} - \Gamma_{yy}^{g1} \quad \Gamma_{xy}^{g1} \quad \Gamma_{xx}^{g2} - \Gamma_{yy}^{g2} \quad \Gamma_{xy}^{g2} \quad \Gamma_{xx}^{g3} - \Gamma_{yy}^{g3} \quad \Gamma_{xy}^{g3}\right)'
$$

$$
M = \begin{bmatrix}
(c_{11}^1)^2 - (c_{21}^1)^2 & 2(c_{12}^1 c_{11}^1 - c_{21}^1 c_{22}^1) & 2(c_{11}^1 c_{13}^1 - c_{21}^1 c_{23}^1) & (c_{12}^1)^2 - (c_{22}^1)^2 & 2(c_{12}^1 c_{13}^1 - c_{22}^1 c_{23}^1) & (c_{13}^1)^2 - (c_{23}^1)^2 \\
c_{11}^1 c_{21}^1 & c_{21}^1 c_{12}^1 + c_{22}^1 c_{11}^1 & c_{21}^1 c_{13}^1 + c_{23}^1 c_{11}^1 & c_{22}^1 c_{12}^1 & c_{22}^1 c_{13}^1 + c_{23}^1 c_{12}^1 & c_{13}^1 c_{23}^1 \\
(c_{11}^2)^2 - (c_{21}^2)^2 & 2(c_{12}^2 c_{11}^2 - c_{21}^2 c_{22}^2) & 2(c_{11}^2 c_{13}^2 - c_{21}^2 c_{23}^2) & (c_{12}^2)^2 - (c_{22}^2)^2 & 2(c_{12}^2 c_{13}^2 - c_{22}^2 c_{23}^2) & (c_{13}^2)^2 - (c_{23}^2)^2 \\
c_{11}^2 c_{21}^2 & c_{21}^2 c_{12}^2 + c_{22}^2 c_{11}^2 & c_{21}^2 c_{13}^2 + c_{23}^2 c_{11}^2 & c_{22}^2 c_{12}^2 & c_{22}^2 c_{13}^2 + c_{23}^2 c_{12}^2 & c_{13}^2 c_{23}^2 \\
(c_{11}^3)^2 - (c_{21}^3)^2 & 2(c_{12}^3 c_{11}^3 - c_{21}^3 c_{22}^3) & 2(c_{11}^3 c_{13}^3 - c_{21}^3 c_{23}^3) & (c_{12}^3)^2 - (c_{22}^3)^2 & 2(c_{12}^3 c_{13}^3 - c_{22}^3 c_{23}^3) & (c_{13}^3)^2 - (c_{23}^3)^2 \\
c_{11}^3 c_{21}^3 & c_{21}^3 c_{12}^3 + c_{22}^3 c_{11}^3 & c_{21}^3 c_{13}^3 + c_{23}^3 c_{11}^3 & c_{22}^3 c_{12}^3 & c_{22}^3 c_{13}^3 + c_{23}^3 c_{12}^3 & c_{13}^3 c_{23}^3
\end{bmatrix} \tag{7}
$$

$$
P = \left(\Gamma_{xx}^p \quad \Gamma_{xy}^p \quad \Gamma_{xz}^p \quad \Gamma_{yy}^p \quad \Gamma_{yz}^p \quad \Gamma_{zz}^p\right)'
$$

Substituting the installation angles of GGIs into equation (7), it can be seen that when there is no parallel between the three GGIs, the following conclusion is established:

$$
Rank(M) = 5 < 6 \tag{8}
$$

Thus, there is no unique solution to the equation (6). The result also explains why there is no redundancy in the FGGM using the Lockheed Martin's FTG.

4. Simulation

In order to verify the above conclusion, the simulation is carried out according to the following steps: first, setting up the installation angles of GGIs; second, calculate the direction cosine matrix C_p^{gi} and M ; third, calculate the rank of M , and check whether it is 6 and then we know that whether the redundant component is measurable.

The steps of the simulation are shown in figure 4

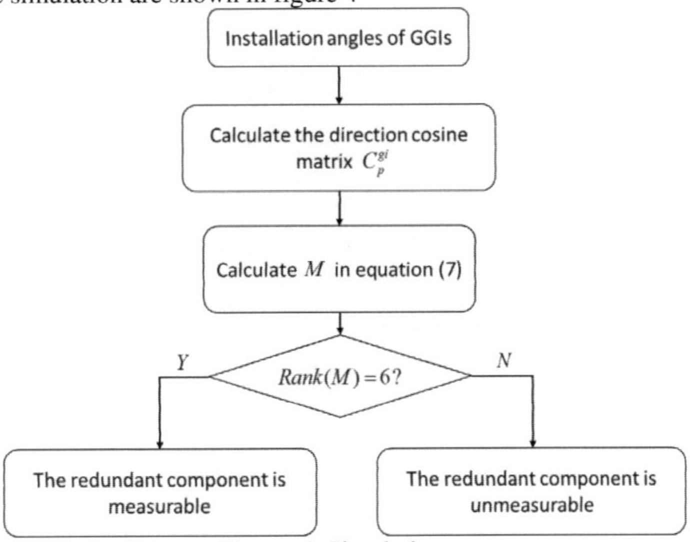

Figure 4. Simulation steps.

The installation angle of the three GGIs are selected as shown in table 1.

Table 1. The installation angle of the three GGIs.

	$\theta_i(rad)$	$\psi_i(rad)$	$\phi_i(rad)$
GGI-1	$\frac{\pi}{10}$	$\frac{\pi}{5}$	$\frac{3\pi}{10}$
GGI-2	$\frac{\pi}{4}$	$\frac{17\pi}{36}$	$\frac{5\pi}{36}$
GGI-3	$\frac{\pi}{12}$	$\frac{5\pi}{12}$	$\frac{\pi}{5}$

Then, the rank of M in equation (7) is calculated as follows:

$$rank(M)=5 \tag{9}$$

Thus, equation (6) has no unique solution, and the measurement of redundancy can not be achieved.

5. Conclusion

In this paper, we have demonstrated the following conclusion: in the FGGM, redundant signal is non-measurable, the installation angles of GGIs have no influence on FGGM. That is, regardless of how the installation angle is selected, no redundant information will be obtained. This conclusion of the study shows that in the FGGM, there is no relationship between the measurement of the gravity gradient components and the installation angle of the three GGIs. Therefore, when installing GGIs, the only conditions that need to be satisfied is:

$$\begin{vmatrix} \theta_1 & \psi_1 & \phi_1 \\ \theta_2 & \psi_2 & \phi_2 \\ \theta_3 & \psi_3 & \phi_3 \end{vmatrix} \neq 0 \tag{10}$$

Reference

[1] Zheleznyak LK, Koneshov VN, Mikhailov PS. Experimental Determination of the Vertical Gravity Gradient below the Sea Level. Izvestiya-Physics of the Solid Earth. 2016;52(6):866-8.

[2] Yuan Y, Gao JY, Chen LN. Advantages of horizontal directional Theta method to detect the edges of full tensor gravity gradient data. Journal of Applied Geophysics. 2016;130:53-61.

[3] You KH, DiCarlo D, Flemings PB. Impact of gravity on hydrate saturation in gas-rich environments. Water Resour Res. 2016;52(2):1265-85.

[4] Tscherning CC, Herceg M. Using ground gravity to improve ice mass change estimation from GOCE gravity gradients in mid-west Greenland. Stud Geophys Geod. 2016;60(1):56-68.

[5] Zheng W, Wang ZK, Ding YW, Li ZW. Accurate Establishment of Error Models for the Satellite Gravity Gradiometry Recovery and Requirements Analysis for the Future GOCE Follow-On Mission. Acta Geophysica. 2016;64(3):732-54.

[6] Braga MA, Endo I, Galbiatti HF, Carlos DU. 3D full tensor gradiometry and Falcon Systems data analysis for iron ore exploration: Bau Mine, Quadrilatero Ferrifero, Minas Gerais, Brazil. Geophysics. 2014;79(5):B213-B20.

[7] Zhou WN, Nan ZY, Li JY. Self-Constrained Euler Deconvolution Using Potential Field Data of Different Altitudes. Pure and Applied Geophysics. 2016;173(6):2073-85.

[8] Zhao L, Liu F, Yuan H. The Design and Analyse for Accelerometer-based Gravity Gradiometer. Guo S, Fukuda T, Yo Y, Yu H, editors2009. 2409-14 p.

IOP Conf. Series: Materials Science and Engineering **339** (2018) 012038 doi:10.1088/1757-899X/339/1/012038

A PBOM configuration and management method based on templates

Kai Guo, Lihong Qiao, and Yifan Qie

School of Mechanical Engineering and Automation, Beihang University, 37 Xueyuan Road, Haidian District, Beijing 100191, China

E-mail: lhqiao@buaa.edu.cn

Abstract. The design of Process Bill of Materials （PBOM） holds a hinge position in the process of product development. The requirements of PBOM configuration design and management for complex products are analysed in this paper, which include the reuse technique of configuration procedure and urgent management need of huge quantity of product family PBOM data. Based on the analysis, the function framework of PBOM configuration and management has been established. Configuration templates and modules are defined in the framework to support the customization and the reuse of configuration process. The configuration process of a detection sensor PBOM is shown as an illustration case in the end. The rapid and agile PBOM configuration and management can be achieved utilizing template-based method, which has a vital significance to improve the development efficiency for complex products.

1. Introduction

Manufacturing process technique keeps an important position in modern manufacture industry, which is closely related to the process of design and fabricating. Bill of Materials (BOM) is regarded as the core of the organization of product information by many manufacturing enterprises. It is efficient to acquire, organize and apply various data and information of products from BOM. PBOM connects Engineering Bill of Materials (EBOM) and Manufacturing Bill of Materials (MBOM). And it can be converted from EBOM and then converted to MBOM[1-4].

The configuration method is a reasonable combination process to meet the individual needs of customers[5]. Researches on product configuration methods have been carried out widely and deeply, including product configuration methods based on feature matching, instances-based product configuration methods, and rule-based product configuration methods and so on[6]. Feature matching based method realizes the product configuration scheme to meet feature requirements of customers, in which main structure of component information is established[7]. The approach of configuration cases retrieval, matching, and adjustment are used to obtain the solution to meet the demand in the process of instance-based product configuration method[8]. Rule-based product configuration method can make logic explicit, which is widely used in knowledge expression field[9]. In order to improve the reusability of the configuration design, the process description and configuration management have become popular research points. Product configuration template that can be constructed by functional module units is able to effectively support the management of configuration process[10].

Currently the research of product configuration management is mainly focused on the classification of configuration modules, the expression of configuration rules and methods of results evaluation. The PBOM configuration and management method based on templates is described in the perspective of customization in this paper.

Nowadays, market demand is diversified, personalized and uncertain. It is increasingly becoming the pursuit goal for enterprises to respond to market demand with the shortest time of product development and the lowest cost. PBOM plays an important role in the process of product development. As product family, especially complex product family needs to be constantly varied and updated. The original PBOM cannot meet the requirements easily, thus it is necessary to redesign the PBOM. In order to reuse the original PBOM resources, improve the efficiency of design and development, and implement the rapid generation of PBOM, the configuration method aimed at PBOM reconstitution can be utilized and mined. On this basis, the configuration and management implementation technology of complex products is researched.

In order to achieve rapid and effective allocation, new requirements are added to the description of the configuration process and configuration object, including:

(a) Highly reusable configuration process. Some components have strict process specification. Therefore, the standard process design flow could be reusable.

(b) Simplified and time-saving configuration process. The configuration process requires simplification. Meanwhile existing instances data, related standards, and specifications need to be fully utilized.

(c) Efficient management of huge data. The quantity of complex products, accumulated instances, and product family PBOM is enormous, and the constraints between configuration objects become more and more complex, which makes configuration become a heavy engineering.

On account of the requirements of PBOM configuration and management mentioned above for complex product, it is significant to structure a configuration and management function system which plunk for PBOM configuration solving, configuration process reuse and configuration process easy implementation to support PBOM design, accumulation and achievement application.

2. PBOM configuration and management method

According to PBOM configuration requirements for complex products, the configuration method of data view based on product process structure is adopted. Existing PBOM data and manufacturing process knowledge are managed and organized in this method to accomplish PBOM design structure quickly and accurately.

2.1. A PBOM configuration and management framework

To meet the requirements of complex product PBOM configuration and management, the configuration management system of PBOM has been set up from the view of PBOM related information management and configuration procedure management. PBOM instance information, standard process parameters and configuration design procedure are organically combined in the framework.

As shown in Figure 1, configuration and management framework consists of PBOM configuration process management and configuration database management. Configuration process management mainly involves requirement features, configuration customization, configuration execution and configuration results management.

(i) Requirement features

Requirement features describe a technical parameter and/or a design essential factor of the product. They are associated with the nodes of the PBOM and act as entry parameters for the product or component PBOM configuration process.

(ii) Configuration customization

To make configuration process and data possessing controllability and reusability, configuration template is established to apply to configuration process, and data definition is actualized in the template.

(iii) Configuration execution

In this step, existing PBOM instance property parameters, standard process parameters and configuration rules are extracted from the configuring base database so as to determine the allocation of configuration process parameters, and ultimately to obtain new PBOM configuration instances.

Configuration execution is composed of a series of ordered configuration activities, and the order of activities is defined in the configuration template and configuration module.

(iv) Configuration results

As the result of configuration process, PBOM structure is managed through a unified defined BOM view. PBOM structure consists of assemblies, subassemblies, machined parts, process parts, dummy parts, fixtures, moulds and etcetera, which are generated by configuration. Newly generated process components due to requirement change are managed through creating BOM view version.

(v) Configuration database

The configuration database is the basic data source that supports the configuration process. The database provides configuration object information and configuration rules constraints for configuration process. It mainly includes requirement feature database, PBOM instance database, standard process parameters database, etc.

2.2. PBOM configuration template construction

The purpose of template customization is to make a unified description of configuration object and configuration process, including configuration procedure and procedure nodes. Configuration template is an abstract expression of associated set of configuration activities. Configuration activities take configuration modules as base operation objects, and configuration template is structured through combination and association among different modules.

2.2.1. Configuration template definition.

Template is a form of business process model, which contains data association and activity methods. And it is the foundation of business process definition and management. PBOM configure template is an abstract description of the configuration process, where configuration modules are associated and connected according to the configuration precedence order. It can be represented as following tuple:

PBOMCT={ Attributes, Modules, Process, Constrains }, in which

Attributes is a set of properties which describes configuration process. Such as the ID of PBOM, name, etc.

Modules describes the configuration objects and is the nodes in the configuration template. Module defines the information of configurable component object, such as assembly, subassembly, machined part, process part, dummy part, fixture, mould, process parameters, rules, etc.

Process represents the relations between configuration process, organizations, resources, and activities.

Constrains describes the constraint relations of configuration object, such as prodromic, subsequent and mutex relationship.

As shown in Figure 2, PBOM function model consists of assembly hierarchy relationships between parts, components, fixtures, moulds and so on. Activity process model describes the precedence relationships between activities that implement product process design and configuration. Process features and hierarchical relationships of products or components drive the generation of activity process. The activity is aimed at the realization of the product process. As to the business configuration procedure of PBOM in product development process, configuration templates are relational integration of product structure, component process information and configuration process activities. Template construction is the realization of data mapping and process mapping. Different combinations of parts and components are produced through the collocation of templates, modules and submodules in the PBOM configuration template structure according to certain configuration rules. And the PBOM configuration template is a dynamic structure, so as to express serialized PBOM of product through version management.

2.2.2. Configuration module construction.

Configuration module is the node in the configuration template. It is a structural unit with interchangeability, versatility, independent structure, specific function and standard interface[11]. The configuration module is a classification module for the PBOM configuration. And configuration module is a composable object that can be configured as a single entity. The module has the function to centrally organize and manage process components, fixtures, moulds, and other process information in the configuration process.

First of all, it is necessary for configuration module to be provided with the configurability of process components of product family in the configuration process. As the functional and structural unit of modular configuration, module is characterized with low coupling and high cohesion. In the process of configuration and management of PBOM, configuration template corresponds to the whole PBOM, and configuration module corresponds to the decomposed PBOM combination, such as components, fixtures, moulds and standard process parameters. And module can further be decomposed into sub-modules.

The information model of configuration module is established as shown in Figure 3 in order to make a further description of the relationship between the relevant information that constitutes the configuration module. The model is centered on the PBOM configuration module class, and associated with configuration process components class, configuration fixture and mould class, and configuration process parameters class. The PBOM correlation class has the unified definition of BOM view to make it easy to manage PBOM process and results. Association class is used to represent the relationship between configuration modules, and to express prodromic and subsequent module object in the definition process of configuration template.

When a variety of market demand appear or market demand changes rapidly, module classification focuses on fast variant design. If the requirements are modified, functional modules that are associated with requirements also need to be varied. However, the original loosely coupled modules should be retained as much as possible. Classification and decomposition is actualized from top to bottom according to product functionality structure and meanwhile functional modules keep independent. The constraint relation between configuration modules should be maintained as little as possible in module construction phase.

An inspection sensor is used as an example to build the configuration template in this paper. The directional graph representation of the configuration template of the sensor is presented as shown in Figure 4. The different nodes correspond to the configuration module of process parts and assembly in the directional graph. The information of configuration process parts, components and process parameters is defined in the module. This PBOM configuration process improves the development efficiency of the inspection sensor.

3. Conclusion

Based on the demand analysis about complex products, the system for PBOM configuration and management has been established. Configuration templates and modules are defined in the framework to support customization and reuse of configuration process. The PBOM configuration of an inspection sensor is shown as an illustration case. The rapid and agile configuration and management of PBOM can be achieved by utilizing this template-based method, which has a vital significance to improve the development efficiency of complex products.

References

[1] Guo J F, Qiao L H and Liu W A 2005 *Comput. Integra. Manuf. Sys.* **11** 1301-06
[2] Lee J H, Kim S H and Lee K 2012 *Comput-Aided. Des.* **44** 253-73
[3] Liu X B, Huang X W, Ma Y, Meng Q N and Meng Y Y 2002 *Comput. Integra. Manuf. Sys.* **8** 983-87
[4] Chung Y K and Fischer G W 1994 *Computers. Ind. Engng.* **26** 321-39
[5] Brown D C 1998 *Ai. Edam.* **12** 301-05
[6] Lu H, Yue T, Ali S and Zhang L 2016 *Inform. Software. Tech.* **72** 68-89
[7] Schuh G, Rudolf S and Riesener M 2014 *Procedia CIRP.* **17** 290-95
[8] Tseng H E, Chang C C and Chang S H 2005 *Expert. Syst. Appl.* **29** 913-25
[9] Zhang J S 2006 *J. Huazhong. Univ. Sci. Techno. : Nat. Sci. Ed.* **34** 108-10
[10] Zhou C J and Lin Z H 2007 *Comput. Integra. Manuf. Sys.* **13** 845-49, 876

[11] Qi Z Y, Wang J Z and Han X M 2007 *Chin. J. Mech. Eng.* **43** 87-94

Acknowledgement
This work is supported by research grant A0320131002. The authors would also like to thank Beijing Municipal Education Commission (Build a Project) for its support.

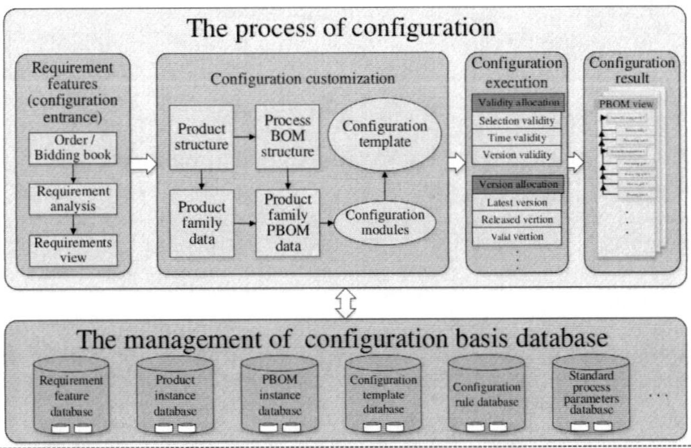

Figure 1. Configuration and management framework of PBOM.

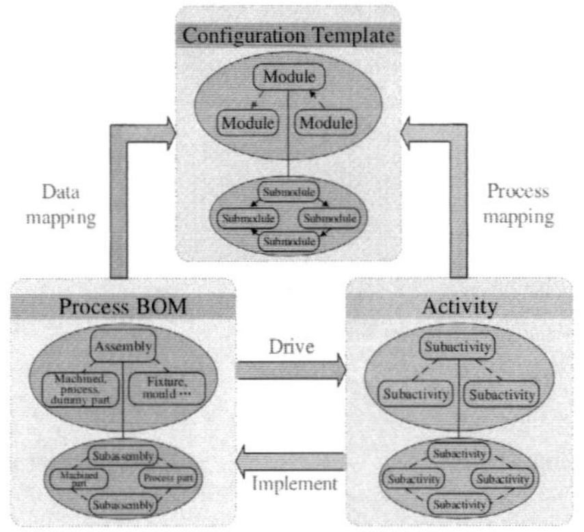

Figure 2. Schematic diagram of configuration template, PBOM and activity mapping.

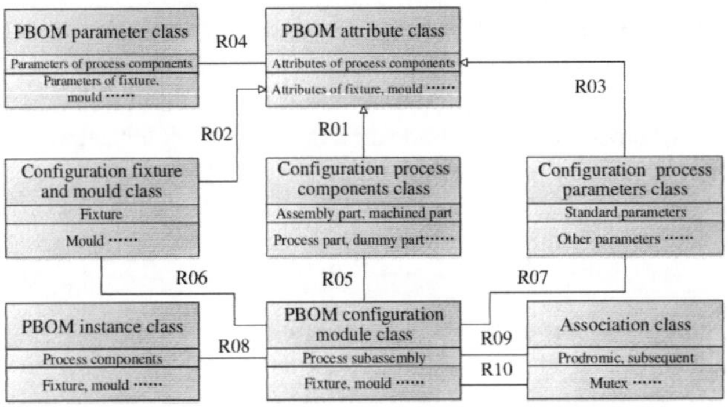

Figure 3. Information model of configuration module.

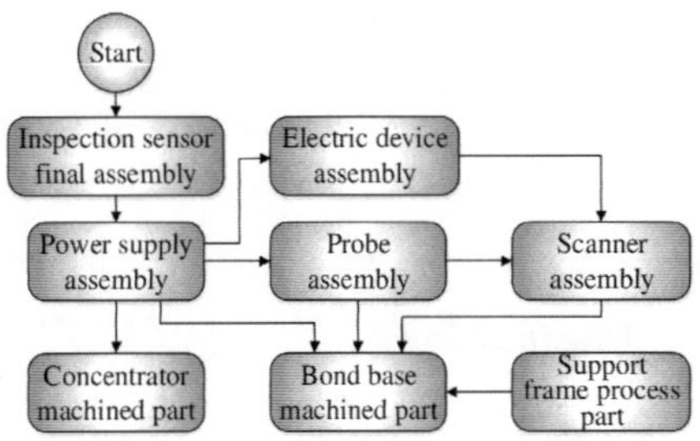

Figure 4. Directional graph representation of a sensor configuration template.

Fault detection of Tennessee Eastman process based on topological features and SVM

Huiyang Zhao [1,2,*] , Yanzhu Hu [1] , Xinbo Ai [1] , Yu Hu [1] and Zhen Meng [1]

[1] Beijing Key Laboratory of Work Safety Intelligent Monitoring, Beijing University of Posts and Telecommunications, Beijing 100876, China

[2] School of Information Engineering, Xuchang University, Xuchang 461000, China

Email: zhaohy@bupt.edu.cn

Abstract. Fault detection in industrial process is a popular research topic. Although the distributed control system(DCS) has been introduced to monitor the state of industrial process, it still cannot satisfy all the requirements for fault detection of all the industrial systems. In this paper, we proposed a novel method based on topological features and support vector machine(SVM), for fault detection of industrial process. The proposed method takes global information of measured variables into account by complex network model and predicts whether a system has generated some faults or not by SVM. The proposed method can be divided into four steps, i.e. network construction, network analysis, model training and model testing respectively. Finally, we apply the model to Tennessee Eastman process(TEP). The results show that this method works well and can be a useful supplement for fault detection of industrial process.

1. Introduction

The problem of fault detection and diagnosis is an essential part of industrial process. It is very important for production. The methods of fault detection can be broadly classified into two general categories: model-based methods and data-driven methods. Generally speaking, the model-based methods need a priori knowledge or a specific mathematic model. However, it is very difficult for some industrial systems. Different from the model-based methods, the data-driven methods are only dependent on the monitored process variables. Thus, the data-driven methods have been extensively studied and developed over the past few decades. We will review some data-driven methods as follows.

KNN(K-nearest neighbors) is a data-driven method used in fault detection. Xiong et al. [1] proposed an information fusion fault diagnosis method that is based on a static discounting factor and combines KNN with dimensionless indicators. Wang et al. [2] proposed a novel fault diagnosis method derived using KNN reconstruction on maximize reduce index (MRI) sensors. Tennessee Eastman (TE) process was provided to demonstrate that the proposed approach can identify the responsible variables for the multiple sensors fault.

Support vector machine (SVM)[3] is also an approach used in fault detection. Wu et al. [4] provided a combined measure of the original SVM and PCA (principle component analysis) to carry out the fault classification, and compared its result with what is based on SVM-RFE (Recursive Feature Elimination) method. Mahadevan and Shah [5] proposed a new approach for fault detection and diagnosis based on One-Class Support Vector Machines. Gao and Hou [6] propose a multi-class support vector machine based on process supervision and fault diagnosis scheme to predict the status of the TE process.

Of course, there are many more data-driven methods for fault detection and we have not mentioned above, such as Decision tree classifier[7], Bayesian classifier [8], neural network[9], random forest[10], PLS (partial least squares)[11], and so on.

However, conventional data-based methods have two problems. One problem is dimension reduction for large-scale industrial processes and the other is that this kind of methods do not consider the relationships among different measured variables and the effect of single measured variables on the overall performance of the system [12].

To solve the problems described above, we proposed a novel method, integrated complex network theory with support vector machine and apply it to Tennessee Eastman process(TEP) which is a well-known typical industrial process.

2. Methods

The main principle of the proposed method is described in figure 1.

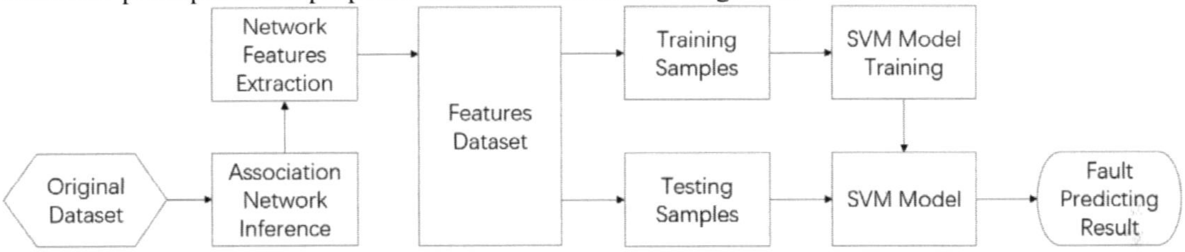

Figure 1. The main principle of the proposed method for fault detection.

From figure 1, we can see the detail of the proposed method. The start of the proposed method is original dataset, generally in form of multivariate time series. Then, the association network is inferred with SSPSTESGC [13, 14]. From the resulted network, some topological features are extracted and analyzed. Next, we will generate a features dataset which are composed of those topological features. This features dataset is considered as the sample of our selected classifier model which is SVM. Fault detection is a problem of classification. Thus, the features dataset is divided into training samples and testing samples. The training samples are used to train a SVM model. With the SVM model and testing samples, we can predict whether some faults have occurred in the observed system. The procedures of fault detection with the proposed method are described in detail as follows.

2.1. Network inference

Some essential methods for association network inference have been proposed for association networks inference, such as correlation [15], information theory [16], Granger causality (GC) [17], neural network [18], and so on.

We have proposed SSPSTESGC for association networks inference in [13, 14] and the good performance have been proved by experiments. Thus, it is selected as a part of the proposed method in this paper. The main principles of SSPSTESGC are the small-shuffled surrogate (SSS) method proposed in [19-21] and the partial symbolic transfer entropy(PSTE) shown in [22].

The primary task for inferring a network is how to define the relations between nodes. PSTE has been proved properly for this work. It is defined conditioning on the set of the remaining time series $z=\{v_3, v_4, \cdots, v_n\}$ and represented by Equation (1).

$$PSTE_{v_2 \to v_1} = \sum p(\hat{v}_{1,t+\tau}, \hat{v}_{1,t}, \hat{v}_{2,t}, \hat{z}_t) \log \frac{p(\hat{v}_{1,t+\tau} \mid \hat{v}_{1,t}, \hat{v}_{2,t}, \hat{z}_t)}{p(\hat{v}_{1,t+\tau} \mid \hat{v}_{1,t}, \hat{z}_t)} \tag{1}$$

where the rank vector $\hat{z}_t$ is defined as the concatenation of the rank vectors for each of the embedding vectors of the time series in z. The partial symbolic transfer entropy can eliminate some of the indirect correlation and remain the pure or direct information flow between v_2 and v_1.

Partial Symbolic Transfer Entropy Spectrum(PSTES) is defined as follows:

The PSTES between time series Y and X is composed of their many partial symbolic transfer entropy curves drawn in a rectangular coordinate system. The horizontal axis represents different time delays and the vertical axis represents transfer entropy. One of the transfer entropy curves is resulted from original data and other curves are resulted from shuffled data.

2.2. Network analysis

Many topological measures have been proposed for complex network studying. Topological measures can be divided in two groups, i.e., measures at global network level and measures at local node level [23, 24], corresponding to the measurable element. Since the observed object in our study is the network as a whole, only those graph-level measures will be selected. In other words, node-level such as the degree of a certain node will not be taken into account. The topological features selected in the proposed method include average degree(AD), diameter(DIA), average path length(APL), density(DEN), clustering coefficient(CLU), degree centralization(DC), closeness centralization(CC), betweenness centralization(BC) and eigenvector centralization(EC).

2.3. Fault detection

After the step of network analysis, we will fit a model of fault detection by SVM. The topological features dataset is randomly split into training and testing groups. The ratio of training dataset is 70% and the ratio of testing dataset is 30%. When fitting an SVM model, a parameter should be specified, i.e. kernel function.

3. Results

3.1. The description of TEP

The TE process is a widely used realistic simulation program for chemical plants and has been widely accepted as a benchmark for control and monitoring studies [25, 26]. The process consists of five major transformation units: the reactor, the product condenser, the vapor-liquid separator, the recycle compressor and the product stripper.

In our experiment, 22 process measurements [27] are selected as original dataset. The TE process contains 21 preprogrammed faults[28] which are described. These faults are divided into six types, i.e. step, random variation, slow drift, sticking, unknown and constant position. 22 data sets (1 normal data set and 21 fault data sets) are generated from the TE process. With a sampling interval of 3 minutes, 960 observations are generated for each data set. All faults are introduced from the 161th observation.

3.2. Fault detection of TEP

3.2.1. Association network construction. In this section, the method SSPSTESGC [13, 14] will be used to infer the association network. The result of applying SSPSTESGC method is called PSTE spectrum between each pair of variables, such as two figures shown in figure 2 and figure 3. The value of horizontal axis is time delay and the value of vertical axis is PSTE. According the rule whether there is a relation from one variable to the other variable, we can conclude that *X7* is influenced by *X6* from figure 2 and *X2* is not influenced by *X1* from figure 3.

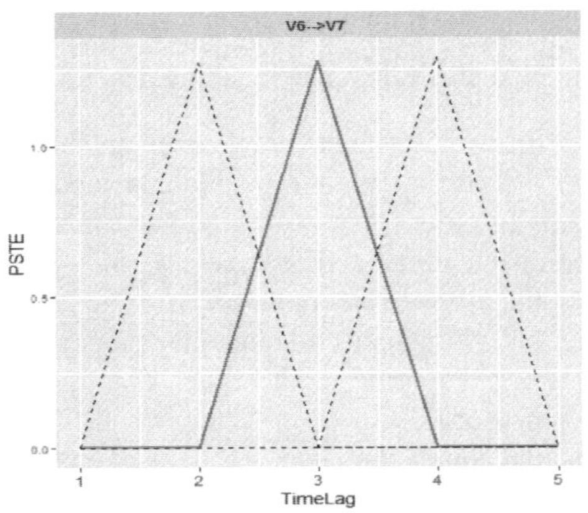

Figure 2. PSTE Spectrum from variable *V6* to *V7*.

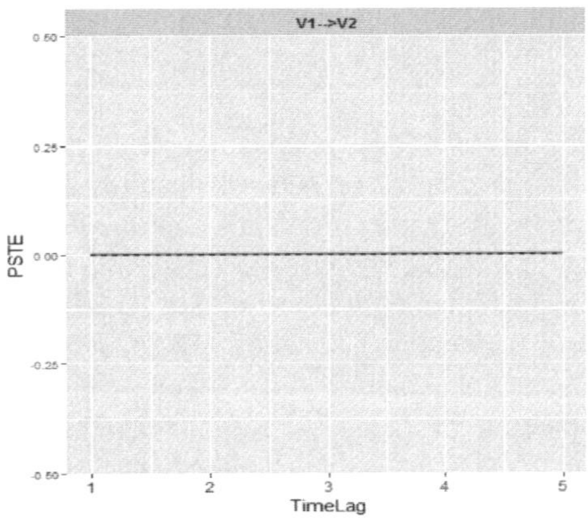

Figure 3. PSTE Spectrum from variable *V1* to *V2*.

According to the PSTE spectrums and the rule of network construction, the network at each time can be inferred. For example, the network with normal state is shown in figure 4 and the network with fault 1 is shown in figure 5. There are some differences between the two networks. For instance, the core vertex with maximal degree is *V6* in figure 4. In figure 5, the network changes with the occurrence of fault 1. The vertex with maximal degree changes to be *V11*. Additionally, there are some other differences between two networks. The degree of some vertex changes to be larger, such as *V7*, *V16* and *V22*. However, the degree of *V21* changes to be smaller. Moreover, some vertex changes from isolate vertex to normal vertex, such as *V2* and *V5*. However, some vertex changes from normal vertex to isolate vertex, such as *V1* and *V9*.

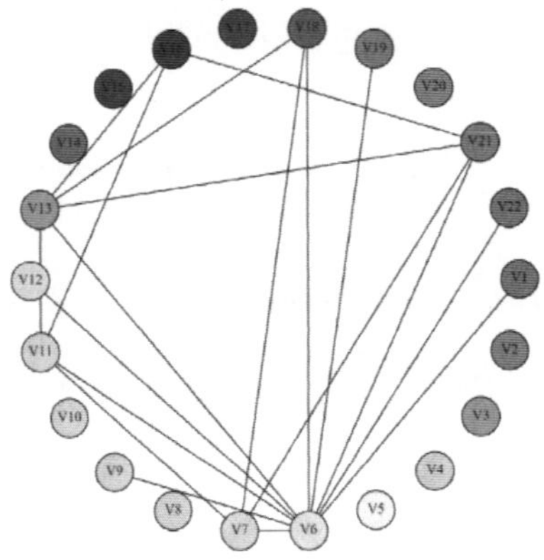

Figure 4. The network with normal state at time point 181.

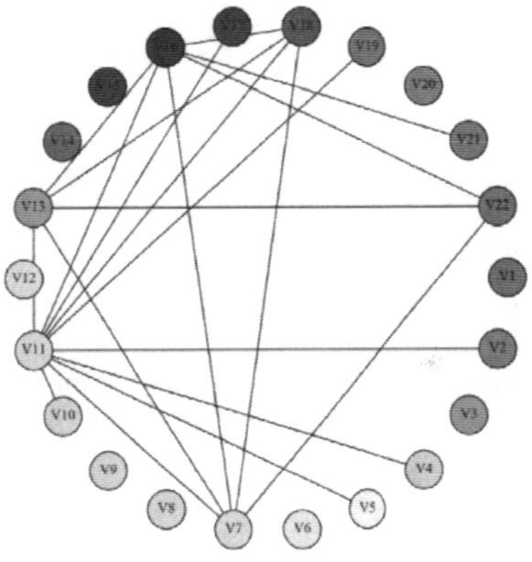

Figure 5. The network with fault 1 at time point 181.

3.2.2. Network analysis with complex network theory. Network features are calculated for each network and part of the results are shown in Table 1. The abbreviations SS represents the state of the system. This column has 22 different values, i.e. N (normal state), F1 (fault 1), F2 (fault 2) and so on.

Table 1. The topological features extracted from association networks.

ID	AD	DIA	APL	DEN	CLU	DC	CC	BC	EC	SS
1	1	2	1.56	0.02	0.29	0.12	0.02	0.06	0.9	N
2	1.27	2	1.48	0.03	0.44	0.14	0.02	0.05	0.89	N
3	1.09	2	1.54	0.03	0.38	0.12	0.02	0.05	0.89	N
4	1.73	4	2.18	0.04	0.12	0.23	0.05	0.21	0.87	F1
5	2.73	4	2.2	0.06	0.18	0.46	0.07	0.39	0.86	F1
6	2.36	4	2.07	0.06	0.43	0.29	0.06	0.26	0.82	F1
7	0.64	3	1.67	0.02	0	0.06	0.02	0.02	0.86	F2

8	0.64	5	2.29	0.02	0	0.06	0.02	0.04	0.88	F2
9	0.55	3	1.55	0.01	0	0.06	0.02	0.02	0.9	F2

3.2.3. Fault detection and evaluation. In this section, we will carry out fault detection by applying the SVM algorithm to the network features extracted above. With the fitted model, we can predict the system state is either normal or abnormal on testing dataset. In order to assess the performance of the proposed method, we introduce into three measures, i.e. precision, recall and accuracy.

The results of fault detection are shown in table 2. With the proposed method, precision is 0.84, the recall is 0.99 and the accuracy is 0.89. As a comparison, the performances of some other methods are also shown in table 2 As a whole, the proposed method is superior to the other five methods. Compared to precision, we pay more attention to the recall measure.

Table 2. The model assessment and comparison.

ID	Method	Precision	Recall	Accuracy
1	NFSVM	0.84	0.99	0.89
2	Naive Bayes	0.77	0.89	0.79
3	KNN	0.78	0.92	0.81
4	Decision Tree	0.79	0.95	0.84
5	Random Forest	0.81	0.96	0.86
6	Neural Network	0.79	0.96	0.85

4. Conclusions

In this paper, based on the complex network theory and SVM, we have proposed a novel method for fault detection of industrial systems. We applied the proposed method to TEP and the performance of proposed method is evaluated by three measures. As a result, the method makes a good performance on fault detection. However, there are still some topics that are worth studying in future.

References

[1] Xiong J, Zhang Q, Sun G, Zhu X, Liu M and Li Z 2016 An information fusion fault diagnosis method based on dimensionless indicators with static discounting factor and KNN *IEEE Sensors Journal* **16** 2060-9

[2] Wang G, Liu J and Li Y 2015 Fault diagnosis using kNN reconstruction on MRI variables *Journal of Chemometrics* **29** 399-410

[3] Cortes C and Vapnik V 1995 Support-vector networks *Machine Learning* **20** 273-97

[4] Wu F, Yin S and Karimi H R 2014 Fault detection and diagnosis in process data using support vector machines *Journal of Applied Mathematics* **2014**

[5] Mahadevan S and Shah S L 2009 Fault detection and diagnosis in process data using one-class support vector machines *Journal of process control* **19** 1627-39

[6] Gao X and Hou J 2016 An improved SVM integrated GS-PCA fault diagnosis approach of Tennessee Eastman process *Neurocomputing* **174** 906-11

[7] Ye F, Zhang Z, Chakrabarty K and Gu X 2016 Adaptive board-level functional fault diagnosis using incremental decision trees *IEEE Transactions on Computer-Aided Design of Integrated Circuits and Systems* **35** 323-36

[8] Duda R O, Hart P E and Stork D G 2012 *Pattern classification*: John Wiley & Sons)

[9] Eslamloueyan R 2011 Designing a hierarchical neural network based on fuzzy clustering for fault diagnosis of the Tennessee–Eastman process *Applied soft computing* **11** 1407-15

[10] Cerrada M, Zurita G, Cabrera D, Sánchez R-V, Artés M and Li C 2016 Fault diagnosis in spur gears based on genetic algorithm and random forest *Mechanical Systems and Signal Processing* **70** 87-103

[11] Khan A A, Moyne J R and Tilbury D M 2008 Virtual metrology and feedback control for semiconductor manufacturing processes using recursive partial least squares *Journal of Process Control* **18** 961-74

[12] Cai E, liu D, Liang L and Xu G 2015 Monitoring of chemical industrial processes using integrated complex network theory with PCA *Chemometrics and Intelligent Laboratory Systems* **140** 22-35
[13] Hu Y, Zhao H and Ai X 2016 Inferring Weighted Directed Association Networks from Multivariate Time Series with the Small-Shuffle Symbolic Transfer Entropy Spectrum Method *Entropy* **18**
[14] Hu Y, Zhao H and Ai X 2016 Inferring Weighted Directed Association Network from Multivariate Time Series with a Synthetic Method of Partial Symbolic Transfer Entropy Spectrum and Granger Causality *PLOS ONE* **11** e0166084
[15] Guo X, Zhang Y, Hu W, Tan H and Wang X 2014 Inferring Nonlinear Gene Regulatory Networks from Gene Expression Data Based on Distance Correlation *PLoS ONE* **9** e87446
[16] Margolin A A, Wang K, Lim W K, Kustagi M, Nemenman I and Califano A 2006 Reverse engineering cellular networks *Nat. Protocols* **1** 662-71
[17] Schiatti L, Nollo G, Rossato G and Faes L 2015 Extended Granger causality: a new tool to identify the structure of physiological networks *Physiological Measurement* **36** 827
[18] Mahdevar G, Nowzaridalini A and Sadeghi M 2013 Inferring gene correlation networks from transcription factor binding sites *Genes & Genetic Systems* **88** 301-9
[19] Small M 2005 *Applied Nonlinear Time Series Analysis: Applications in Physics, Physiology and Finance*
[20] Nakamura T, Hirata Y and Small M 2006 Testing for correlation structures in short-term variabilities with long-term trends of multivariate time series *Physical Review E* **74** 041114
[21] Nakamura T, Tanizawa T and Small M 2016 Constructing networks from a dynamical system perspective for multivariate nonlinear time series *Phys Rev E* **93** 032323
[22] Ai X 2014 Inferring a Drive-Response Network from Time Series of Topological Measures in Complex Networks with Transfer Entropy *Entropy* **16** 5753-76
[23] Sun X and Wandelt S 2014 Network similarity analysis of air navigation route systems *Transportation Research Part E: Logistics and Transportation Review* **70** 416-34
[24] Durek P and Walther D 2008 The integrated analysis of metabolic and protein interaction networks reveals novel molecular organizing principles *BMC Systems Biology* **2** 100
[25] Maurya M R, Rengaswamy R and Venkatasubramanian V 2007 Fault diagnosis using dynamic trend analysis: A review and recent developments *Engineering Applications of artificial intelligence* **20** 133-46
[26] Golshan M, Pishvaie M R and Boozarjomehry R B 2008 Stochastic and global real time optimization of Tennessee Eastman challenge problem *Engineering Applications of Artificial Intelligence* **21** 215-28
[27] Cai E, Liang L and Xu G 2015 Monitoring of chemical industrial processes using integrated complex network theory with PCA *Chemometrics and Intelligent Laboratory Systems* **140** 22-35
[28] Chiang L H, Russell E L and Braatz R D 2000 *Fault detection and diagnosis in industrial systems*: Springer Science & Business Media)

Acknowledgments
We thank the anonymous reviewers for their valuable comments and suggestions to improve the quality of the paper. This paper is supported by the National Natural Science Foundation of China (61503034), by the National Key Research and Development Program of China (2016YFC0701309-01), the National Natural Science Foundation of China (61627816), by the Key Scientific Research Project of Henan Province Universities (15B520031), and by the Xuchang Science and Technology Program (1502098). We thank all the fund organizations for offering us the studying conditions.

Institute of Physics
Dirac House, Temple Back
Bristol BS1 6BE UK

ISSN: 1757-8981
ISBN 978-1-5108-6115-2

CURRAN ASSOCIATES INC.
proceedings.com

1st International Conference on Big Data and Cloud Computing (ICoBiC 2017)

Journal of Physics: Conference Series Volume 1018

Kuching, Malaysia
25 – 27 November 2017

Institute of Physics